ENGINEERING
INVESTMENT
DECISIONS
planning under uncertainty

ENGINEERING INVESTMENT DECISIONS

planning under uncertainty

l.m. rose

Privatdozent

Systems Engineering Group, Technische-Chemisches Laboratorium
Eidgenössische Technische Hochschule, Zürich.
Switzerland

ELSEVIER SCIENTIFIC PUBLISHING COMPANY
335 Jan van Galenstraat
P.O. Box 211, Amsterdam, The Netherlands

Distributors for the United States and Canada:

ELSEVIER/NORTH-HOLLAND INC.
52, Vanderbilt Avenue
New York, N.Y. 10017

ISBN: 0-444-41522-X

Printed in The Netherlands

To my Mother

PREFACE

The essence of engineering is to design equipment that, once built, will function. As engineering predictive methods improve, the equipment should not merely function, but function more reliably, meet its specifications closely, require less maintenance and supervision and involve lower cost.

Engineering research is usually aimed at producing an improved predictive method that will enable equipment to be designed to closer tolerances which, in turn, result in still lower cost. In the design of manufacturing equipment, sophisticated engineering techniques can now produce plants with capacities within ± 5 to 10% of the requested output.

It is usually a source of annoyance to engineers to find, that having gone to great lengths to achieve this degree of accuracy for their designs, the planning that led to the choice of the process and capacity was made very superficially, and the decisions were taken after very little analysis. There seems to be little point in employing sophisticated engineering techniques unless improved planning techniques are also used.

Admittedly, since planning depends on forecasting, which is always associated with uncertainty, one's initial reaction is not to work in depth with the forecasts since they will change anyway; but methods for forecasting, probability analysis, Decision Theory and Risk Analysis which accept that the figures will change are available in the literature. Why are they rarely used? This may be partly because they are not well known and partly because they have been rarely worked up from theory into directly applicable procedures.

The aim of this book is to collect together all the techniques that are involved in the planning for investment in engineering plant. The available literature is diverse, and has not been presented suitably in a form for planners -- who are often engineers -- to appreciate. Further to this, the book presents practical methods, reinforced by numerous examples, for analysing decisions concerning plant investment. These methods attempt

to use all available quantitative information in the analysis, so that all
that can be done with the available information is done. By using these
procedures the least that can be achieved is that the engineers can feel
that the planner has tried to determine the best design basis. The best
that can be achieved is that a more economic, more flexible, less risky
plant is built.

Chapters I, II, and III present the relevant Economic, Probability
and Decision Theory and these are linked by Chapter IV to the remainder
of the book which is concerned with the application of these theories to
engineering investment decision making.

Chapter I is concerned with the economic evaluation. Though there
are many books already published in this area, it is necessary to repeat
the subject in this one chapter as it does form the cornerstone of all
planning and Decision Analysis. The chapter emphasises the selection of
alternatives as opposed to the simple economic evaluation of a single
proposal, and describes a number of very different methods, showing their
advantages and disadvantages.

Chapter II is concerned with uncertainty, -- its sources, and the
techniques by which it can be handled. A survey of forecasting methods is
also given because the forecast is usually the source of the major uncertainty. This chapter describes the concept of probability and surveys
the type of distribution that can be used to represent uncertain data, and
the way in which probability distributions can be combined.

Chapter III is concerned with those parts of Decision Theory that
are applicable to decisions concerning engineering investment. Decision
Trees are described in some detail, including their use for determining the
value of better information. Also, decision criteria in the face of uncertain information are discussed, particular emphasis being laid on the
theoretically attractive, though rarely applied, Utility Function.

Chapter IV describes what is being looked for in decisions concerning
plant investment. In particular it separates planning into two distinct
stages:

      1)   The selection of one alternative out of those
          proposed (Planning Stage)

      2)   A rigorous economic evaluation of the chosen proposal
          (Evaluation Stage)

Chapter V adopts the general techniques described in the first three
chapters to the particular requirement of the engineering planning

environment. Methods are developed for carrying out a search for the best combination of investments, and for a Monte-Carlo simulation procedure suited to engineering investment with uncertain demand.

The search method summarises the results as a table which clearly shows the decision-maker the various possible outcomes of this decision. The simulation attacks the problem of correlation between demand, price and year, which is normally inadequately treated in the literature.

Proposals are made for using a general purpose planning program as a focal point of planning a department in the same way that Process Flowsheet Programs have become the focal point of large process engineering departments. The components of such a program (PLADE) are described.

Chapter VI concerns itself with the selection of one out of a number of basically different ways of making the investment. (i.e. process selection):

- how deep must the risk analysis be to make a
  realistic comparison?

- how important is the choice of decision criterion?

- when is one satisfied with the result?

and

- what is the value of obtaining better information

These questions are discussed with particular reference to four examples which are developed in this chapter and run through the remainder of the book.

Chapter VII is concerned with selecting the capacity for the investment, after the basic principle has been established. Demand normally grows in an uncertain fashion out to the horizon. The optimum size is the outcome of an economic balance, after which further reductions in size may be made to provide some insurance against possible high losses. The chapter derives optimum plant sizes for a number of idealised situations, and follows these with a number of more practical cases. Though the idealised cases provide useful generalisations regarding the importance of individual factors on optimum plant size, the size for each project is best chosen by individual economic optimisations. This can be done conveniently using a planning computer program.

The final adjustment of plant capacity is to decide on the "Safety Factor" to be employed in the design to account for engineering uncertainty. This can be determined by an economic balance, again most conveniently by a computer program, as shown by an example.

After the basic principles and the capacity for the investment have been chosen, the last step is the detailed economic evaluation of the proposal. Chapter VIII is concerned with the factors that most influence the accuracy of this economic prediction. The importance of inflation, data uncertainty and the allowance for future policy changes are shown by examples. The chapter then discusses factors concerning uncertainty. The importance of correlation between demand/price and year, and the accuracy required for the various "subjective judgements" in order to get meaningful results are discussed, and numerical examples given. The chapter ends with comments on the interpretation of the probability distribution of profitability for decision making purposes.

Chapter IX is a postscript to show that, although all the examples and discussion have centered around the process Industries, most engineering decisions have equivalent features and therefore can be treated by similar methods. Production Engineering, Civil Engineering, Energy Generation and Transport are areas which should benefit equally with Process Engineering from improved planning techniques.

Appendix I is the "Planning Manual" for the PLADE planning computer program and Appendix II contains normal probability tables, and discount tables.

The book should be useful within planning departments and Operation Research groups of most engineering companies, because it is a review of the literature in the area, because it contains many numerical "idealised" examples which planners themselves never have time to make and, because it demonstrates -- and details -- a computer program for carrying out studies more completely than has previously been possible.

As this book is one of the few to summarise those parts of Probability Analysis and Decision Theory most useful in an engineering context it should find application as a text in Production Engineering, Operations Research and Management Science courses and in the broader type of chemical, mechanical, electrical and civil engineering courses that include this type of material.

It would be wrong to write a book in Zurich on Investment without mention of the Gnomes and the methods used so successfully by them. To cover this approach, traditional Swiss nursery rhymes are given in each chapter which are more or less appropriate to the matter under discussion, since these might contain their secret.

I am grateful to many people for help I have received, both in the subject matter and in the preparation of this book. I would like to thank

Prof. D.W.T. Rippin for his interest in the area and for the provision of such a comprehensive library and literature coverage, Prof. J. Myhre for the development of the Dynamic Programming algorithm which formed the starting point of the study, and Dr. O.H.D. Walter for the efficient programming of that algorithm.

For assistance in the preparation of the natural book I am very grateful to Mrs. J. Wegmann, Mrs. M. Bryant and Mrs. J. Johns for their careful typing, my Wife, Jane, for correcting the style to make it readable, and Benita Rose for assistance in some of the computations.

L.M. ROSE
Z urich 1976.

CONTENTS

CHAPTER 1

ECONOMIC EVALUATION

> Ich und due sind Brüederli,
> schaffe tüe mer liederli,
> ich und du händ Gäld im Sack,
> die andere nu Schnupftaback.

1.1 WHY EVALUATE?

Investment proposals involve spending comparatively large sums of money initially, in the hope that this will result in the receipt of even larger sums of money sometime in the future.

Economic Evaluation is the name given to the calculation made before the investment is actually made to predict whether the investment will be "worthwhile" or not. One first measure of "worthwhile" is to determine whether the income following the investment is greater than the capital sum originally invested. If it is not, then clearly the investment is not "worthwhile". If it is, then is the extra money earned above that initially spent large enough to merit the effort and risks associated with the project? Could more money be earned by putting it to another use? Sometimes one has a number of ways of spending different amounts of capital to make a product. How does one choose which way should be used? Sometimes one has a number of different possible products and limited capital. Which should be chosen?

The Economic Evaluation helps answer all these questions. It is inconceivable that large capital expenditures could be made without an accompanying Economic Evaluation. Even if the final decision were contrary to the recommendation of the Evaluation, the evaluation would at least be made.

The objective of the analysis is usually summarized as "maximizing the profit" obtained from any capital investment. "Maximum Profit" for a company can mean maximizing the returns to the shareholders. If this were the objective--ensuring that those people with money get the absolute maximum money for having some in the first place--then the whole subject would be of minor importance. There is, however, a much more important reason for

"Maximizing Profit": money is really a common unit by which most things
can be measured. The Profit is therefore a measure of whether carrying out
a project is worthwhile. Now since there is a shortage of capital in the
world, there has to be some selection process, and one based on accepting
the most profitable investments--in order to make the most of a scarce
resource--is very sensible.

It is very necessary for a firm, industry and even country to carry out
economic evaluations and act on them accordingly. Without proper investments
in our industrial society it would no longer be possible to remain competitive, to the detriment of all.

Those readers to whom the word "Profit" is abhorrent should therefore
have patience because the arguments put forth throughout this book apply
equally well to a profit-making or a non-profit-making society. Exactly
who receives this profit is a question for the politicians, not the engineer,
even though engineers may sometimes feel that they are not making a good job
of it.

1.2 NON-ECONOMIC FACTORS

Any decisions regarding the expenditure of capital therefore should be
preceded by an Economic Evaluation. However, this is never the whole story
and "Non-Economic" factors should also be taken into consideration before
the decision is made.

Sometimes these factors are irrefutable, they override any economic
considerations. These may involve health hazard, pollution or matters of
National Interest. Such matters are made clear by a country's legal system
which introduces laws to cover them. The legal system is a great help in
these areas because without it individual firms would never feel justified
in, for instance, expensive pollution control, for fear of no longer being
competitive. When everyone has to play to the same rules, restrictions are
accepted with good grace.

Sometimes these "Non-Economic" factors are indisputable advantages
associated with some investment but are virtually impossible to equate with
a monetary value. For example, one proposed new road scheme may fit better
into the landscape than another, but this advantage may be impossible to
evaluate in monetary terms. In such cases, selection of alternatives can

be made using a method involving simple points systems rather than using an
economic criterion. This is discussed at more length in Chapter 3 on
Decision Theory.

It is in many ways more satisfactory to reduce all benefits to a monetary value to enable a choice of alternatives to be made. One way of doing
this is to employ the technique of Cost-Benefit analysis. This technique
attempts to put a monetary cost on every criterion, so that there is no
need to revert to subjective points schemes (6). The method itself cannot
resolve difficulty in the selection of alternatives, but its advantage is
that it enables a more rational discussion to be held than would otherwise
be possible. For example, by assuming a cost for a road accident, and
using accident statistics, a cost due to accidents for each alternative can
be determined and compared with the capital cost of each proposal. A
quantitative selection based on safety and cost can be made. There can still
be discussions--has a reasonable cost of an accident been assumed?--but this
is easier to discuss than discussing "safety versus cost" in general.

The technique is very well developed, and various theorems enable
values to be put on all the so-called "Qualities of Life". If one accepts
the theory of the market economy, then the way the average man spends his
money is his quantitative evaluation of his various preferences. The cost
of noise for instance can be estimated from the effect of noise on house
prices on the free market. This can be multiplied by a number of people
affected by any proposal to produce a cost equivalent to the reduction of
"Quality of Life" due to the noise.

The cost of wasted time is also estimable and the method is subtle
enough to use different values for different sections of the community. If
society pays the average air traveller more than the average city dweller,
then it is accepted that the former's time is more valuable, and hence
different values should be used in any evaluation comparing time wasted by
the two groups. This method however is not aimed at comparisons based
only on money but on such "satisfaction". Hence it includes the concept of
Utility, which is dealt with in Chapter 4, where a poor person's money is
given a much higher value for "Utility" than a rich person's.

The method has been developed for community projects and it has to
recognize the different interests within a community. Each project will
affect each section in different ways. For example, one road scheme will

have the following groups as different interested parties:

      --the council providing the funds

      --the road users

      --the road construction company

      --the local inhabitants

    The equivalent cost to each group can be determined as well as the total costs and benefits. Any accepted project should have an acceptable benefit to the community as a whole and not to only a single group. The "acceptable benefit" can be evaluated as any normal evaluation, usually Net Present Worth as discussed in the next section.

    Cost Benefit Analysis is designed for analysis of large community projects, its application to private industry is rather limited because the economic system is so structured that industrial evaluations are only based on the interest of the private company, with society's requirements being introduced as constraints imposed by the legal system. However, the technique does show that virtually all advantages and disadvantages can be converted into an equivalent cost. It would therefore not be surprising to find that a modified form could be introduced into private industrial decision-making to structure rational discussions and decisions.

    Another type of "Non-Economic Factor" is the introduction of uncertainty. Data may be used in a financial evaluation, but because the data is uncertain, the assumptions made in the financial analysis are open to question-- "if sales develop as forecast"--"if no technical difficulties are encountered." A manager reserves the right under "Non-Economic Factors" to query these assumptions and override a purely economic assessment because of intuition, based on his experience, that a particular assumption may not hold. It is right that such a possibility exists, but it should only be applied when the factors cannot be quantified. Apart from the egoist, no manager should be happy injecting such qualitative judgements unless there is no alternative. In fact many of these can be quantified by probability calculations and risk analysis, and the major part of this book is devoted to the description of methods of making such analyses. The decision has still to be made by the manager after these analyses, but he can arrive at a decision from much more quantative data.

1.3 CASH FLOW

1.3a Types of Cash Flow

An economic assessment consists of listing the flows of cash into and out of the firm as a result of a particular project, and from this pattern of flows, determining how attractive the project is.

Cash paid out by the firm, the cash "outflow," can be conveniently subdivided into the following types:

a) Investment Capital

b) Working Capital

c) Yearly Fixed Costs

d) Variable Costs

e) Taxes

Under the heading "Investment Capital" come all the monies paid out as non-recurring payments which are considered as capital investments by the tax authorities. The usual profit tax rules permit such capital to be allowed against profits over a period of years before taxation is calculated. When tax laws become complicated, this group has to be further divided to account for "buildings" or "machinery" for which different tax rules some-times apply.

"Working Capital" consists of all monies involved filling the time gap between expense and income in a fully operational business. Raw materials have to be paid for before the product is sold, and so the business must employ a quantity of money which it recovers only at the end of the project. Working Capital therefore is money spent during the project which is fully recoverable at the end of the project. It can be calculated from the value of the raw material and product held at normal stock levels, plus the value of product delivered but not yet paid for, minus the value of raw materials received but not yet paid for.

In valuing the stored product for Working Capital purposes, marginal cost of production, rather than the sales price, has to be taken. This is the real cash outflow associated with the storage.

Working Capital differs from investment capital involved in a project, in that it is a relatively safe investment, since it involves saleable material and expected payments. One could imagine raising a loan to cover this cost at normal interest rates, with no increased rates to cover the

risk. For accurate estimation of working capital it is necessary to have
discussions with company accountants, both to set the value of the product
and the rate of interest required from this money.

Working Capital should not be neglected in evaluations because it can
sometimes be a considerable expense, although it is fairly normal to use
the same discount rates as for the other monies in the evaluation. This
simplification however should always be examined with each new project,
since projects involving expensive material and cheap plants may give a
false picture because the money at risk (the capital cost of the plant),
for which the evaluation is being made, may sometimes be swamped by the
working capital.

The third group, "Yearly Fixed Costs" is all the monies paid out each
year independently of the production rate. This includes the majority of
the manpower costs, the majority of maintenance costs, all management and
supervision costs, general overheads such as building taxes, insurance, and
a part of water and energy costs.

The fourth group, "Variable Costs" involves payments that are directly
proportional to production rate. The clearest example is raw material costs
which, assuming plant efficiency remains constant, are directly proportional
to plant output. Part of the water and energy costs are also directly
proportional to output. There may also be a part of the plant manpower employed in proportion to the output--in multi-pupose batch plants or packing
departments for instance. Some maintenance may also be considered proportional to output.

This variable cost is quoted in money units per unit of product made.
Hence the yearly cash outflow associated with variable cost is the

(units produced in that year) x (variable cost/unit)

Variable costs are therefore <u>marginal</u> production costs/unit.

Taxes are also represented as cash outflow from the firm. The quantity
of tax paid is calculated from the difference in the annual cash-in minus
the cash-outflow according to the laws of the particular country concerned.
It is usual to remove a part of the original capital expense--depreciation--
before determining the tax outflow. The rules governing the amount of
depreciation allowable are also carefully laid down by laws. These laws
are frequently changed since they represent a form of control by which

government can attract investment when they feel an incentive is necessary
on economic, social or regional grounds.

The cash received by the firm, the cash-inflows, can be divided into
three:

a) sales income

b) scrap income

c) working capital income.

The sales income is, of course, the major reason for the project's
existence. It is accurately calculated as the sum of

$$(sales\ quantity)\ \ \ x\ \ \ (price)$$

for each sale , since it is unusual for the product to be sold at a single
price. However, it is adequate for project evaluation purposes to evaluate
the sales income as

$$(total\ sales\ quantity)\ \ \ x\ \ \ (mean\ selling\ price)$$

There is often difficulty in deciding the mean selling price,--does
it include delivery, or packing, or the sales office expenses? The price
used has to be defined very carefully because this definition must agree
with the line drawn for the cash-outflow,--do these include distribution
and selling costs? Both costs and selling price must have the same basis.

Scrap Income represents the cash-inflows that occur whenever a plant
is shut down for good. The simplest situation is where the plant is dismantled and the equipment is sold as second-hand equipment, or as scrap
metal. The scrap income is then this income, less the expense of demolition
and transport. It is frequently the case that the demolition and transport
costs equal the income from selling the dismantled plant--often a contractor
will remove it free of charge! This means that more often than not, the
scrap income from the plant is zero.

A more complicated situation exists when the plant is put to another
use; to make another product for instance. One approach here is to include
this second product in the evaluation, but this tends to complicate the
evaluation and make the analysis of alternatives cumbersome and difficult.
It is often better to separate the two projects completely by employing a
transfer price, a hypothetical value that is credited to one project and
debited to the second to represent the sale of the plant. This transfer
price in the simplest case can represent the cost of building an equivalent

plant for the second project. However, this is often unacceptable because
it means that one project gains all from the coincidence that a suitable
plant is available, and the second project gains nothing. When such
conflicts exist, the right approach is to consider both products in the
single evaluation, but the convenient and adequate approach may be to agree
on a transfer price which is between zero and the cost of a new plant,
depending on the relative incentive to each project for the transfer to be
made. This transfer price then becomes the scrap income in the analysis.

It is convenient to assume a definite project lifetime for the evaluation; 15 years is common in capital intensive industries. However, with
some products, plant lifetimes are more than 15 years. Hence, some evaluations result in plants still existing at the end of the evaluation, and
there is every reason to believe that they will still be producing after
this time. This plant therefore has some value. Some approximate value
can be chosen to represent this, and this value used in the evaluation as
a scrap income to credit the project with this remaining useful plant.

This method is not to be recommended because of the difficulty of
arriving at a value for the plant. Errors introduced are not usually
serious because of discounting, but since it is comparatively easy to extend
the evaluation period, as will be shown in Chapter 5, it seems wrong ever
to use scrap income to convey the concept of further useful plant life.

Working Capital income is simply the recovery of the money laid out
during the project as working capital, which should equal the total
working capital put into the project.

Of the eight cash flows described here, six completely define any
project. These are:

   a) Investment capital
   b) Yearly fixed costs
   c) Variable costs / Unit output
   d) Sales income
   e) Scrap income
   f) Working Capital

This is a very important concept because it enables any investment to
be characterized and evaluated from only these standard items. It enables
standard evaluation and selection programmes to be written and is the
biggest single contribution to an organized approach to investment analysis.

This, of course, is a simplification, particularly to the accountant who is much more interested in detail and accuracy. He may worry about the subdivision of capital and other costs; he might not assume that working capital income equals working capital outlay. However, this book discusses investment analysis for engineers and since these details will not have a significant effect on the ranking of alternative proposals, or the acceptance of a project, there is no reason to introduce this detail into economic analyses for engineering purposes.

1.4   METHODS OF INVESTMENT APPRAISAL

1.4a   Analysis Without Discounting

Pay Back Time Method

The crudest measure to determine whether a project is worth carrying out is to see if the total cash inflow is greater than the total cash outflow, when considering the whole life of the project. Usually a "project" consists of a heavy initial outflow, followed by a steady cash inflow until and beyond the point where the initial outlay has been recovered.

The simplest common method of investment appraisal which is based on this straightforward cash flow consideration is the Pay Back Time (PB). The Pay Back Time is the time taken from the start of the project before the total cash flows for the project become positive. Expressed mathematically one can say:

$$PB \quad = \quad n \tag{1.1}$$

where n satisfies the equation

$$I - \sum_{m=1}^{n} \left((F_{in})_m - (F_{out})_m \right) = 0 \tag{1.2}$$

where I is the single initial capital investment and F_{in} and F_{out} are the respective annual cash flows.

This is a fairly commonly used method: it is easy to understand; it indicates how risky the project is; it gives a measure of how long there will be a strain on the liquidity of a firm. These factors are all particularly important for small companies and this is where it is most commonly used.

Rate of Interest Method

A second fairly simple method of assessment is to look at a "typical" year's operation, and decide whether cash inflows are greater than outflows, and if so, how much greater compared with the original investment. This method is the Rate of Interest Method (ROI). The concept is to keep the capital intact by paying it back over the life of the project, so that at the end of the project the capital is available for re-use and during the project an annual profit has been made. The annual profit is therefore the cash inflow less the cash outflow, less that year's capital repayment. This capital repayment is the original capital divided by the project life. The annual profit is then expressed as a percentage of the capital employed which is the ROI, sometimes referred to as the Rate of Return:

$$ROI \ = \ \left(\sum F_{in} \ - \ \sum F_{out} \ - \ \frac{I}{L} \right) \frac{100}{I}$$ (1.3)

where   F represents the sum of the yearly cash flows,

I is the capital investment

and   L is the life of the project (years).

This analysis has to be made for an "average" year of a project.

1.4b   The Introduction of Discounting

There is a time value attached to money. Given a chance of having £100 now, or in one year's time, it is obviously worth more if collected now. This advantage can be quantified by assuming we put it in a bank immediately, at an annual rate of interest of r. In one year it will have earned £100r interest and we will own £100 (1+r). Conversely, if we had £100/(1+r) today, we would have £100 out of the bank in a year's time. The amount of money now equivalent to a future sum is called the "Present Value" or "Present Worth" of that sum. The present worth of £100 next year is £100/(1+r). Similarly, using compound interest, assuming interest is paid at the end of each year, the £100 now will be worth £100(1+r)(1+r) in two year's time, or conversely, the Present Worth of £100 in two year's time is £100/(1+r)^2.

In general, the Present Worth of any sum S_n, flowing in n year's time is:
$$PW \ = \ S_n(1+r)^{-n}$$

The above discussion has assumed that all transfers and interest calculations are made at fixed time intervals--discrete discounting has been
assumed. If one considers continuous discounting whereby the interest is
calculated continually and continually added to the capital sum, then our
derivation would be slightly different. At any time x, the total sum in the
bank, S, is the original sum plus the interest payments. The rate of
increase of this sum ∂S due to interest payments in time x to x + ∂x would
be:

$$\partial S = Sr \, \partial x \tag{1.5}$$

hence, for an initial sum S_1, after a time n years, this final sum S_n is
given by:

$$\int_{S_1}^{S_n} \frac{dS}{S} = \int_0^n r dx \tag{1.6}$$

therefore $\qquad\qquad \ln\left(\frac{S_n}{S_1}\right) = rn \tag{1.7}$

$$S_n = S_1 \, e^{rn} \tag{1.8}$$

Us,ing this approach, our NPW S_1, of the sum S_n at year n is given by:

$$PW = S_n e^{-rn} \tag{1.9}$$

This equation results in a slightly different net present worth, as shown
by Table 1.1.

It is usual for economic evaluations to use the discrete discounting
formula. The continuous formula, which could be argued is theoretically
more correct, is sometimes applied in mathematical derivations.

Engineering Investment Decisions

Table 1.1

Comparison of Discrete and Continuous Discounting Formula

(Interest rate 15%)

| n | NPW of £100 received in n years time calculated by, | | Difference |
	(a) Discrete discounting formula Equation (1.4)	(b) Continuous discounting formula Equation (1.9)	£
1	86.96	86.07	0.9
5	49.72	47.24	2.5
10	24.72	22.31	2.4
20	6.11	4.98	1.1

Net Present Worth Method

With the aid of Present Worth we can now make quantitative comparisons of sums of money spent at different times. We can now answer difficult questions such as "Do you prefer £100 today to £105 in one year's time?" The answer is "Yes" if you know of a bank which has an interest rate greater than 5%, and "No" if you feel you cannot find a 5% interest rate.

For a complicated investment project, with cash flowing in and out throughout its life, we can reduce each cash-flow (or sum of cash-flows within one year) to Present Worth, sum all these Present Worths and so decide what the whole project is equivalent to in terms of a sum of money now. If this is negative, then the project is certainly not attractive. If the result is near zero, then at least the project has earned an interest rate of r on the monies involved--which by itself may be adequate grounds for the project. However, if the result is a positive sum, this is the present day equivalent of the money earned by the project after it has paid interest at a rate r on the monies involved.

This basically is the Net Present Worth method, where the Net Present Worth (NPW), sometimes referred to as the Net Present Value (NPV), is defined by:

$$\text{NPW} = \sum_{n=0}^{L} \left(\left((F_{in})_n - (F_{out})_n \right) (1+r)^{-n} \right) \qquad (1.10)$$

where F_n represents the total cash flows for year n, including capital, and L is the life of the project.

This summation is expressed in equations as starting from n=0. In practice there is considerable flexibility allowed for the definition of "the present" for NPW calculations. It may actually be the present moment in time at which the calculations is being made. In this case the summation of all cash flows start at year 0. Often it is convenient to define "the present" as the moment production starts--which may be 2-5 years later than the time at which the calculations are made. In this case, the summations of cash flows involves flows that will be spent before n=0. The summation might start at n=-2 or -5 for example. This starting point should be chosen to include all cash flows yet to occur, concerned with the project. Cash already spent on the project is completely excluded. It is irrecoverable money and should in no way influence the spending of the future money.

This definition assumes that all cash transfers occur at instances in time. Furthermore it is normally assumed that these instances correspond to period-ends (usually year-end). The use of non-integer values for n is unusual, although this simplification can lead to appreciable errors for sums transfering in earlier years. In particular, large errors can result if the timing of the capital out-flow is too inaccurate, and therefore use of non-integer values of n for this flow should always be considered.

When identical cash flows occur every year--for example the benefit B accrued by a project may be considered to be the same every year--then the NPW for this flow (based on a discrete cash inflows at the end of each year) is:

$$\text{NPW} = \sum_{n=1}^{n=L} \frac{B}{(1+r)^n} = B \sum_{n=1}^{n=L} \frac{1}{(1+r)^n} \qquad (1.11)$$

It is sometimes helpful to express this series formula, which is tedious to evaluate by hand, by the following formula, which can be shown to be equivalent to equation (1.11):

$$\text{NPW} = B \sum_{n=1}^{n=L} \frac{1}{(1+r)^n} = \frac{B}{r}\left(1 - \frac{1}{(1+r)^L}\right) \qquad (1.12)$$

The NPW method calculates the total amount flowing into an organization as a result of a project, after paying for all the capital used at an acceptable interest rate.

Discounted Cash Flow Method

A particularly interesting solution to equation (1.5) is the value of r when the NPW is zero. This is the interest rate earned by all the monies involved in the project. It can therefore be compared with other available interest rates, to determine whether the project is better or worse than alternative more straightforward uses of the capital such as simply lending it at a guaranteed fixed rate of interest.

This method is favoured by economists and was derived from economic theory in the late 19th century (11). It lays particular emphasis on the profitability per unit of capital invested, and since this is a yardstick often used in economic circles--and generally considered an index of good management--it probably explains why it is such a widely used criterion. It has a number of names: the interest rate when the NPW is zero can be called the "DCF rate of Return" (DCF), the "Internal Rate of Return (IRR), or simply, amongst economists, the "Yield."

The Cost of Capital

The discount rate r has to be set for the NPW evaluation, or used for comparison of the result with the DCF calculation. This rate is referred to as the Cost of Capital. Generally, the capital comes from private investors who have money they wish to save. They lend this money through the money market to those requiring capital who offer the best terms. This competitive situation means that a firm raising money has to offer terms comparable with the mean market rate and as it is a true market governing this rate, it fluctuates considerably, for reasons known only to itself.

When considering a long-term capital investment, we are interested in using a long-term cost of capital. No-one knows what this will be; all we can ever do is to look at past figures.

Taxation must be properly taken into account. The investor is only interested in his interest receipts after tax. Hence, the cost of capital should refer to the interest rates after tax payments.

Inflation has a considerable effect on interest rates. In market terms,

under inflation, everybody wants to convert money into material goods which
are likely to keep their value, and this demand for capital causes its cost
to rise. It is normal to correct for inflation by quoting the cost of
capital at the interest rate less the inflation rate. That is, the interest
rate is calculated for a constant value money analysis--sometimes referred
to as the "real" interest rate. In fact, it is reasonable for the investor
to keep the purchasing power of his investment and then receive an interest
from it.

The cost of capital we are looking for is therefore the past long-term
performance of companies, as the profit they have achieved after removal of
tax, and after correcting for inflation.

This figure is in the 7-9% range (11). It depends on the particular
industry, and is dependent on whether the funds are retained profits, newly-
raised share issues, or short or long-term loans. The optimum method of
raising capital for projects considers the cost of capital from alternative
sources (11), but it is well beyond the scope of this book.

This figure of 7-9% is lower than that used by most companies in their
analysis, where 12-15% have been found to be necessary criteria to achieve
reasonable performance. The difference can be explained by the approximate
method of evaluation which is normally used. Normally, inflation is ignored.
In the past this has been 2-3% per year. Hence the 7-9% becomes 9-12% to
compare with the 12-15%. The remaining difference of about 3% is probably
associated with "risk". It is unusual to carry out a careful risk analysis;
it is more usual to treat risk by demanding a higher rate of interest on the
capital involved. This in fact is formalized in a "Venture Worth" evaluation
method where the extra interest rate is considered to be the premium for an
insurance to cover the risk.

This general approach leads to firms setting multiple standards for
their Cost of Capital. For instance, 7% for safe manpower reduction studies,
12% for expansion of basic products and 15% for new products. This is a
complete confusion of cost of capital, inflation and risk. Should a product
be wrongly categorized, it will either become a lost opportunity or a
profitless investment.

The aim of this book is to present methods of rigorously analysing
these contributions separately. Therefore the variable r will always refer
to the cost of capital alone, after allowing for inflation. It should be

agreed with the companies' accountants after they have clearly understood
its definition, and should be in the 7-9% range.

The Disadvantages of Discounting

The concept of discounting is so universally accepted in the Western
World, where money is expected to make money, that it becomes difficult to
work and co-operate with countries that do no accept this premise. For
instance, governments in developing countries often claim that Western
"help" is "exploitation" because much more money is taken from their country
than is invested in it. Discounting inevitably produces this situation,
even when the "grounds" for investment are philanthropic--with the investor
doing no more than covering the "Cost of his Capital". This makes industrial
cooperation with the developing countries difficult.

Discounting can also be seen to have its immature aspects (14)--an
attitude of 'think of today far more than tomorrow'. When social planning
is involved this can lead to very wrong approaches. For example, prior to
1972, the oil companies' attitude was to extract and sell as much oil as
possible, as quickly as possible. A discounted evaluation immediately shows
that oil sold today is more profitable than the same oil sold next year.
This is exactly opposite to society's interests which should involve conser-
vation for the future.

One could say that such situations can be recognised, and further constraints put on the discounted evaluation. However, with discounting so
widely accepted by Western economists, it is likely that it is sometimes
applied when it should not be.

1.4c Less Common Methods of Evaluation

There are over 20 "methods" of evaluation for capital investment, all
of which are serious proposals which have their own particular advantages.
They may have been proposed because they are particularly simple, or emphasize the importance of capital, or emphasize the importance of profit after
interest repayments, or emphasize risk, or they may be modifications to
established methods to overcome shortcomings. The subject is further confused by the many names for the same method and the multitude of minor corrections that can be applied, each resulting in a new name.

In the following paragraphs a further dozen or so methods will be
briefly reviewed. These are additional to the four most common methods
already described. This review shows the major problem in this field--the
difficulty in setting up an agreed objective for capital investment. Because no single objective can be agreed, a large number of alternative objective functions have grown up.

Benefit/Cost Ratio

A disadvantage of the NPW method is that the profitability is not referred to the capital required to produce it. The project may have a high
NPW but can still be unattractive because it requires a very high capital
outlay. DCF is one method which relates the NPW to the capital involved
but it has two disadvantages. Firstly, the function can, mathematically,
have two or more solutions under certain conditions. This causes confusion
and means, under some circumstances, that the method is not applicable (11).
Secondly, a high DCF is a hypothetical analysis unless the monies recovered
in the project are invested in another project of equal profitability.

These disadvantages can be overcome using the less well known Benefit/
Cost ratio (BCR) method. In this case the ratio of the NPW of the benefits
(usually yearly profits) to the discounted cost (usually the capital investment costs) is calculated. The discount factor used is the estimated
cost of capital.

$$BCR = \frac{\sum_{n=0}^{L} \left(\left((F_{in})_n - (F_{out})_n \right) (1+r)^{-n} \right)}{\sum_{n=0}^{L} \left(I_n (1+r)^{-n} \right)} \qquad (1.13)$$

where F_{in} and F_{out} in this instance represent the yearly cash flows, excluding capital, and I_n is the capital investment in year n.

When this ratio is less than 1, the project does not even cover the
cost of the capital. When it is greater than 1 it earns more than the cost
of capital. In fact, (BCR-1) is the NPW of the project expressed as a
fraction of the discounted capital involved. Hence the magnitude of the
profitability can be judged at a glance to those accustomed to dealing with

the capital cost of projects and so the ratio is particularly handy for
engineers.

Written mathematically:

$$(BCR-1) = \frac{\sum\limits_{n=0}^{L} \left(\left((F_{in})_n - (F_{out})_n - I_n \right) (1+r)^{-n} \right)}{\sum\limits_{n=0}^{L} \left(I_n (1+r)^{-n} \right)}$$

$$= \frac{NPW}{(Total\ Discounted\ Capital)} \qquad (1.14)$$

The BCR therefore is a measure related to the capital employed. It
has only one value, since NPW for any set of cash flows can only have one
value.

Equivalent Annual Value

Not only is NPW difficult to appreciate because it bears no relationship to the capital investment; it is difficult to appreciate its significance when viewed from the financial side of the business. Company finances are carried out on a yearly basis, and comparative figures should therefore also be on a yearly basis. The Equivalent Annual Value (EAV) is the
equivalent constant annual sum that a project will contribute to a firm's
finances, after the capital has been paid for, that is:

$$\sum\limits_{n=0}^{L} (EAV) (1+r)^{-n} = NPW \qquad (1.15)$$

therefore

$$EAV = NPW / \left(\sum\limits_{n=0}^{L} (1+r)^{-n} \right) \qquad (1.16)$$

This enables a manager to appreciate at a glance the importance of
the profits involved, in relation to the company's annual balance sheet.
Conversely, the engineer finds the EAV somewhat less significant than the
Benefits/Cost Ratio.

Terminal Value

In certain investments the terminal value (TV), that is the Compounded Worth at the end of the lifetime of the project, is more significant than the Present Worth--the discounted worth, referred to the beginning of the project. This applies, for example, to the annuity type of investment, where money is invested yearly to provide a single sum at the end of the project. For a yearly sum, S_n, invested for L years, the Terminal Value is given by:

$$TV = \sum_{n=0}^{L} \left(S_n \ (1+r)^{(L-n)} \right) \tag{1.17}$$

The relation between Terminal Value and Net Present Worth is quite straightforward:

$$NPW = \sum_{n=0}^{L} \left(S_n (1+r)^{-n} \right) = TV(1+r)^{-L} \tag{1.18}$$

The TV of a constant sum S, invested for L years is given by:

$$TV = S \frac{\{(1+r)^L - 1\}}{r} \tag{1.19}$$

The Annual Cost and Capital Charge Methods

These methods are derivatives from the Rate of Interest method and not based on discounting, but on investigating the flows in a typical year by subtracting yealry costs including a yearly cost for the capital involved, so giving a yearly profit. As with the Equivalent Annual Value, an annual profit is a particularly convenient value for management to consider. The problem is always to determine the annual charge due to the capital.

Firstly there is the cost of capital; this has already been discussed for the NPW method. Secondly, there is the recovery of this capital. Discounting assumes that this money is entirely recovered in profit; non-discounting methods use the concept of depreciation--each year money is returned from the project to the company's capital balance,--so that at the end of the project the capital has been exactly recovered. Hence a yearly capital

depreciation charge must be removed from the project as a constant sum each
year. In the simpler methods, such as the Rate of Interest Method, this
sum is the capital divided by the project life. In the more developed
methods, such as the Annual Capital Charge or Annuity methods, the sum is
that value which provides a Terminal Value equal to the total capital involved.

The Annual Cost Method (11)simply compares the annual cost of alternatives assuming the income to be fixed. The most complicated variant, the
Sinking Fund Return Method, develops the rate of interest on capital after
it has subtracted a "sinking fund depreciation" at a different rate of interest related to normal cash borrowing rates.

The Minimum Revenue Requirements Discipline (MRRD)

This method is essentially a development of the Annual Cost Method.
Any proposal has certain costs that must be covered by revenue--normally
yearly cash flows, plus the cost of capital, plus the annual depreciation
charge--in the form of an annuity. This is a minimum revenue requirement
for all projects under consideration, a constraint for every investment
which should not be confused with the profitability of the project. The
figure that is interesting, the one that affects the performance of the firm,
is the difference between this Minimum Revenue and the total project revenue.
It is this difference that should therefore be compared for the different
alternatives. To emphasize the objectivity of the method, this difference
is then expressed as an improvement in the yield of the company's shares.

The method is commercially oriented and does not involve discounting
(apart from the depreciation annuity) but is based on a mean annual consideration. Levelling of cash flows to an equivalent annual flow, and strict
definitions of all the terms, means that the application of the method is
complex, requiring a book of 600 pages to explain it. (8)

The Extended Yield (11) and Modified Pay-Back Methods(1)

The DCF and PB methods are so well understood and established that
there have been attempts to rectify their shortcomings by modifications.

For example, the DCF method does not give useful results when cash
flows are initially predominantly positive, with a large negative flow
appearing late in the project--for instance, an open cast mining project

with a high landscaping cost at the end of the project. DCF methods are
also ambiguous when there is more than one capital flow in the life of the
project, in that there is more than one discount rate that gives a zero NPW.
These disadvantages can be overcome by using the <u>Extended Yield Method</u>. If,
during the DCF calculation the net cash flow becomes positive to account
for a later negative cash flow, then this positive quantity is considered
to be transferred to a sinking fund, appreciating at the firm's cost of
capital, to pay off this later negative cash flow. Hence the project evaluation itself never has positive net cash flow situations before a capital
outlay. There is then no problem with the multiple solutions of the DCF
method.

The modified pay-back method--<u>Equivalent Maximum Investment Period</u>
(EMIP) is an attempt to extend the pay-back method to projects with an un-
steady cash flow with more than one capital outlay. Figure 1.1 gives a
typical break-even chart, with a large initial outlay followed by a steady,
but not constant, recovery as the project earns money.

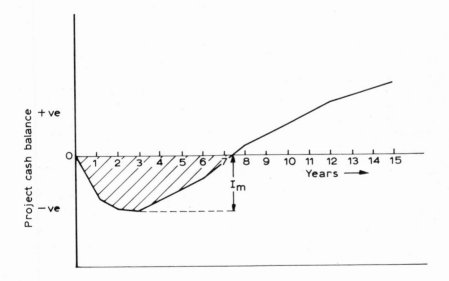

Figure 1.1 The EMPI Method

The above pay-back is 7.4 years from the first investment, but it is
only 4.4 years from the last investment. The EMIP method removes this

ambiguity by determining a single figure that can be considered to be the
number of years required to pay back the maximum negative cash flow balance
given the same performance. This figure is determined by dividing the total
shaded area in Figure 1.1 by I_m. This EMIP is used in place of pay-back
where flows are irregular.

Other Methods

        To show that the list is still not exhausted we can name Capitalized
Cost, MAPI formula, Avoided Cost or Test Year methods described by Jeynes
(8) and Postponability, or the Annual Return on Capital Employed, described
by Merrett and Sykes (11).

        This list of methods is not intended to be exhaustive but only to give
some idea of the number of attempts that have been made in this field.
Each method is a serious attempt at meeting a need and the reason for describing them here is to give the reader a feel for the inconclusive nature
of the subject.

1.5 COMPARISON OF METHODS

        When there is a need to choose between two projects we use a method to
evaluate the economics of them both and, if there is nothing to choose
between the Non-quantitative Factors of the project, then the choice is
made according to the outcome of the economic evaluations. If all the
economic criteria resulted in the same preference order there would be no
problem. However, this is not the case. In fact, it is often possible to
choose different methods of evaluation and get different preference orders.
There have been numerous papers published in the literature showing cases
where NPW and DCF conflict or DCF and ROI conflict, but alas, no paper yet
which shows they all conflict! This subject is discussed interminably and
readers wishing to do so can read 1000 pages on the debate in the literature (2, 8, 11), where most authors come up with a definite recommendation--
DCF, NPW, MRRD or EAV.

        Space here is better spent in analysing the difficulties rather than
making any further recommendations.

        The problem arises because all these evaluation methods are merely
rules of thumb or guide lines. They do not clearly define the real objective

and then evaluate according to that criterion.

The objective obviously concerns the future of the firm, but there is still confusion on what the objectives of the firm are. It may be to maximize benefits to shareholders, or simply to stay in existence. For "staying in existence" simply a well-controlled cash flow to maintain liquidity may be the primary objective. "Satisfying the shareholders" involves dividend and capital growth criteria.

Jeynes, being American, is quite confident that the objective is maximum benefit to shareholders. This enables the MRRD method to be applied with confidence. This is a particularly interesting method because it is an attempt at identifying the real objective first, and only then devising an evaluation method. However, the MRRD method is too cumbersome for engineers to use for their individual problems.

A further reason for the endless debate is that some methods produce a more easily understood result than others, even though the less easily understood result may be more suitable. Here is a meeting place of disciplines; economics, accountancy, management, and engineering. All have different outlooks which are best served by different "rules of thumb." The economist is particularly interested in the efficient use of capital-- DCF is often his choice. The accountant is interested in the behaviour of the firm, for which the effect on the yearly accounts is most meaningful. He therefore prefers time averaged yearly evaluations, without aversion to discounting--Minimum Annual Cost, Annual Capital Charge, Equivalent Annual Value or MRRD are most meaningful here. The manager will be rather conservative and less interested in technical accuracy. Rate of Interest, Payback, or, for the more up-to-date, DCF appeal to him. The engineer is averse to inaccuracy, and so he prefers the methodical accuracy of discounting methods to the averaging methods. He prefers the simplicity of handling NPW, which is additive, to the non-additive DCF, although since the engineer is primarily occupied with capital cost the relation of DCF to capital is particularly significant to him. Hence he has some respect for DCF but Benefits/ Cost Ratio really combines the best of NPW and DCF for him.

The characteristics of the project have a further influence on the most suitable "rules of thumb." Some methods are fitting only for single capital investments--Pay-back and Rate of Return; some methods are suitable only when data for the average year can be realistically defined, Annual

Capital Charge and MRRD for example. It is interesting to note that these
are reported as being used mainly in the Nationalized Industries (11) and
Public Utility Companies (8) where a fairly constant yearly performance can
be expected. Their application to, for instance, a short-lived, high growth
pharmaceutical product would be most inappropriate. In such a case, Payback
or a discounted method would be preferred.

If a project has an unusual cash flow profile, several expenditures or
high expenditures towards the end of its life, the DCF cannot be used, and
a NPW method or the Extended Yield would be appropriate.

If there is a real limitation on capital, then DCF, Benefits/Cost or
Annual Capital Change will be maximizing the company's performance within
the capital constraint. If capital is really no problem, as may often be
the case in a big firm, there is no point in employing methods which use
the ratio of profit to capital--NPW or EAV are better as unconstrained
wealth maximisers. In fact, if any other factor was constraining, one
would be quite justified in creating a new criterion to meet the siutation.
For example, a labour shortage which limits the number of plant expansion
possibles would be adequate grounds for judging projects on a NPW/man
employed basis.

Since this book is for engineers, the major discussion will concern
NPW. It is a widely accepted criterion very suitable for many types of
project and its particular additive properties make it very suitable for
the manipulation required in risk analysis and selection of alternatives.

Despite all the discussion on the subject, there has been no quantitative work to show how badly a firm would perform if it used the wrong
evaluation criteria. Possibly, when projects are so close that they are
in the region where different criteria give different results, the overall
effect on the firm in making the wrong decision may be hardly noticable.
In other words--it might not matter!

1.6 THE INCLUSION OF TAX

1.6a Why Involve Tax?

As described earlier in Section 1.3, one of the cash flows that every
company has is the payment of tax on profit. The benefit that a project is
to a firm depends on its after tax cash flow. Before tax is paid on the

yearly cash-inflow minus outflow it is allowed to remove a portion of the
capital cost from the flows and pay tax only on the remainder. Hence two
apparently similar projects based on cash flows before tax will no longer
be similar after tax if the capital involved for the project is different.
Hence the true preference order to the firm can only be known with certainty
when taxes are included in the evaluation.

Taxes are sometimes used by governments to direct a particular policy--
to favour projects involving investment, or the creation of employment in
depressed areas for instance. This is often done by modifying the way in
which allowances can be removed from projects before calculating the taxes.
These measures constitute gifts by the government for particular projects
and they therefore should be evaluated before any decision is made. These
measures vary with country, district and time and so there is no point in
recording any particular regulation here. However, details of any scheme
should be obtained from government departments before any important decisions
are made, since it is the intention that the effect of the regulation should
be large enough to be significant in any project evaluation.

1.6b The Tax Calculation

The tax calculation involves the concept of "depreciation" and "capital"
as do the time averaged methods, i.e. the ROI, Minimum Annual Cost and Annual
Capital Charge methods. The discounting methods on the other hand do not
treat capital differently from other cash flows nor involve a yearly sum to
recover the capital called depreciation. However, these concepts have still
to be introduced into discounting methods in order to calculate the tax
liability.

For tax purposes the annual taxable profit can be considered to be:

Yearly Cash Inflow - Yearly Cash Outflow - Depreciation

Having obtained this taxable profit, the tax is simply a percentage of
this. For example, in England this is around 50%,--in Switzerland 25%.

The cash-outflow here excludes capital. The depreciation is a fraction
of the capital involved. Over the project life the sum of the yearly
depreciation must equal the capital invested.

The Straight Line Depreciation Method

A common method is to allow straight line depreciation over a certain period, for example, 10 years. The depreciation is then calculated as one tenth of the capital investment, every year for 10 years. After this period the depreciation for tax becomes zero.

For evaluations for engineering decision purposes there is no need to calculate the profit before tax or tax payable and so the net yearly cash inflow after tax (S_n) can be reduced to

$$S_n = (F_{in_n} - F_{out_n}) - (F_{in_n} - F_{out_n} - D_n) \, T/100$$
$$= (F_{in_n} - F_{out_n}) \, (1-T/100) + D_n \, T/100 \tag{1.20}$$

where D_n represents the depreciation allowance for year n and T is the % tax rate on profits.

For our example of the 10 year straight-line depreciation case,

$$D_n = C/10 \tag{1.21}$$

until n = 10, then $D_n = 0$. If there is more than one capital investment involved, then

$$D_n = \sum_{i=(10-n)}^{n} C_i/10 \tag{1.22}$$

This can sometimes be inconvenient to calculate because it is necessary to keep a record of the previous ten years' capital cost to evaluate the overall cash flow for just one year. This difficulty can be overcome by using a Modified Capital Cost in the investment calculations. The tax allowances on depreciation are effectively credited against the capital employed. A capital sum C, with discount rate r will have a total tax allowance credit, discounted to the time the capital was spent, of

$$\sum_{n=1}^{m} \frac{CT}{100m} \frac{1}{(1+r)^n} \tag{1.23}$$

where m is the number of years over which the capital is allowed to depreciate and the first tax allowance is assumed to be received one year after the expenditure of the capital. The capital cost was therefore equivalent to:

$$C - \frac{CT}{100m} \sum_{n=1}^{m} \frac{1}{(1+r)^n}$$

$$= C\left(1 - \frac{T}{100m} \sum_{n=1}^{m} \frac{1}{(1+r)^n}\right) = FC \qquad (1.24)$$

where F is the <u>Capital Tax Factor</u>.

We can therefore use this value in place of Capital Cost, C, and neglect the tax allowance on the depreciation in the yearly calculations. Since, as shown by equation (1.12),

$$\sum_{n=1}^{m} \frac{1}{(1+r)^n}$$

can be written directly as

$$\left(1 - \frac{1}{(1+r)^m}\right) \frac{1}{r} \qquad (1.25)$$

the capital tax factor can be alternatively expressed as:

$$F = 1 - \frac{T}{100mr}\left(1 - \frac{1}{(1+r)^m}\right) \qquad (1.26)$$

This derivation has assumed the first value of n is 1. This means the first tax rebate is received 12 months after the capital is spent. The mechanism for recovering the tax is usually such that there is a two year delay before the first rebate is received. Reworking the above derivation with n = 2 to m + 1 given the following expression in place of (1.26).

$$F = 1 - \frac{T}{(1+r)100mr}\left(1 - \frac{1}{(1+r)^m}\right) \qquad (1.27)$$

In situations where it is inconvenient to keep a history of the capital investments to make the evaluation, this modified capital approach can be used. This is analogous to the accountant's use of Tax Tables to simplify his taxation calculations.

All the formula so far quoted assume that every year there is a profit greater than the depreciation allowance. This is an acceptable assumption because profits are calculated on a company-wide basis, so if the project concerned does not make a profit then the tax allowance will be taken from the profits made elsewhere within the firm.

The Reducing Balance Depreciation Method

A second common method of calculating the depreciation is to depreciate annually a constant fraction of the outstanding undepreciated value. In this case there is always some depreciation to set against taxes because the equipment is never completely written off until the last year of the project. For the last year of the project the depreciation is the total outstanding amount less any scrap income. The depreciation in the first year is therefore Cd, the second year $Cd(1-d)$ and in the nth year $Cd(1-d)^{n-1}$.

The outstanding undepreciated capital at the end of the project (L years) is $C(1-d)^{L-1}$. Discounting these values to the beginning of year 1 gives the following expression for the total tax credits for an investment C at the end of year 0:

$$\frac{CdT}{100(1+r)} + \frac{CdT(1-d)}{100(1+r)^2} \cdot \cdot \cdot + \cdot \frac{CdT(1-d)^{L-2}}{100(1+r)^{L-1}} + \frac{T[C(1-d)^{L-1}-Z]}{100(1+r)^L} \qquad (1.28)$$

where Z is the scrap income.

This can be simplified to give the equivalent expression to equation (1.25) for the factor with which to modify the capital investment, but now with the depreciation being calculated by the reducing balance method:

$$C\left[1 - \frac{Td}{100} \sum_{n=1}^{L-1} \left(\frac{(1-d)^{n-1}}{(1+r)^n} - \frac{T}{100} \frac{[(1-d)^{L-1}-Z]}{(1+r)^L} \right) \right] \qquad (1.29)$$

English law allows the firm to choose between straight-line and Reducing Balance for some investment items. The Reducing Balance obviously has the advantage of giving credit earlier. Other items, for example, buildings, must use the straight-line method. A project involving a mixed system could very easily be evaluated using the two expressions, (1.25) and (1.29) to modify the appropriate capital costs. However, it would be surprising to find that this degree of accuracy is necessary for project selection purposes.

1.7 THE EFFECT OF INFLATION

For the last 40 years inflation has always been present. The rate is usually between 2-10% per year, but it has always been positive. This means that any investment analysis that assumes money will flow over a long period of time must clearly define its treatment of inflation.

It is useful to define two types of money--constant value money and current money. Constant value money means that the money units represent equal purchasing power whenever they occur, that is, inflation-corrected; current money refers to normal money units which decline in purchasing power with time under inflationary conditions.

Evaluations are normally made using constant value money--that is, one assumes that when there is inflation, prices will be increased so that they cover increased costs.

It is possible to relate constant value money to current money quite simply. For a given annual rate of inflation, f, £M at the end of year 1 equals £M/(1+f) at the beginning of year 1; likewise at the end of year n, £M current money is equivalent to $£\frac{M}{(1+f)^n}$ constant value money. Hence the present worth of £M current money at the end of year n is $£\frac{M}{(1+r)^n(1+f)^n}$ constant value money.

Consider a DCF calculation entirely in current money: The interest rate earned by the project is r, given by:

$$\sum_{n=1}^{L} \frac{M_n}{(1+r)^n} = 0 \qquad (1.30)$$

Now consider the same project with monies reduced to constant value money: The DCF rate, r' is given by:

$$\sum_{n=1}^{L} \frac{1}{(1+r')^n} \frac{M_n}{(1+f)^n} = 0 \qquad (1.31)$$

Hence the relation between the rate of return given by constant value money (r') and the current money (r) is given by:

$$(1 + r) = (1 + r')(1 + f)$$

When r' and f are small numbers, this can be further simplified to:

$$(1 + r) = (1 + r' + f)$$

or

$$r = (r' + f) \qquad (1.32)$$

Inflation and interest rates are generally less than 10% per annum, and when this is so, the maximum difference between (1+r')×(1+f) and (1+r'+f) is 1.21-1.20=0.01. This simplification is therefore justified.

1.7a Choice of Discount Rate in Inflationary Conditions

Interest rates are quoted to investors in current money, i.e. always
r is quoted. However, investors generally expect to cover any inflationary
trends and still have an acceptable interest rate. In other words, they
are looking for a steady constant value money interest rate, r', whatever
the inflation rate is. The current interest rate (r) therefore tends to
fluctuate with the inflation rate, but the constant value money rate (r')
is more stable, and is the real cost of capital, as discussed in Section 1.4b.

Hence, if the calculation is made with constant value money the discounting should be done with the normal cost of capital (r'). If the cash
flows throughout are in current money terms, the discount rate r should be
r' + f, that is, the cost of capital plus the inflation rate.

In some projects, contracts without inflation clauses are agreed, fixing
a price for a number of years. This is fixed in current money and so a
calculation in constant value money must have a decreasing income from such
constant price sales in inflationary conditions. Some projects may involve
imported material where different inflation rates are to be expected.
These can be reduced to constant value money with the appropriate difference
in rates. By clearly stating the basis for the calculation, usually constant
value money, it is then obvious how to handle the various cash sources and
how to choose an appropriate discount rate.

Tax allowance on depreciation has a reduced benefit when constant
value money evaluations are carried out because depreciation is generally
defined in current money terms. Hence, in inflationary conditions, one
never gets the full tax credit one would expect in constant value money
terms. This point is referred to further in Chapter 5.

1.8 THE LIFE OF THE PROJECT

Projects are normally assumed to have a definite lifetime. This may
be 5 years for new products with an uncertain future, 15 years for an established heavy chemical, or 40 years for a public investment such as a road
or a power station. This lifetime is necessary for time-average methods of
analysis to establish the yearly capital repayment charge. It is necessary
in the discounting methods to determine over how many years to accumulate
the income. There are two particular problems associated with the lifetime

of a project: The first problem concerns the comparison of investments with
different lives. All the evaluations are capable of showing 2 projects as
being equally attractive, though one might have a 3-year life and the other
a 30-year life. In such a case, the 3-year project is to be preferred
because there is a further 27 years in which to increase the return in other
ventures. It follows therefore, that when a 3-year project appears worse
than a 30-year one, a rational choice cannot be made without enquiring what
the profitability of the remaining 27 years will be with the same resources
that have been freed by the closure of the 3-year project. Hence, a proper
comparison requires both evaluations to be made for the same lifetime, with
the return from the future projects in the 3-year case being included in
the evaluation of the 3-year project itself. This point is clearly seen
when one is comparing alternative methods of capacity expansion. It is
possible to put in a cheap expansion to satisfy the next two years as a
"single project." One could replace the plant by one three times the size
to cater for the next 15 years as an alternative project. A choice between
these alternatives is impossible without first knowing what would be done
after the cheaper extension for the later 13 years. This point is referred
to in progressively more detail throughout the book.

The second problem concerning the life of the project concerns the way
in which the project is ended. It is generally assumed that the plant is
"scrapped" with an appropriate scrap value being credited to the project.
However, in practice, scrap values are usually zero and, furthermore, plants
are generally not "scrapped" at the end of projects--they continue to operate
until replaced by new processes. They are replaced only when the new processes can justify the closure of the old plant. Hence it is reasonable, in
many cases, to assume the plant operates beyond its "project life." It is
more realistic to consider extended project lives than to allow for a ficti-
ous scrap value. This point is referred to in much more detail in Chapter 5.

1.9 COMPARISON OF ALTERNATIVES

Evaluations are made either to determine whether a project is profitable
enough to be carried out by the company, or to choose which should be selected for development out of a series of possibilities.

When choosing from a set of possibilities, it is important to determine
whether they are "mutually exclusive." That is, whether by choosing one

project, the others are automatically rejected. Projects that are mutually
exclusive are alternative ways of producing the same product, or alternative
ways of using some limited resource--be it raw material, capital, site area
in a works, or manpower. They are truly <u>alternatives</u> and this word will be
used to mean "mutually exclusive possibilities."

For projects that are not mutually exclusive, process selection is a
comparatively simple matter. In this case, one is only comparing the project with a profitability standard defined by the company. This could be
that the NPW at a given discount rate should be positive, or the DCF should
be greater than 10%, or a criterion for any other profitability measure set
by the management. If any project is better than the standard, it is accepted; if worse, rejected. A particularly useful point is that the accept/
reject decision produced by the different criteria is consistent. In particular, the same select/reject decision is obtained from any project, whether
one chooses NPW or DCF (1). Although these criteria give different rankings
to a series of projects, they give the same result when every project is
subjected to a 'Go/No-go' test against fixed criteria.

Mutually exclusive alternatives are a much more difficult problem, and
this is the case that the engineer normally has to deal with. Not the least
of the problems is that the different methods of economic evaluation in
this case do choose different alternatives as the most attractive. This
problem can best be overcome if we can re-organise the choice from alternatives to be in the form of 'Go/No-go' economic decisions. This can be done
as follows: If we re-phrase the question to ask if the "extra profit" in a
more expensive alternative justifies the "extra capital" then we have converted the problem into a non-mutually exclusive one and we have overcome
the problem of different evaluation criteria producing different results.
In fact, the whole evaluation procedure is a series of comparisons against
the standard. Firstly, which of the projects are acceptable as an investment evaluated as a whole, and, of those that are, which have additional
costs over the lowest capital case--the "base case"--that are justified
when considered as a difference case. However, these comparisons can only
be done in pairs and so when a number of alternatives are being considered
this method becomes cumbersome because of the large number of combinations
to be evaluated.

This comparison with a "base case" also has the advantage of checking

that the individual parts of a large investment are in themselves worthwhile.
Every engineer knows that one can hide a life-long ambition, such as an
architecturally designed amenity block, inside a profitable project with the
whole still looking attractive. It is of course impossible to evaluate
every capital addition as an alternative but one should carry out a rough
calculation at each stage. Commonly, the more expensive case consists of
the addition of marginally profitable items and so a story as given by
Table 1.2 is obtained.

TABLE 1.2
Comparison of Two Alternatives

Project	Capital	Yearly Cash Flow 5 year Project	NPW at 10%	DCF
A	1,000,000	335,000	270,000	20
B	1,500,000	475,000	300,000	17.7
B-A	500,000	140,000	30,000	12.3

By inspecting the A and B results in this table it is impossible to
resolve whether A or B should be preferred since this is a case where NPW
and DCF give contradictory results. However, the advantage of B over A can
be determined from the B-A evaluation. Would (B-A) be acceptable as a project? This is less attractive than A alone but if it is still above management criteria for acceptance then it should be accepted.

This evaluation by difference is, however, often impossible in engineering work because of the number of alternatives involved. It is also
difficult to implement when one is looking for a maximum profitability where
some variables--such as operating conditions--can be varied in a continuous
rather than a discrete fashion. One cannot choose values for continuous
variables using such a method. However, comparison with a base case is a
good discipline and it is an important principle in decision analysis.
Every project is actually a difference between "reject project" and "accept
project." The base case is "reject project," which often does not result
in zero cash flows.

For engineering design work, selection of alternatives is usually done by comparison with NPW; it is easy to manipulate mathematically and on pure economic grounds can be defended as well as any other single criterion. DCF is sometimes used although it is more tedious to evaluate. For very rough screening calculations the Rate of Return can be used; this is simple enough to be done mentally in the course of a discussion or meeting if necessary.

1.9a A Simple Example

Consider a simple investment which involves the capital of £100,000 and makes a product which realizes an annual sales income of £50,000. The fixed costs are £10,000 and the variable costs amount to £20,000 giving an annual cash flow of £20,000.

The plant is expected to run for 10 years at constant output, after which time the plant will be shut down and have no scrap value. It is necessary to calculate the NPW of such a proposal using a 10% discount rate.

This can be done by hand in the form of a table, calculating for each year the discount factor, cash flow before tax, depreciation allowance for tax, cash flow after tax, and discounted cash flow after tax which can be summarized to give the NPW of the project.

Assuming the capital is spent one year before the project operates, there is straight-line depreciation for 10 years, and the tax rate is 50%; then the whole calculation is given by Table 1.3.

This example is very simple because there is only one capital outflow and there are no cash-outflows other than the one constant yearly amount.

There are numerous techniques one can apply with such a simple case to make the evaluation even quicker. Since the yearly amounts are constant, one can evaluate the cash flow after tax without carrying out the whole calculation for every year. Since

$$\sum_{n=1}^{10} \frac{(C-D)\ (1-t) + D}{(1+r)^n} = [(C-D)\ (1-t) + D\ ] \sum_{n=1}^{10} \frac{1}{(1+r)^n} \qquad (1.33)$$

When C and D are the same for each year, one need only multiply the one evaluation $((C-D)(1-t)+D)$ by the sum of the ten discount factors. To make this even simpler the discount factors are usually available from Discount Tables. Discount Tables are included at the end of this book in Appendix II. Alternatively, it is possible to use the Annuity Tables of Appendix II which give directly

$$\sum_{n=1}^{L} \frac{1}{(1+r)^n}$$

Since NPW's are additive, that is, one can write

$$NPW_{(1+2)} = NPW_{(1)} + NPW_{(2)} \qquad (1.34)$$

it is possible to divide the flows into a constant yearly part, plus occasional separate flows. This means only the occasional separate flow must be calculated completely, with the major flows being treated by the short-cut method above.

TABLE 1.3

Simple Hand Calculation of NPW

(units: £1,000)

Discount rate = 10%
Tax (t)   = 50%
Straight-line depreciation
over 10 years.

Year	(a) Discount Factor	(b) Cash Flow before tax	(c) Depreciation	d=(b-c)t+c Cash flow after tax	e=da Discounted cash flow
-1	1.10	-100		-100	-110.0
0	1.0	0		0	0
+1	0.9091	+ 20	10	15	13.64
2	0.8264	+ 20	10	15	12.40
3	0.7513	+ 20	10	15	11.27
4	0.6830	+ 20	10	15	10.25
5	0.6209	+ 20	10	15	9.31
6	0.5645	+ 20	10	15	8.47
7	0.5132	+ 20	10	15	7.70
8	0.4665	+ 20	10	15	7.00
9	0.4241	+ 20	10	15	6.36
10	0.3855	+ 20	10	15	5.78

NPW of project   -17.82

This additive property is very useful when one is considering minor changes to a basic study. Once a major study has been carried out,

alternatives can be evaluated by determining the NPW of the difference, then
subtracting this from the NPW of base-case.

The calculation of DCF rate of return by hand consists of the same NPW
calculation carried out a number of times using different discount rates
until the rate is found where the NPW becomes zero. In practice this means
three or four attempts to get a positive and negative NPW with discount
rates within say 2 discount %. Then a linear interpolation will give a
reasonably accurate value for the DCF.

1.9b A Complex Example

Consider now an example more like the normal investment proposal. Let
us consider a proposal for an extension to an existing growing product.
Assume there is a 20,000 ton/yr. plant already existing but the sales are
expected to rise to 46,000 tons per year over the next 15 years, according
to Fig.1.2. Owing to competition the price commanded by the product is
expected to fall in constant money terms according to Figure 1.3.

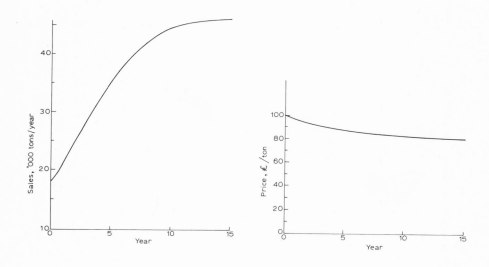

Figure 1.2   Figure 1.3

Sales Forecast - A Complex Example   Yearly Price Forecast-A Complex Example

It is proposed to build a second plant of 26,000 tons/year capacity. The new plant will be more efficient than the old, with a lower variable cost--the appropriate data being given by Table 1.4.

Table 1.4

Data for Complex Example

	Existing Plant	New Plant
Variable cost/ton	40	35
Fixed cost/year	100,000	80,000
Scrap Value	0	400,000
Capacity, tons/year	20,000	26,000
Capital	3,000,000	4,000,000
to be spent in:	already spent	year -1

The new plant will cost £4,000,000 and will be ready to operate at the beginning of year 1. The capital can be considered to be spent as one sum 1 year earlier than start up. Tax is 25% with 10 years straight-line depreciation. The project expects injections of working capital in years 5 and 10. The NPW at 15% discount rate is required to determine whether the investment in a second plant is justified.

The calculation is shown on Table 1.5. A point to note is that we are investigating a difference in operation between the existing and the existing plus new plants, and because the new plant is more efficient than the old, the new plant should be operated fully before the old is used.

Referring to Table 1.5 from left to right:

The Discount Factor	=	$(1/(1+r)^n)$ - can be obtained from discount tables in Appendix II.
The Capital		is £4 million spent at the end of year -1 and the scrap value of the new plant is credited in year 15.
Working Capital		is £100,000 in year 5 and again in year 10, which is all recovered in year 15.
Sales		taken from the forecast graph, Fig. 1.2.
Price		taken from the forecast graph, Fig. 1.3.
Sales Income	=	Sales x Price for each year.

TABLE 1.5

A Complex Example
(£'000)

Year	Discount Factor at 15%	Capital	Working Capital	Sales	Price £ per ton	Sales Income	Variable Cost (existing + new-existing alone)	Fixed Cost	Annual Cash Flow before Tax	Annual Cash Flow after Tax	Discounted Annual Cash Flow after Tax
-1	1.15	-4000		0	-	0	0	0	-	0	-4098.1
0	1.00			0	-	0	0	0	0	0	0
1	0.8696			19,500	99	0	+97.5	-80	17.5	+13.1	11.4
2	0.7561			22,600	96	249.6	+ 9.0	-80	178.6	134.0	101.3
3	0.6575			26,600	93	613.8	-134.0	-80	399.8	299.8	197.1
4	0.5718			30,400	91	946.4	-286.0	-80	580.4	435.3	248.9
5	0.4972		-100	33,800	90	1242.0	-422.0	-80	740.0	555.0	226.2
6	0.4323			36,800	88	1478.4	-542.0	-80	856.4	642.3	227.8
7	0.3759			39,200	87	1670.6	-638.0	-80	952.4	714.3	268.5
8	0.3269			41,200	86	1823.2	-718.0	-80	1025.2	768.9	251.4
9	0.2842		-100	42,800	86	1960.8	-782.0	-80	1098.8	824.1	234.2
10	0.2472			44,000	86	2064.0	-830.0	-80	1154.0	865.5	189.3
11	0.2149			45,000	86	2150.0	-870.0	-80	1200.0	900.0	193.4
12	0.1869			45,400	85	2159.0	-886.0	-80	1193.0	894.7	167.2
13	0.1625			45,700	85	2184.5	-898.0	-80	1206.5	904.9	147.0
14	0.1413			45,800	85	2193.0	-902.0	-80	1211.0	908.2	128.3
15	0.1229	+ 400	+200	46,000	85	2210.0	-910.0	-80	1620.0	1215.0	173.9

NPW after tax = -1282

NOTES

1) Capital tax factor (F) = $\dfrac{1-t}{100Mr(1+r)} \left(1 - \dfrac{1}{(1+r)^m}\right)$ = 0.8909

2) No tax on working capital flows

3) 25 % tax rate

4) 10 years straight-line depreciation

Variable Cost

This must be estimated for each plant separately. To produce a reasonable estimate of expected cost, the preferred working of the efficient
plant must be allowed for. Hence it is necessary firstly to work the most
efficient plant, either to its maximum or to the sales demand, whichever is
limiting, and secondly to work the next efficient plant to its maximum or
until the sales are again limiting. However, since we are in fact evaluating
a difference case to an existing plus new, we must subtract the variable
costs that would have been incurred with the existing plant alone. From
this analysis the total differential variable cost can be entered in
. Table 1.5.

Fixed Cost

Again the difference in fixed cost is required, which is simply the
fixed cost of the new plant.

Annual Cash Flow Before Tax (S_n).

This is obtained by adding all the cash flows each year from the
previous columns, excluding capital and working capital.

Annual Cash Flow After Tax

It is convenient to credit the capital investment with the total discounted worth of the tax allowance, and then tax the remaining cash flows
without reference to depreciation allowance. This can be done by calculating the capital tax factor given by equation (1.27). It could alternatively be obtained from Tax Tables (12). In our case the factor is 0.8909.
The remaining sums, excluding working capital, are subject to full tax.
Hence the annual discounted cash flow after tax in year n is:

$$C_n (1 - t)/(1 + r)^n - 0.8909 \text{ (Capital)}_n - \text{(Working Capital)}_n$$

This is given in the last column. Note that the income from scrap is a
taxable income but the income from working capital is not, because working
capital is merely a re-allocation of a company's resources within the firm.
Summation of this column gives the overall project NPW of the addition of
the second plant. In this example the value is -£1,282,000, that is, the

project should not be accepted if the company requires the 15% return on its capital. The whole calculation is rather tedious. If the DCF rate is taken as the criterion, this calculation would have to be repeated a number of times, which adds to the tedium. As will be described in later chapters, project evaluations are usually repeated for many alternatives, and so it is sensible, and usual, for project evaluations to be done by computer.

Besides the mere calculation effort it will require, a second reason for using a computer is that when a number of alternatives are being compared, it is essential that identical assumptions are made in all cases. It is not so important that the assumptions be exactly right since all evaluations are only an estimate of future performance: it does not require the same accuracy as, for instance, the commercial accounting activities of a firm. Hence there is some latitude in the exactness of the assumptions made, but none in the reproducibility of a calculation. Such a programme is presented later in this book.

1.10 UNCERTAINTY AND RISK

If all the important data for investment evaluation were accurately known we should now be at the end of the story. The investment decision could be taken by simply looking at the appropriate evaluation criteria. However, the data in the calculation are no more than estimates of what might happen in the future. No data is known with certainty--except possibly, fixed price contracts. Sales, costs, prices, and inflation can all vary from the estimated values. A good decision therefore remains acceptable, even when these estimates are found to be in error. A good decision also allows latitude for later corrective decisions to reduce losses should initial estimates prove to be very wrong.

This is the main theme of this book, the introduction is now over and the following chapters are concerned with the source of the uncertainties and techniques for allowing for these uncertainties in investment appraisal.

1.11 BIBLIOGRAPHY FOR CHAPTER I

1. Allen, D.H., Economic Evaluation of Projects, Inst. of Chem. Engrs., London, 1972.

2. Betts, G.G., Investment Appraisal, Chem. Engr., 78,(Feb. 1972).

3. Cadman, M.H., Business Economics, Macmillan, London, 1968.

4. Chambers, J.C., Mullick, S.K.and Smith, D.D., How to Choose the Right Forecasting Technique, Harvard Business Review, 49, 45,(July-August, 1971).

5. De Garmo, E.P., Engineering Economy (4th Ed.), Macmillan Co., London, 1967.

6. Frost, M.J., Value for Money--The Techniques of Cost Benefit Analysis, Gower Press, London, 1971.

7. Grant, E.I.W.G., Principles of Engineering Economy (5th Ed.), Ronald Press, New York, 1970.

8. Jeynes, P.H., Profitability and Economic Choice, Iowa State University Press, Ames, Iowa, 1968.

9. Lawson, G.H., Windle, D.W., Capital Budgeting and the Use of DCF Criteria in the Corporation Tax Regime, Oliver and Boyd, London, 1967.

10. Malloy, J.B., Risk Analysis of Chemical Plants, Chem. Eng. Prog., 10, 67, (1971).

11. Merrett, A.J. and Sykes A., The Finance and Analysis of Capital Projects, Longmans, London, 1963.

12. Merrett, A.J. and Sykes, A., Capital Budgeting and Company Finance, Longmans, London, 1966.

13. NEDC, Investment Appraisal, H.M. Stationary Office, London, 1971.

14. Schumacher, E.F., Small is Beautiful, Blond and Briggs Ltd., London, 1973.

15. Wilkinson, J.K., Watson, F.A. and Holland, F.A., Introduction to Process Economics, Wiley, New York, 1974.

CHAPTER II

UNCERTAINTY AND FORECASTING

Was?
Liebi Frau Bas,
wenns rägnet, so wird s nass,
wenns schneit, so wird s wyss,
wenns frürt, so git s Ys.

2.1 INTRODUCTION

The factor in common with all economic evaluations is the concern with predicting future performance. It is necessary to use predictions of future sales, future prices and future costs as data. Even in those cases where long-term contracts are signed fixing a price, this price is usually fixed in current money terms, whereas analyses are best done in terms of constant value money. Hence, one has still to predict inflation rates before the "real" price is known.

The future is never certain and so forecasts cannot be expected to be accurate. It therefore follows that any economic evaluation will never be exact. It will have an accuracy which can be related to the accuracy of the individual forecasts used as data for the evaluation. The accuracy of an evaluation is of considerable importance in making a decision; if there is a reasonable chance of an extremely high loss, due to the possibility of the sales not developing, the decision maker should know this before he makes his decision.

Forecasting, economic evaluation and decision analysis are therefore closely related subjects, yet it is surprising how rarely they are treated together. The salesman, statistician, economist and O.R. man need to join forces to help the engineer, who has to base all his work on decisions concerning capacity and process selection, often made after a very superficial analysis.

2.2 THE SOURCES OF UNCERTAINTY

All the data used in economic evaluations are subject to risk or uncertainty. To convince the reader of this, each data item required by the
evaluation is discussed seperately in the following pages. Two distinct
types of uncertainty are referred to in decision theory literature: risk is
used to describe situations for which there is probability data. Past data
can be used to predict the chance of a road accident, or age at death or
the tossing of a coin. Uncertainty refers to situations where there is no
suitable past data on which to base an estimate of the chance of an occurrence. The sales of a new product for instance, would be described as
being "uncertain."

An economic analysis can be performed once one has the following data
available:

     (a)   Sales volume by year

     (b)   Sales price by year

     (c)   Plant capacity

     (d)   Investment cost

     (e)   Variable cost per unit produced

     (f)   Annual fixed cost

     (g)   Plant scrap value

2.2a Sales Volume

This is undoubtedly the most important data in the evaluation and also
the most difficult to forecast. The sales forecast determines the size of
the plant that is built. If the sales forecast is too high, the resulting
plant will never be operated at full capacity, thus ensuring that the project would make no profit. If the forecast is too low, a small plant will
be built which will result in higher cost per unit product which may result
in the product being uncompetitively priced. The volume sold will depend
both on the amount used by the community, and the proportion of the market
captured by the company. Both of these are difficult to forecast.

Some products are elastic with respect to price. This means if the
price is reduced, much more can be sold--which further adds to the difficulty of forecasting. Both consumer goods and replacement products (plastic
to replace metals, for instance) are elastic.

Some products are inelastic--chemical intermediates for instance. No-
one is likely to use much more sulphuric acid just because its price is
reduced. These items are easier to forecast when the company is in a monopolistic position. However, in a competitive situation even sulphuric acid
is an elastic commodity for each individual firm, since a change in price
of only a few per cent with respect to the competitor's price can produce
wide variations in sales.

Some products are basically stable with steady growth rates linked to
the country's national Gross National Product (GNP). Forecasts for such
products are no more difficult than forecasting the GNP! (11) Some products
are completely new and launched on the market in an attempt to displace an
established product. Their success depends upon market acceptance, the
discovery of secondary effects and the retaliation for the producers of the
established product. New products are most difficult to forecast.

In an attempt to generalize, one can say that all products' sales
follow a curve of the form shown in Figure 2.1, known as the Gomperz Curve,
with the five sections to the life of the project. It is sometimes referred

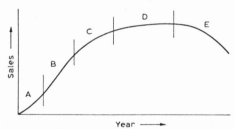

Figure 2.1 Generalised Sales Curve for Any Product

to as the S growth curve with a final decline. Time span A is the slow
acceptance of the new product, B represents a comparatively rapid growth
period, moving into C when the market becomes saturated, and then the
steady period D where sales follow the GNP. Finally, there is the decay
period where the product is eventually replaced by something new.

This curve by itself is of little value because no indication of turn-
ing points are given, but at least it gives the basis from which one can
begin to forecast. The subject of forecasting is discussed in much more
detail in Section 2.3.

2.2b <u>Sales Price</u>

    After sales volume, the next most important data to be forecast is the
selling price. However, there is an element of stability in the selling
price which makes forecasting less important than sales volume forecasting.
The price of a new product is normally determined by "what the market can
stand," that is, that the new product is attractively priced compared with
the existing method. For example, a new insecticide must be so priced that
it is profitable for the farmer to buy it rather than accept the damage done
by the particular pest. However, with time, particularly as patent protection runs out, the pricing structure is much more determined by "manufacturing cost plus an acceptable return on capital," as competitive forces remove opportunities for higher-than-average profits.

    The manufacturing costs reduce slightly with time as process improve-
ments are discovered; this drift is asymtotic, as shown by Figure 2.2 (1,16).

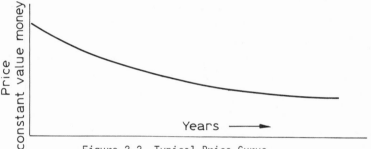

Figure 2.2 Typical Price Curve

    Hence, for an established product, if all manufacturers are using the
same raw materials and all are subject to the same cost fluctuations, prices
will be calculated to produce a reasonable return on investment and one
company's price fluctuations will be paralleled by his competitors--there
will be no change in the competitive position. This is exemplified by the
way in which prices of petroleum products of various companies meticulously
follow each other.

    With a new product the forecast price must be a good estimate of the
price that will prove to be attractive to the customers. Too high an estimate would mean that the lower price charged in reality to obtain acceptable
sales volume would result in much lower profits than forecast. In the
extreme, an incorrect price forecast may result in a project being accepted
where in fact the economics were such that it should have been rejected.

With existing products the important question is the comparison with competitors' prices and costs. If it is clear that the company is equally or favourably situated with respect to transport and raw material costs, compared with his competitors, then present profit levels are likely to be maintained, whatever price fluctuations occur in the future. This balance could be adversely upset by a competitor building a larger plant than its rival company because they would have lower unit costs, or by their discovering some process improvement that reduces their costs. The probability of these occurrences can be minimized by building big plants and by maintaining some form of process activity in the hope of making process improvements parallel to, or ahead of, the competitors.

The price uncertainties in competitive situations are therefore concerned with competitors' costs changing in relation to each other. These changes could be due to new sources of raw material, different labour costs, different plant capacities or process improvements. These changes have to be forecast before one knows whether further investment in that particular product would give an acceptable return.

Such forecasts can be made in constant value money terms as described in Section 1.7 of Chapter 1 and only differential inflationary trends-- between different countries for instance--need be forecast. Long-term price forecasting in current money terms is virtually impossible because of the unpredictability of inflation.

2.2c Plant Capacity

Compared with uncertainties associated with sales and price forecasting, the uncertainties associated with plant capacity are relatively small. These uncertainties arise due to incomplete data at the engineering design stage. It therefore follows that the more that is known about the process, the less will be this uncertainty. Well known processes, such as petroleum refining can define their capacities within $\pm 5\%$, whereas a new process would probably be engineered once sufficient data was available to guarantee a capacity within $\pm 20\%$. This uncertainty is very different from that relating to sales and price; it is much more controllable and further process development can effectively eliminate it. Further research on sales and price may reduce the uncertainty in the commercial data, but it could never be "effectively eliminated."

A comparatively small error in plant capacity can however have a large effect on the profitability of the project. Once can often say a plant has to work at 70% capacity to break even on costs and only when operating at over 70% capacity is there anything left for profit. In such a case, if the plant were 30% under sized, there would be no profit in the project.

Because of the magnified importance of the top range of plant capacity it is common for engineers to oversize by adding safety factors during the design. This means plants often have a high probability of meeting the stated capacity, and have an expected capacity greater than their quoted capacity (16).

Hence, refinery items are generally 5% oversized, and new processes often have a 20% safety factor to increase the probability of making the rated output.

2.2d Investment Cost

This data is very similar to plant capacity, in that it contains uncertainty that can be very nearly eliminated by increasing the engineering effort put into the design. It is normal to have a number of levels of design detail from which a range of estimates with different accuracies can be produced. These range from very quick estimates for research guidance (analogy costing) to the different levels of factorial costing (Lang methods) and to detailed costing from a final plant design and a complete inventory list (detailed costing) (4). The roughest costing may have an error of 50 to 100%, the Lang methods probably between 10 and 30%, and the detailed costing estimates can achieve ±5%. However, the design work necessary to achieve the improved accuracy is very considerable, and such accuracy could not be considered for estimates required at the research stage. Usually, though not always, the final investment decision is made using a ±5% accuracy estimate, but research planning decisions, or investment decisions involving comparisons with projects in the research stage may have to work with estimates no better than ±50%.

Capital cost estimates are normally made about 2 years before the money is spent. Hence they are subject to 2 years inflation, which is an uncertainty that engineering efforts cannot eliminate. However, large companies develop their own estimate of the rate of inflation, aimed at the particular industry involved. This is done by discussions directly with

equipment suppliers and contractors on how they see their near future. Are
their costs likely to remain stable? Do they see pay rises in the near
future? Is there a dropping off in orders which will depress prices in the
construction industry?

Such discussions result in large inflation corrections over a 2-year
period, but suprisingly accurate ones.

2.2e Variable Costs/Unit Produced

This is an important item in any evaluation, and one subject to uncertainty from a number of sources. The variable cost is normally the sum
of the raw material and service costs, and is therefore calculated as the
cost per unit of the raw material or service, multiplied by the usage per
ton of the product. The cost of the raw material or service is subject
to inflationary trends and also to individual cost trends associated with
the individual raw material. For instance, a world shortage could lead to
multifold price increases of certain materials, but new processes or increased magnitudes of production for a raw material could produce equally
spectacular decreases in price.

The usage/ton of the raw material or services is a measure of the
efficiency of the process. With time this efficiency improves as unpredict-
able technical improvements are made to produce reductions in cost.

This discussion is the same as the one in Section 2.2b--the variation
in sales price--one man's sales price is another man's cost, in intermediates manufacture. There are again two cases: existing and new products.
For both types of product the trend will be to pass raw material cost
increases on to the selling price so as to maintain the same profit margin.
Hence, for existing products in a competitive situation when the competitors
all have the same raw material costs, raw material cost variations are
comparatively unimportant. But for new products a cost increase may result
in the product being no longer attractive as compared with the existing
products. This cost increase may make the product uneconomic.

Clearly, the forecasting of sales prices and raw material costs are
so closely linked that they should be made at the same time on exactly the
same basis.

For a new product, or a new process, one must also predict the efficiency that will be achieved on the plant. Efficiencies will be known for

the laboratory scale and possibly for the pilot plant scale, but the same
efficiencies may not be achieveable at full scale because of the effect of
the larger scale. This uncertainty can be reduced at the cost of further
research and development work.

2.2f Fixed Costs

Fixed costs--supervisory wages, maintenance, overheads--are often only
minor cash flows in the investment analysis and so the uncertainty in these
figures is comparatively unimportant.

The uncertainty is proportional to inflation, but using a constant
value money basis the major uncertainty that arises is the accuracy of prediction of the manpower requirement. For example, a plant may be thought to
be operable with 10 men, but in practice 12 men may be found to be required.

In projects with significant labour costs, most of this cost will be
a variable cost since the manpower requirement will change with the output.
It is therefore difficult to find projects where the uncertainty in fixed
costs is important because the fixed costs themselves are of second order.

2.2g Plant Scrap Value

Of all the data discussed, scrap value has probably the greatest uncertainty but the least importance. Normally, one is not even certain when
a plant will be scrapped, and so the prediction of its new use and the
equivalent value of this use is very uncertain.

The most common fate of plants is that they are removed by a scrap
metal dealer for nothing. Hence the scrap value is zero, or the value of
the cleared site if the site value is within the scope of the evaluation.

When a plant can be put to an alternative use, the scrap value is the
cost of providing a new plant for that alternative use. Reasonably certain
values may be estimated when this occurs in the near future--say within 5
years,--but further into the future, the definition of an alternative use
is usually very difficult.

Hence the predicted income that will arise from the scrapping of the
plant is very uncertain. However, because it is usually done at the end of
the project it is so heavily discounted as to be insignificant in the
analysis. Should part of a project be scrapped after only a few years, the

uncertainty associated with the scrap value will be correspondingly reduced.

As a conclusion to section 2.2, Table 2.1 is presented which summarizes typical accuracies associated with the data for investment analysis. The Table is based on data from Baumann (4). It is interesting to note that in this table very few items have a symmetrical distribution; most have a greater pessimistic than optimistic range.

Table 2.1

Range of Uncertainty in Economic Evaluation Data

		Probable Variation from Forecasts over 10-year Plant Life (%)
1.	Cost of capital investment	- 10 to + 25
2.	Sales volume	- 50 to +150
3.	Price of product	- 50 to + 20
4.	Plant maintenance costs	- 10 to +100
5.	Raw material availability and price	- 25 to + 50
6.	Inflation rates	- 10 to +100
7.	Interest rates	- 50 to + 50
8.	Working capital	- 20 to + 50
9.	Scrap value	-100 to + 10

From this Section it can be seen that there is a need to pay particular attention to the prediction of:

(a) Sales volume per year from good demand forecasting

(b) Sales price and raw material costs from good price forecasting

(c) Plant capacity and efficiency from good process development.

Obtaining adequate technical data from process development is the essence of engineering and since this book is written for engineers it will be assumed that the reader must already be an expert in this field. Forecasting however, is generally not understood by engineers, although almost universally decried by them. It is therefore worth discussing the techniques and problems of forecasting in some detail.

2.3 FORECASTING METHODS

Sales forecasting--sometimes referred to as Market Research--is an
extremely important activity in any business. It is generally associated
with the commercial side of the business, and so engineers rarely involve
themselves with the forecasters. The engineers' usual impression is that
forecasting is nothing but guesswork and the forecasters are continually
changing their minds.

Forecasting has to be subjective. Subjective judgements have to be
made by people who know the particular market well. They can read information from a multitude of sources, obtain a feeling for the general state of
the economy, the general competitive situation and how it is likely to
change, the relative importance of the end uses of the product, possible
new products that will enter the market, possible new applications that
will expand the existing market, and so on. Then, together with many past
years performance and forecasting history, they put forward forecasts as
their considered opinions.

This approach is diametrically opposite to the numerical approach of
the engineer and because the whole of the engineer's work has to be based
on forecasted estimates, the antagonism is easy to understand.

There are numerical methods that can be used in forecasting but, as
will be shown, they are only aids and do not replace the subjective judgements that must be made at some point in the analysis.

Forecasts are used for many purposes within a business besides the
comparatively minor one of determining the design capacity for new investments. Short-term forecasts (14) (less than 1 to 3 years), are probably
the most used. Any operating business should know its next quarter's
sales level so that it can control its inventory, production plans and raw
material supplies. The sales forecasts for the next year are necessary to
secure raw materials and to predict the immediate future economic health of
the business in order to spot difficulties ahead of time. In particular,
the management must have an accurate financial picture of the firm to ensure that it has sufficient cash to prevent bankruptcy, to know what level
of investment can be sustained from its own profits and to know whether
and when any future loan must be floated.

Medium-term forecasts (11)(3-15 years), are necessary both for planning
future production capacity, and for planning further financing of the firm.

Long-term forecasts (17)(15-50 years), generally known as Technological Forecasting, are useful for research departments selecting new long-term research topics, and companies wishing to broaden their product base to ensure they remain in existence after their major 'raison d'etre' has passed--the oil companies for example.

Hence there are three types of forecast: short, middle and long term, and though there is a multitude of forecasting techniques in the literature --one can number about 20 from 2 references alone (5, 11)--any one is gener- ally applicable to only one of the three types.

The short-term methods, such as asking regional sales organisations to forecast, are inappropriate for the medium term, and some long-term methods are too inexact for medium term work. In the following pages only the fore- casting methods that are suitable for medium term planning and production capacity planning will be described.

The methods can be divided into groups. Firstly there are the purely Subjective Methods using available data from within the firm. Secondly, there are User Expectation Methods which make an effort to obtain new data from outside the firm. Thirdly there are Statistical Methods for extrapol- ating past data, and finally there are the Modelling Methods where a math- ematical model is constructed to simulate the market.

2.3a Subjective Methods

The Jury of Executive Opinion (1) methods consists of collecting to- gether those in a firm that have considerable experience in the business, usually senior members of the firm, who then give their expectation of the future sales and price levels. They are people who are very well informed about the business, and before the meeting they should have been circulated with relevant and up-to-date information regarding the product and its market.

A development of this is the Delphi Method (17) which is aimed at formalizing the discussion between experts and at removing the domineering effect of the more senior members of the panel. The method consists of the forecasts being produced anonymously by the team members in a first round. In a second round, each member re-forecasts, having seen all the first round forecasts. If he forecasts outside the middle two quartiles he must give his reason. The third round of estimates allows members to be swayed

by these reasons, and this time counter arguments must be given by the out-
liners. These counter arguments may be affected by the team members' fourth
round of estimates. These fourth round estimates are the ones that are
averaged and taken as the final forecast.

2.3b <u>User Expectation Methods</u>

These methods consist of asking the user of the product how much they
will expect to be using in the future.

The simplest case is when a product is an intermediate which is sold
to a comparatively small number of other firms as their raw material. Each
of these firms can be then asked to forecast its future requirements and
these are then added together to give the forecast.

This method is popular in the chemical industry where so many products
fit exactly the description of our simplest case here.

If there are a considerable number of customers, one can take a statistically significant random sample instead of every customer. If the product
is a consumer product, one must ask the distributors and users for their
opinions.

This method has its drawbacks: the informers are outside the company
and it is not important to them whether they are right or not. Personal
interviews are therefore better than postal surveys so that one can estimate how seriously the informants are taking the matter. Personal interviews however make the method expensive and slow.

Finally one must remember the number is still only a forecast: the
product user has simply passed on the forecasts of his product to your firm.
This method is only passing the problem on down the line. Of course, the
consumer should be in a better position to forecast the future sales of his
product since he is in that business, but if he himself has a less developed forecasting department than your firm, it may be that you could estimate
his usage better than he could himself.

Under this same heading of "User Expectation Method" could be classed
the method of <u>Pilot Trials</u> for new consumer products (19, 22). In this
method a small area, such as a town, is distrubuted with the product and
backed by appropriate advertising to develop a typical consumer situation.
The sales level can then be used to determine what a nation-wide demand
would be. The method is supported by market research and a detailed
statistical analysis.

2.3c Time-Series Methods

Time-series methods are methods of extrapolating existing data--an existing time-series--to produce forecasts of future data. Given a steadily increasing data set representing sales against time (Figure 2.3a) there is no problem in its extrapolation. It could be done with a ruler and pencil. However, when random error is present (Figure 2.3b) the "best" line is difficult to draw and the differences between the alternative fitted curves are magnified on extrapolation. At this stage it is useful to have a simple

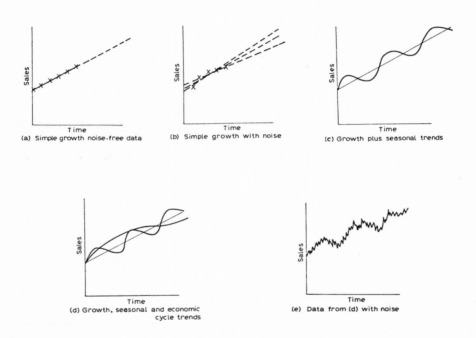

(a) Simple growth noise-free data   (b) Simple growth with noise   (c) Growth plus seasonal trends

(d) Growth, seasonal and economic cycle trends   (e) Data from (d) with noise

Figure 2.3 (a-e) Types of Sales vs. Time Data for Extrapolation

single linear equation to find the best least squares fit to the existing points in place of the ruler and pencil. Since such random noise is inevitably present on commercial data, regression is a useful tool when forecasts are being produced from past data.

In practice, many products undergo seasonable variation; that is, superimposed on the growth curve is a cyclic effect with an amplitude of

one year (See Fig. 2.3c). Furthermore, business also goes in "Stop-Go"
cycles with an amplitude of a few years in phase with world economics.
Hence, on top of the growth curve is the longer cyclic variation. The contribution of the growth and the two cycles produce a curve as shown in
Fig. 2.3d.

A forecaster is trying to define these individual curves in order to
obtain a reliable extrapolation, but all he is presented with are the quart-
erly figures which include the random noise of Fig. 2.3b--the data presented
to him is shown in Figure 2.3e.

The statistical methods of Time-Series Analysis (6,7) can sort out such
data into cyclic variations and simpler trends. If a reasonable fit can be
obtained, it is possible to predict future data, assuming that no new
factors interfere. The method is particularly useful in predicting near-
future data: for instance, next quarter's results for stock and production
requirements, taking account of the expected seasonal and economic cycle.
Extrapolations of more than two years are not recommended (13). Medium
term forecasts are therefore too far ahead for such statistical methods to
be used with much confidence, but they are valuable as secondary, confirma-
tory evidence.

For the prediction of medium term forecasts from past data, a safer
extrapolation is obtained if a characteristic sales curve, rather than an
undefined curve is fitted. Such a curve is shown in Figure 2.1. This
still employs a statistical regression with a single variable, but the form
of the equation is more accurately defined.

2.3d Correlation Methods

The extrapolation and curve-fitting techniques so far mentioned assume
that the future individual contributions to the sales will remain identical
to those of the past. This in practice is unlikely, and so the methods
have distinct limitations. This disadvantage is overcome using Multiple
Linear Regression Techniques or Correlation Techniques because they attempt
to relate a number of important factors to the past sales. Once a successful correlation has been found, predictions can be made, assuming only that
the effect of the individual factors remains the same as the past. This is
more satisfactory than assuming that all conditions, both the effect of and
the proportion of each factor, will also remain constant as must be assumed

for the simpler curve fitting techniques to be applied.

Take as an example the prediction of a future total sales of television sets in a country: The number bought will depend upon

(a) the number of people still without television sets (buying for the first time)

(b) the number of people with old sets (replacement buying)

(c) the average of disposable income after paying for food and housing (the chance of people being able to afford to buy).

A straightforward linear regression can fit the equation.

$$\text{Sales} = B_0 + B_1 a + B_2 b + B_3 c \qquad\qquad (2.1)$$

The best value of B_0 to B_3 found from fitting past data can be used to predict future sales.

Notice that now we need future values of a, b and c before we can predict future sales. This is sometimes an advantage and sometime a disadvantage. If good forecasts are already available for a, b and c then this prediction is being related in a logical fashion to reliable or basic forecasts. If there are no good forecasts of a, b and c then the prediction cannot be good.

These correlation methods can be taken one stage further so that they also forecast the correlating variables a, b and c. For example, the number of first-time buyers is dependent upon the number of people in the marrying age range, d, the number of people already married, e, and the number of single people, f, a fraction of whom will buy their own television set that year. Hence a linear regression can be used to predict c:

$$c = B_5 + B_6 d + B_7 e + B_8 f \qquad\qquad (2.2)$$

In this way the method is using still more fundamental data, population statistics, to predict a value for c where forecasts do not exist.

The correlation method can also be refined to consider trends in "second-order effects." For instance, in our example, b--the number of people already with sets--gave a lead to the number of replacements sets that would be bought. This correlation could then be used for prediction, assuming the effect of the individual factors do not change in the future. But what if sets became more reliable--or if there were a growing tendency for homes to own more than one television set? Obviously the forecast becomes more accurate if we can include these trends. These effects can be included in correlation analysis since we can write

$$b = B_9 + B_{10}g + B_{11}h \qquad (2.3)$$

where g is the ratio (number of televisions)/(number of homes with one or more television set)
and h is the mean life of a set when scrapped.

Both g and h are likely to change with time, that is

$$g = B_{12} + B_{13}t \qquad (2.4)$$

$$\text{and} \quad h = B_{14} + B_{15}t \qquad (2.5)$$

where t is the year.

The forecast now produced has a greater likelihood of being valid if future trends in reliability and growth of multi-TV homes remain similar to those in the past.

Notice that these correlation methods, unlike the Time-Series methods, do not actually produce a forecast, but only enable a forecast to be obtained from other forecasts. Their aim is to convert the problem of one difficult forecast into one of a number of easier forecasts. Easier, either because more information is available, or because the data is of much more general interest and so has already been reliably forecast by other parties.

2.3e Modelling Methods

Mathematical Modelling methods are attempts to describe mathematically the complete system leading to the demand for the product. These equations then relate the forecast to available data. They can be considered as logical progressions from the Correlation Methods described in the previous section. The Correlation Methods try to relate the forecasts to more fundamental data which are important components of the system. The Modelling Method takes this one step further by allowing the effect to be described by equations that have some theoretical basis, and also by widening the system so that all the input data required by the model is readily available. This is in contrast to the Correlation Method which uses empirical equations and still requires forecast information as data input.

Let us continue our Television set forecasting problem as an example. We have already said that the number sold in a year is the sum of the number sold to homes already with television sets plus those sold to homes without sets and that both these can be sub-divided into

      --sets bought as replacement

      --sets bought as second (or more) sets

and

      --sets bought by new couples

      --sets bought by single persons

and to which we could add

      --sets bought by homes that have previously resisted.

     A model of such a system is given by Figure 2.4. The number of people
with and without sets can be calculated from population data, and initial-
ization data which informs the model about the number of homes with a set
and the age distribution of the sets at the present time.

     Such a model will advance year by year predicting the number of marriag-
es and the number of people coming of age. The existing sets will become
one year older and a certain number will be scrapped. Hence the number of
new sets bought can be predicted and this used to update the number of homes
with and without sets and to modify the age distribution of the sets.

     In fact every year produces a complete updating of the old data so that

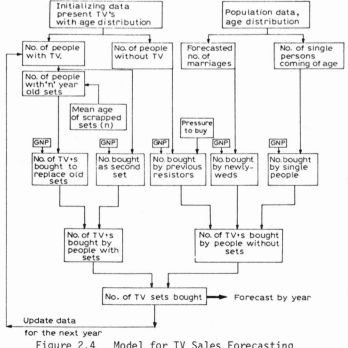

Figure 2.4   Model for TV Sales Forecasting

the model is "simulating" the passage of time, and each year re-calculating
the number of sets purchased. Extra effects can be included. For instance,
a Gross National Product effect (GNP) may be introduced to modify the pur-
chases and this could have a cyclic nature. The number of possible refinements is unlimited--all at the expense of a more complex model and therefore
one more difficult to handle. Of course, the inclusion of more refinements
does not guarantee a better forecast, but it does enable a good qualitative
picture to be obtained.

These models must go through a fitting stage where the model is fitted
to historical data by searching for the best value for the many constants
in the equation. This fitting stage is also the time when the model shows
whether it gives logical results or not (18).

The overriding advantage of such a technique is that the results will
be sensible. A long-term forecast may show each home has on average 1.5
sets. A long-term Time-Series extrapolation might give a figure that is
equivalent to every man, woman and child owing 3 sets, and there would be
no way of knowing this from the Time-Series study alone.

Modelling methods involve much work in their preparation. Each project
requires an individual study, a computer programme and a fitting stage,
which takes months to prepare. It does however present more advantages
than any other method. These can be listed as:

(a) logical switches in the programme enable subjective judgement
     to be investigated, "If this were to happen . . ."

(b) Time-Series extrapolations can be included--economic or season
     trends can be written into the programme

(c) The forecast can be compounded from whatever effects the fore-
     caster feels are important in the way he thinks they contribute
     to the forecast.

(d) the model is based on (fitted to) past data and can be tested
     by predicting other past data before being used for predicting
     future data.

(e) the method can also be used where there is no past data.

Probably the most well-known is Forrester's attempt to forecast world
population and living standards by means of a crude forecasting model (12).

There are continual attempts to produce adequate economic models for
individual countries for governmental use in forecasting the effect of
various government policies on total economics (2,10,18). The success of
such models is debateable because extremely complex systems are being
studied, but, on the less ambitious front, modelling has been successfully

used for forecasting individual product demands by industry. There are
reports of its use in predicting the number of cattle slaughtered (11), in
forecasting the US fertilizer industry (13) and forecasting potato sales
(22).

Modelling techniques are by far the most powerful and promising of the
forecasting methods, but their application is restricted by the time and
expertise required to produce them. However, it can be expected that these
methods will become progressively more common as more generalized techniques
are developed and as the number of models written increases.

We have discussed forecasting methods that are particularly suitable
for middle-term forecasting needed for production capacity planning work.
It is evident that no one method can be expected to give a reliable result.
The golden rule in forecasting is to obtain estimates by using more than
one method. Once different methods are in agreement there is confidence in
the forecast; when there is not, one has to investigate why discrepancies
have occurred.

Particularly important is the incorporation of opinions within the
forecasts. All the mathematical methods assume the future will be the
same as the past. If one thinks otherwise, be it due to fashion, technological change, energy shortage or anti-pollution measures, these can only
be introduced subjectively. The last of these methods, the Modelling
Method enables such subjective judgements to be quantified, but they, none-
theless stem from the man who knows business and not the mathematician.

All the forecasting techniques take time and money to carry out.
Simple techniques may take only one man-week but by the time customer
interviews or market surveys are contemplated the time will be running into
man-months. The mathematical Modelling methods can take man-years of effort
in data collection and computer programming. Obviously, if the same sized
plant is going to be built whatever the forecast--which may be the case if
it is usual to build standard units--money spent on improving the forecast
is wasted.

It is only reasonable to give a forecaster an estimate of the value of
a more accurate forecast before asking him to improve his estimates.
Decision theory describes methods of determining the Value of Information,
and this is described in Chapter 3. Methods described in Chapter 5 show
how the Value of better Information can be obtained from investment
appraisal studies.

2.4 THE DISTRIBUTION

This chapter so far has emphasized that none of the data we use in
investment analysis is known with certainty. It is all subject to inaccuracies of various degrees. We do, however, have some idea of the magnitude
of the inaccuracies since we can imagine the worst and decide how much this
differs from our estimates. Furthermore, we subjectively have information
on the likelihood of the worst or the best happening. Sometimes we can say
they are equally likely, or more often, much less likely than an estimated
middle way. This is all very useful information. The decision concerning
a project would be different if the likelihood of the worst occurring was
1 in 100 or 1 in 3. The decision would again depend on whether "the worst"
represented a catastrophic loss or just a reduced profit. This information
on the range of values and the corresponding likelihood must not be lost
because it is information which is very relevant to the decision.

The most convenient way of storing this information is by means of
probability distributions. A probability distribution is a plot of relative
frequency versus value for a data item that has been sampled a very large
number of times. Figure 2.5 is an example of a probability distribution.

The Y-axis "frequency" has no units because it is plotting relative
frequency. The X-axis gives the range that the data can take and the
height of the curve indicates the relative probability of it occurring.
The area under the curve is probability. Hence, since the total probability
must equal 1.00 for a complete distribution, in place of a scale on the Y-
axis there is the condition that the area under the distribution must be
1.00.

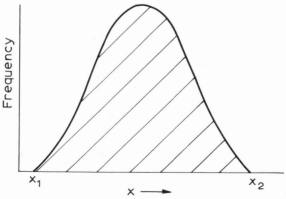

Figure 2.5   A Probability Distribution

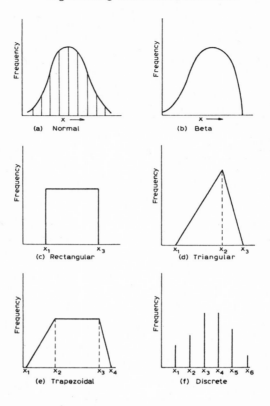

Figure 2.6 (a)-(f)   Types of Probability Distribution

There are a number of <u>types</u> of probability distribution that are described in the literature. These are compared on Figure 2.6.

2.4a The Normal Distribution

If a coin is tossed ten times there is a very small chance that the result will be ten heads, an equally small chance that it will be ten tails, but a much higher chance that it will be 5/5 heads/tails. This distribution is the result of a chain of ten probabilistic events, which can be determined arithmetically and, if done a sufficient number of times, can be experimently verified. The distribution obtained is bell-shaped, as shown by Figure 2.6(a), and in its integer form called the <u>Binomial Distribution</u>.

The Binomial Distribution represents purely chance events, but it has drawbacks in that it is not a continuous function, and the mathematical expression representing it is inconvenient to evaluate.

The formula

$$F(x) = \exp [-(x-\mu)^2/2\sigma^2] / \sigma\sqrt{2\pi} \qquad (2.6)$$

closely fits the Binomial Distribution but has the advantages that it is a continuous function and it is fairly convenient to handle. It is frequently found to closely represent distributions that occur in nature. It is called the Normal Distribution, and it is commonly used in statistical studies. Its characteristics have been well studied, and are available in tables.

Normal distributions are defined by two values, μ the mean value, and σ, the standard deviation. For the purpose of tabulating the characteristic of the distribution, the "Standard Normal Distribution" has been defined as the Normal distribution with a mean of zero, and unity standard of deviation. Hence for the standard Normal distribution:

$$F(x) = \frac{1}{2\pi} e^{-\frac{u^2}{2}} \qquad (2.7)$$

$$\text{where } u = (x-\mu) / \sigma \qquad (2.8)$$

Appendix II contains some characteristics of the Standard Normal Distribution. Statistical text contain other tables such as statistical significance tests $(F, t, \text{and } \chi)$ which are based on the assumption that the random component of any variable is Normally distributed.

Besides the Normal distribution being found to represent natural random processes, it is the distribution which results from the addition of a number of individual distributions even though the individual distributions are not themselves normal. This is known as the Central Limit Theorem.

For recording distributions in investment data, this distribution has two serious drawbacks. Firstly, it is symmetrical and often data has unsymmetrical errors. Secondly it asymtotes to infinity, whereas the data for physical systems have distinct upper and lower limits.

2.4b The Beta Distribution

The Beta distribution, Figure 2.6b, produces a shape which is not symmetrical and one end does not asymtote to infinity. It therefore has

some advantages over the usual Normal distribution for representing
physical systems. Like the Normal distribution, it can be represented by
a single equation thus:

$$\text{relative frequency} = F(x) = t^{(a-1)}(1 - t)^{(b-1)}$$

which can be useful when analytical manipulations are required. Despite
its advantage over the Normal distribution, it is still not appropriate for
investment studies. Pouliquen of the World Bank (21) claims that it emphasizes the best estimate too much. This is also a criticism of the Normal
distribution.   Beattie and Reader (5) claim that it can be successfully
used but it is unnecessarily complex and precise.

2.4c The Rectangular Distribution

The simplest form of distribution is the Rectangular or Uniform distribution (Figure 2.6c). This means that the value of x lies between two
values, x_1 and x_2 on Fig. 2.6c, and within this range all values of x have
equal probabilities.

This distribution however gives no opportunity to say that the middle
of the range is more likely than the extreme values. Since this is usually
so, the Rectangular distribution is seriously deficient.

2.4d The Triangular Distribution

The Triangular distribution (Fig. 2.6d), overcomes the criticism of
the Rectangular distribution by using three points shown as x_1, x_2 and x_3
on Fig. 2.6d. x_1 and x_3 represent the extreme values and x_2 the most
likely value.

This distribution is similar to the Normal and Beta distributions but
it does not asymtote to infinity and is quite adequate for most investment
analysis work. However, it can be criticized for over-emphasizing the most
likely value, as can the Normal and Beta distributions.

2.4e The Trapezoidal Distribution

The Trapezoidal distribution consists of a middle section of uniform
probability, with the two extremes having decreasing probabilities to
definite points beyond which the probability is zero. This is effectively

a combination of the Uniform and Triangular distributions, as shown in
Fig. 2.6e. The Trapezoidal distribution requires four data items for its
definition. x_1 and x_4 are the total range within which the value must lie,
and x_2 and x_3 are the range in which it is most likely to appear. This distribution overcomes the criticism of the earilier forms and, in addition,
it is the most flexible, since by carefully choosing the four data points
it can adequately represent any of the preceding four forms.

2.4f The Discrete Distribution

The Discrete distribution is used to represent a distribution when the
values of the variables can only be expressed as integer--a number of times
a coin head is tossed for instance. It can be expressed as a form of frequency distribution as shown by Figure 2.6a or 2.6f. In this case the
total length on the chart represents unity probability, and not the total
area as with the continuous distributions. Discrete distributions can also
be used to simplify data collection and handling for continuous variables.
Considerable simplification results if a variable is allowed only to have
a limited number of values. This becomes more evident when the values vary
in a further dimension (e.g. time).

When it is desired to express a continuous distribution by a Discrete
distribution it is best converted via the cumulative distribution. See
section 2.4h.

2.4g Other Distributions

The histogram is a commonly used method of representing oddly shaped
distributions such as multi-peaked or skew patterns that one might obtain
in practice from practical data. Large quantities of data can be collected
into a limited number of groups by making each group cover a definite
range of the variable concerned, and then the frequency data for each
group is simply the number of occassions in each range.

This then allows a mass of practical data to be easily expressed and
understood.

There are very many other types of distribution that are possible and
can be used for special cases(9). The Log Normal distribution is used in
place of the Normal when it is essential that no estimates appear less than

zero. Other distributions include the Exponential and the Weibull distributions used in reliability studies.

2.4h Frequency and Cumulative Distributions

The Frequency distribution so far described has the advantage that the
range and relative importance of the values can be clearly seen. However,
data expressed in this way is very difficult to work with because there is
no scale on the Y-axis. This scale is replaced by the definition that the
area under the curve is unity. The probability of any point value on the
curve is zero. One can only talk of a probability of a result between two
values, which is given by:

$$\int_a^b F(x) \, dx \qquad\qquad (2.9)$$

where F(x) is the function representing the distribution.

It is usually more convenient to convert the Frequency distribution
into a Cumulative distribution so that one can read off useful information
directly. The Cumulative distribution is a plot of P(x) versus x where
P (x) is the probability that a randomly chosen value will be below the
value x. It is obtained from the Frequency distribution, since P(x) can be
evaluated as:

$$P(x) = \int_{-\infty}^{x} F(x) \, dx \qquad\qquad (2.10)$$

Hence, P (x) must lie between 0 and 1.0.

From a Cumulative distribution one can read directly "a 70% probability
that the value will be below x_1" or by difference, "a 10% chance that it
will be between x_1 and x_2."

Figures 2.7a to 2.7f give the corresponding Cumulative distribution
for the six types of distribution given in Fig. 2.6a to 2.6f.

Conversion from any continuous distribution to a Discrete distribution
is done by carefully drawing the steps symetrically over the cumulative
distribution plot of the distribution to be converted, as shown by Figure
2.7 (f).

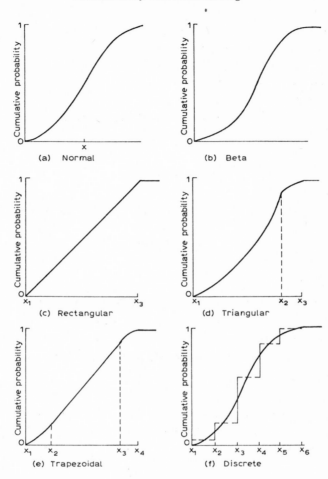

Figure 2.7 (a)-(f)   Types of Cumulative Distribution

2.5   COMBINING PROBABILITIES

2.5a   Direct Combination of Probabilities

Since we are developing arguments for using probabilistic rather than deterministic data, we must look at the methods of working with probabilities.

The most important characteristics of any distribution are the Expected Value and its Variance.

The Expected Value $(E(x))$ is the sum of the values weighted by the probability. For a discrete distribution with n stages

$$E(x) \quad = \quad \sum_{i=1}^{n} P(x_i) \, x_i \qquad\qquad (2.11)$$

where $P(x_1)$ represents the probability of the occurrence x_1,
and for continuous distributions

$$E(x) \quad = \quad \int_{-\infty}^{+\infty} x \, P(x) \, dx \qquad\qquad (2.12)$$

The Variance $V(x)$ is the sum of squares deviation from the Expected
Value.

$$V(x) \quad = \quad \sum_{i=1}^{n} P(x_i) \, [x_i - E(x)]^2 \qquad\qquad (2.13)$$

Often more meaningful is the standard deviation σ, the square root of
the variance.

A third characteristic of a distribution is its Most Likely value.
This is the value of x with the highest probability, it need not necessarily
be the same as $E(x)$.

When combining two events, x and y, with known probabilities, if the
two probabilities are independent, that is, the occurrence of one happening
does not alter the probability of the other occurring, the probability that
two particular events both occur is:

$$P(x,y) \quad = \quad P(x) \times P(y) \qquad\qquad (2.14)$$

As an example: Consider that we have two probability levels for demand, d_1
and d_2, and also two for sales price, P_1 and P_2, and we wish to calculate
the sales realisation. We have only four possible outcomes:

$$D_1 \, , \quad P_1$$
$$D_1 \, , \quad P_2$$
$$D_2 \, , \quad P_1$$
$$D_2 \, , \quad P_2$$

The probability of each of these occurrences is given by equation
(2.14), and so equation (2.11) can calculate the Expected Value and equation
(2.13) the Variance. Table 2.2 gives a numerical example:

TABLE 2.2

Sales Realisation from Probabilistic Demand and Sales

Demand	Probability	Price	Probability	Realisation Worth	Probability	
1) 150	0.5	20	0.5	150x20=3000	0.5x0.5=0.25	$(D_1\ P_1)$
				150x40=6000	0.5x0.5=0.25	$(D_1\ P_2)$
2) 100	0.5	40	0.5	100x20=2000	0.5x0.5=0.25	$(D_2\ P_1)$
				100x40=4000	0.5x0.5=0.25	$(D_2\ P_2)$
Mean:125		30		3750		

Expected Value $E(x)=3750$
Variance $V(x)=2187500$
Standard Deviation $\sigma = 1480$

      Hence in this way we can combine two probabilistic sets of data to produce an Expected Value and a distribution for the product of the two sets.

      If the occurrences had not been independent, but the demand had been dependent on the price, the combination of probabilities would be given by:

$$P(x,y)\ =\ P(x) \times P(y|x) \hspace{3cm} (2.15)$$

with $P(y|x)$ meaning the probability of y, given x has occurred. This now has introduced correlation between variables. This is dealt with later in this chapter (Section 2.5e). The combination of probabilities is treated in more detail in Section 3.4e in connection with decision trees.

      Let us take this principle further to a case in which we are combining many distributions. Consider that we need to manipulate four sets of data, each with a distribution, and to represent each distribution we need 10 discrete values. This requires a total of 10^4 evaluations. Even though this permutation of combinations can be achieved with a computer programme (26) this method of handling probabilistic data is limited to situations where only a few data items are probabilistic. Where more than 4 or 5 data items are probabilistic the number of evaluations beomes too high for the method to be practicable.

2.5b Simulation

Stocastic or Monte Carlo simulation has been developed to handle situations where a complete permutation of evaluations leads to excessive calculations.

The method is based on the fact that an adequate estimate of the distribution and Expected Value can be obtained by evaluating only a fraction of the number of times required for a complete set of combinations. Furthermore, the data for these evaluations can be picked in a random fashion; there is no need to develop specific combination patterns.

The method consists of expressing all the probabilistic data in terms of cumulative probabilities, and then by using a 0-1 random number generator obtain a number which is taken as the probability on the cumulative probability graph. Hence, the corresponding value of the variable can be read from the graph and used in the calculation. A similar process is carried out for each probabilistic variable using a new random number each time, and when values for all the probabilistic data have been generated, the evaluation can be made.

This whole procedure is repeated 100 to 1000 times, each time "stimulating" a random choice of variables. The results can be expressed as a frequency distribution, from which the Expected value can be calculated by equation (2.11).

The number of times the simulation must be carried out depends on the confidence limit required for the final result. It is useful to measure this as a tolerance on the Expected Value.

Consider that we require a probability of (1-a) that the mean value is within an error ±e of the true value. Then the number of simulations required (n) is given by (25):

$$n = \left(\frac{Z\sigma}{e}\right)^2 \tag{2.16}$$

where Z is the number of standard deviation to achieve the required upper tail of a/2, which can be obtained from statistical tables (see Appendix II). 95% confidence limits are normally taken for quoting errors, hence a/2 = 0.025 and Z = 1.96. σ is the standard deviation for the simulated variable. This is not known until some simulations have been done. However, this presents no difficulty in the simulation programme: it can be arranged to simulate a certain number of times, obtain a preliminary value of σ, evaluate n with this preliminary value for a required e by equation (2.16), and proceed with the simulation for that number of evaluations.

Mathematical texts generally describe the Monte-Carlo simulation as an integration method. It is used to evaluate integrals and in fact the determination of the Expected Value from an equation whose components have distributions is a form of integration.

As a integration method it is very unfashionable. It requires enormous amounts of computing to obtain the degree of accuracy required by mathematicians, particularly since the improvement in accuracy is proportional to the square root of the number of iterations.

The problem with the Monte-Carlo method is to increase its efficiency to obtain better accuracy with fewer iterations. Most mathematical theory of the Monte-Carlo method is concerned with various variance reduction methods (19) to improve the efficiency.

The methods basically attempt to remove as much variation from the data as possible by applying mathematical theory (Controlled Variate), isolating and analytically describing the source of the variations (Regression), arranging the sampling to compensate for the variations (Stratified Sampling and Importance Sampling) or attempting to find expressions that compensate one variation with another (Antithetic Variates).

These methods produce efficiency increases of a factor of 10 to 1000, sometimes more. However they are not generally applicable, and one always has the problem of finding a variance reduction technique that is suitable for the particular problem.

2.5c Combining Variances and Expected Values

A third method of combining probabilities is to obtain an Expected Value and a Variance for a system directly from the Expected Values and Variances of the individual components. Frequently these two values alone are adequate to describe a system, but if a complete distribution is required, this can be obtained from normal statistical tables (Appendix II), assuming that the distribution is normal. This is usually a reasonable assumption because combined distributions tend toward the Normal Distribution, even if the individual components are not normal (Central Limit Theorem).

This method requires the knowledge of some combinatorial rules of probability calculus and these are given below. Its main disadvantage is that it can be very tedious to work through all these rules for all but the simplest expressions. The basic rules required for combining distributions are as follows:

a) For a linear function:

$$y = ax \qquad\qquad (2.17)$$

the Expected value is given by:

$$E(y) = a\,E(x) \qquad\qquad (2.18)$$

and the variance by:

$$V(y) = a^2\,V(x) \qquad\qquad (2.19)$$

Note that in equation (2.17), y represents only different units (kg. to tons for example). If y represents a number of independent observations of x, it must be treated as the sum of a number of variables, as follows

b) For a sum of two variables x_1 and x_2:

$$y = a_1\,x_1 + a_2\,x_2 \qquad\qquad (2.20)$$

the Expected is given by:

$$E(y) = a_1\,E(x_1) + a_2\,E(x_2) \qquad\qquad (2.21)$$

and the variance by:

$$V(y) = a_1^2\,V(x_1) + a_2^2\,V(x_2) \qquad\qquad (2.22)$$

when there is no correlation between the variables. More generally, when there is correlation between x_1 and x_2 with the correlation co-efficient being r,

$$V(y) = a_1^2\,V(x_1) + a_2^2\,V(x_2) + 2a_1 a_2 r\,V(x_1)\,V(x_2) \qquad\qquad (2.23)$$

Correlation is discussed in the next section.

c) For a product of two variables

$$y = x_1\,x_2 \qquad\qquad (2.24)$$

tne Expected is given by:

$$E(y) = E(x_1)\,E(x_2) \qquad\qquad (2.25)$$

and the variance by:

$$V(y) = [E(x_1)]^2\,V(x_2) + [E(x_2)]^2\,V(x_1) + V(x_1)\,V(x_2) \qquad\qquad (2.26)$$

assuming there is no correlation between x_1 and x_2.

These rules are sufficient to aggregate distributions for simple relation-ships.

For more details of these rules and the effect of correlation the reader should refer to a more statistical text (23).

We have now identified three ways of combining probability distribu-tions; all these have their applications, as will be seen in later chapters.

2.5d Analytical Combinations

A fourth way of combining probabilities is algebraically to combine the equations described in the distributions of the variables to obtain equations for the combined distributions. This requires that the distribu-tions have known and continuous equations describing them, such as the Normal distribution, and that the system is simple enough for the resulting equations to be manageable. These methods find application in simple situ-ations such as quality control, but find little application to investment analysis (25).

2.5e Correlation

The methods so far discussed for the combination of probabilities have assumed that there is an equal chance of high and low probabilities combining as high and high. The distributions are independent, they are not correlated.

Let us consider the Sales/Price example introduced in Table 2.2. Let us now assume that the higher sales are the result of the lower price and vice versa. That is, the two distributions are related--they are complete-ly correlated. We can now calculate the expected realisations again, but limiting the way in which the combinations are made. This is done in Table 2.3.

Writing this in terms of conditional probabilities, the four outcomes are:

$$P(P_1, D_1) = P(P_1) \times P(D_1 \mid P_1)$$
$$P(P_2, D_1) = P(P_2) \times P(D_1 \mid P_2)$$
$$P(P_1, D_2) = P(P_1) \times P(D_2 \mid P_1)$$
$$P(P_2, D_2) = P(P_2) \times P(D_2 \mid P_2)$$

$P(P_1)$ and $P(P_2)$ are both 0.5, but because of the complete dependence of Demand on Price:

$$P(D_1 \mid P_1) \quad = \quad 1.0$$

$$P(D_1 \mid P_2) \quad = \quad 0.0$$

$$P(D_2 \mid P_1) \quad = \quad 0.0$$

$$P(D_2 \mid P_2) \quad = \quad 1.0$$

Hence Table 2.3 has effectively only two outcomes: $P(P_1,D_1)$ and $p(P_2,D_2)$.

TABLE 2.3

Sales Realisations assuming Complete Negative Correlation

Demand	Probability	Price	Probability	Realisation Worth	Probability
150	0.5	20	0.5	150x20=3000	0.5
100	0.5	40	0.5	100x40=4000	0.5
Mean: 125		30		3500	

Expected Value $E(x)$ = 3500

Variance $V(x)$ = 250000

Standard Deviation σ = 500

Notice that the mean is similar to the uncorrelated case (Table 2.2), but the distribution is very different. In the correlated case the only realisations are 3000 and 4000; there is no chance of a small realisation, whereas Table 2.2 gives a 25% chance of a return of only 2000. Notice also now:

$$E(Realisation) \neq E(Demand) \times E(Price)$$

Now consider the market situation to be yet again different. Assume that if a competitive product appears on the market, the sales will be low and the price will also have to be cut. Now if this was the reason for the probability distribution, then the high must combine with only the high, and the low with the low. This is again complete correlation but with the opposite sign. This is, in fact, a positive correlation, whereas in the previous example correlation was negative. Now let us look again at the realisations shown as Table 2.4:

TABLE 2.4

Sales Realisations assuming Complete Positive Correlation

Demand	Probability	Price	Probability	Realisation Worth	Probability
150	0.5	20	0.5	150x40=6000	0.5
100	0.5	40	0.5	100x20=2000	0.5
Mean: 125		30		4000	

Expected Value	$E(x)$ = 4000
Variance	$V(x)$ = 4000000
Standard Deviation	σ = 2000

In this case the mean value is again similar to the other cases but
the distribution is very wide, with a 50% chance of obtaining a realisation
of only 2000.

This example shows three main characteristics of correlation. Firstly,
the effect on the mean result is relatively small so that until one moves
from deterministic to probabilistic evaluations, it is of no importance.
Secondly, it has an extremely important effect on the result of any risk
analysis. Finally, its presence has to be estimated subjectively by knowing the reasons for the distributions; it cannot be determined by mathematical rules.

Two excellent papers by the World Bank (21,23) emphasize the importance of correlation in actual case studies. It is particularly important
because the decision following a risk analysis is often dependent on the
magnitude and probability of the lowest realisation, rather than on the
mean realisation. Correlation therefore strongly affects the very data
upon which the decision is based.

2.6 OBTAINING PROBABILISTIC DATA

The difference between Risk and Uncertainty has already been described.
Risk infers the distribution is known, either from past events which have
enabled data to be collected, or from a knowledge of the system-eg it may be
a purely random occurrence with all events having an equal probability.

If any of our data falls into this category then we are lucky enough to have data to hand.

In such a case we can resort to standard statistical techniques to determine the distribution from the data. This in itself can be a fairly involved subject. One should for instance estimate the reliability or confidence limits of a distribution obtained from a limited quantity of data (24). This becomes a problem of sample size and introduces the concept of the probability of a probability being correct.

Standard regression computer packages are a help since they determine the standard deviation, give the extent of correlation between the variables, and also compare the distribution with the Normal distribution (8).

However, more often than not we do not have past data; we are faced with Uncertain data, with no historical or theoretical grounds for calculating the probability distribution. This does not mean that we have absolutely no idea of the relative probabilities of the various outcomes; there are very often firm subjective opinions about these, held by people with years of experience of similar situations--the "experts." We have therefore to convert the opinions of experts into probability distributions.

Some mathematicians cannot accept that the concept of probability can be applied to opinions because the usual definition of probability is: the limit of the ratio of the number of the occurrences to the number of trials, as the number of trials increases without limit (24,25).

This number either has been, or can be, established by experiment. The idea of applying this probability to an event that has not occurred, and will probably only occur once, requires a broader definition of 'probability.' This is a subjective or Bayesian definition: the probability is the degree of a person's belief that an event will occur (25).

The conflict between these two definitions can be highlighted by considering the statement that "the process has a 70% probability of working" in the context of having adequate technical data to design it. If this process was built an infinite number of times, according to the first definition, the plant should work in 70% of the cases. In fact it would either work in every case or not work in every case, depending on the value of the technical data used in the design. That is, the probability would be either zero or 1, but never 0.7 (24).

The conflict can be resolved by defining an Expert Opinion as: An informed assessment based on experience of previous technically similar systems whose final outcomes were known.

Consider, for example, a new process for making a dyestuff. The Expert may have seen a hundred similar processes for different types of dyestuff in the past and he remembers that about 70% of them were success- ful. He is therefore justified by either the first or the second definition of probability, in attributing a probability of 0.7 to the chance of a successful outcome for this new process.

The way an Expert develops opinions is by sub-dividing any proposal into constituent parts and then by passing a judgement from his experience of previous histories on the constituent parts. In this way he appears to be attributing probabilities to unique events. The arguments, and the methods developed throughout this book, assume that Experts can give val- uable opinions which can be expressed as meaningful probabilities.

There are no exotic techniques developed for obtaining subjective probability distributions from Experts for uncertain data. The first rule is to discuss the matter with the appropriate expert--the originator of the data--and he may be able to give his considered opinion directly in the form of probabilities. When this is not possible, one can arrive at prob- abilities by an iterative process in a conversation between the expert and analyst (21).

A first estimate is attempted and normalised. For example, a five- level discrete situation may give:

	1	2	3	4	5
"First probability estimate:	0.2	0.2	0.2	0.2	0.2

Q.Analyst: do the extremes have probabilities equal to the middle probab- ilities?

A.Expert : No, say 3 times more probable for the middle ranges.

	1	2	3	4	5
1st Revision :	0.08	0.28	0.28	0.28	0.08

Q.Analyst: Have all 3 centre values an equal probability?

A.Expert : No, level 3 is 50% more likely than either of the other 2.

	1	2	3	4	5
2nd Revision :	0.08	0.24	0.36	0.24	0.08

Q.Analyst: If the value were not to be in the middle section is it more likely to be above or below the middle?

A.Expert : More likely above, say twice as likely.

3rd Revision	1	2	3	4	5
	0.06	0.16	0.36	0.32	0.10

Q.Analyst: Is the higher extreme twice as likely as the lower?

A.Expert : No, they have about the same likelihood.

4th Revision	1	2	3	4	5
	0.08	0.16	0.36	0.32	0.08

Q.Analyst: This has produced a distribution with the 4th level having a
similar probability to the middle, is that reasonable?

A.Expert : No, the middle level is the most likely. The above distribution
has too big a difference between levels 2 and 4, and not enough
difference between 3 and 4. Give way on the earlier criteria;
I would give my final estimate as:

Final Probability Estimate	1	2	3	4	5
	0.08	0.15	0.39	0.30	0.08"

This process of checking that the estimate looks reasonable when the
questions are posed in a variety of ways is referred to as Testing for
Consistency (24).

Some experts may be unwilling to give estimates because they do not
have a good feel for numbers. In such cases the discussion can be held
with visual aids by segmenting a circle and adjusting the size of the segments in a similar iterative fashion.

When probabilities have to be estimated for continuous functions rather
than discrete possibilities, a different technique has to be employed.
Dividing the continuous functions into a number of discrete levels would be
a possibility, but it is not very satisfactory because in a discrete set
the estimator is helped by knowing the reason for each set. For example,
the lowest is when the product is unsuitable; the next lowest occurs if a
competitor enters the market--and so on. The condition that leads to each
discrete level enables the probability to be estimated. A continuous
function divided into discrete sets does not have this advantage. It is
more usual to leave it as a continuous function and estimate:

(a) what type of distribution it is likely to be, and

(b) sufficient data to characterize that distribution.

The different distributions, their application, and the amount of data
required to define them have already been described in Section 2.4. For

instance, if a distribution was thought to arise from purely random occurrences, a Normal distribution would be suitable. Normal distribution
requires a mean (μ) and a standard deviation (σ) to be estimated. The standard deviation can be obtained by estimating the range in which there is a
95% chance (subjectively estimated) that the value will lie. Statistical
tables, (Appendix II), show this total <u>range</u> is equivalent to 3.9σ.

Hence, the subjective judgements required are:

(a) that the distribution will be normal

(b) the "one chance in 40" that it will be higher than . . .
     and a one chance in 40 that it will be lower than . . .

It is sometimes difficult to estimate the one chance in 40 range to
define the Normal distribution, but much more practical to estimate a value
e where the likelihood of the estimate occurring between (μ - e) and (μ + e)
is equal to the likelihood of it being outside this range(25). This value
is the 50% confidence limit and the range is equal to 1.34 standard
deviations. A final consistency check is that a value of ±3 standard
deviations should be considered "extremely unlikely."

As described in Section 2.4e, the Trapezoidal distribution is a particularly flexible one. The distribution has five regions, estimated by
four data points (Figure 2.6e). The centre of the distribution between
x_2 and x_3 is the region where the value is most likely to occur and it is
given equal probability of occurring anywhere in that range. Outside this
section are (x_1 - x_2) and (x_3 and x_4) where the value could occur but with
decreasing probability, and outside x_1 or x_4 there is zero chance that the
value would ever occur. The distribution is defined by estimating x_1 to
x_4 which are comparatively easily defined. Values x_2 and x_3 are the values
which define the range where an estimate would cause "no surprise" whatso-
ever. All values in this range are equally <u>credible</u>. The value x_1 defines
the lower limit, beyond which any value would cause "complete surprise,"
as would any value above x_4. This approach of relative surprise is described by Allen (3) in his presentation of credibilities as an aid to
subjectively estimating distributions.

The reasons for estimating the distribution of individual data items
when one is really only interested in the distribution of a final function
is because it is possible to estimate subjectively the distribution of
individual components but not possible to estimate the distribution of a
final function. This same principle may be useful in obtaining an estimate

of the distribution of the individual data items. It may help in estimating their probabilities by further breaking them down into even finer
detail. Consider the problem of obtaining a distribution for an annual
fixed cost which is composed of labour costs, electrical power costs and
maintenance costs: A distribution for such a fixed cost would be extremely
difficult to estimate subjectively, but distributions for labour costs,
power costs and maintenance costs would be comparatively simple to estimate
separately.

At this level the most useful method of aggregating such detailed
distributions is by the Combination of Variances and Expected Values, as
described in Section 2.5c. Since the straight-forward combination of a
number of distributions tends towards the Normal distribution, even though
the comparative distributions may not be normal, it is reasonable to obtain
a combined variance and use Normal tables to determine the whole distribution.

The reason for breaking data down, or disaggregating, is therefore to
enable the distribution to be estimated. This introduces further difficulties discussed in the next Section (2.7) concerned with correlation. Since
correlation difficulties can be overcome by combining or aggregating data,
there is a limit beyond which this aggregation causes more problems than it
solves.

2.7 SEQUENTIAL STOCASTIC PROCESSES

Market development is probabilistic, but this does not imply that, if
the forecast shows the various levels to have equal probability, then, year
by year there is an equal probability of the sales being at all levels.
The chance of the demands staying on the same level as the previous year or
dropping or increasing one level are much higher than the sales changing
to extreme levels in adjacent years. In other words, next year's level is
determined by the state this year. In the long term all levels can have
equal probabilities--so there is no conflict with the forecasters' original
information--but how it attains this final state is not defined by the
probabilistic forecast. There is clearly a further characteristic of the
market which defines the probability of changing from one state to another
which has so far not been discussed.

Any stagewise stocastic process, with the state in the next stage being dependent only on the state in the proceeding stage, is called a Markov chain. Markov chains are defined by giving the set of probabilities of transferring from level (i) to level (j) (p_{ij}) where i represents the present state (time n) and (j) represents the state of the next stage (time n+1). For a system with N levels we therefore require a N x N Transition Matrix to define the process.

For example, take a 4 level sales forecast, and assume that past experience has shown that there is a 50% probability that the sales next year will be the same as this, and 25% probability that they will rise one level, or fall one level and there is no chance of changing two or more levels in one time period. Our Transition Matrix would appear as Table 2.5.

TABLE 2.5

Transition Matrix - probability of being in any
level next year depending on the level this year

		Next years state (j)			
		1	2	3	4
This years	1	0.67	0.33	0	0
state	2	0.25	0.50	0.25	0
(i)	3	0	0.25	0.50	0.25
	4	0	0	0.33	0.67

Note the special adjustments necessary to our definition at the extreme levels, because we know that the system must always have a level to go to next year. i.e.

$$\sum_{j=1}^{N} P_{ij} = 1 \qquad (2.27)$$

$$\text{for} \quad i = 1 \text{ to } N$$

Once the Transition Matrix has been defined, it is possible to calculate the probability of the system being in each state in any year, given the state in the first year, as shown below for our example.

Consider the market begins (year 1) in level 2, from Table 2.5. We can say that the probability of the market being in levels 1-4 in year 2 is

0.25, 0.50, 0.25, 0 respectively. These are the probabilities of the market being in these levels in year 2--denoted normally by $P_1(2)$ to $P_4(2)$. Notice the difference between these and the elements of the Transition Matrix, P_{ij} which denotes the probability of Transition.

Proceeding a further year, we can determine the probability of the market being in each level $P_1(3)$ to $P_4(3)$, as the sum of the probabilities of transferring into these levels from all other levels. The probability of having transferred into j from i is $P_i(1) \times P_{ij}$, i.e. the probability of being in state i multiplied by the probability of transferring from i to j.

In general we have N equations of the form:

$$P_j(n+1) = \sum_{i=1}^{N} P_i(n)\, P_{ij} \qquad (2.28)$$

where j=1 to N, with N being the total number of states per year. In our example

$$P_1(3) = 0.2925, \quad P_1(4) = 0.2947, \quad P_1(5) = 0.2866$$

$$P_2(3) = 0.3950, \quad P_2(4) = 0.3565, \quad P_2(5) = 0.3366$$

$$P_3(3) = 0.2500, \quad P_3(4) = 0.2444, \quad P_3(5) = 0.2458$$

$$P_4(3) = 0.0625, \quad P_4(4) = 0.1044, \quad P_4(5) = 0.1310$$

It can be shown that for such a system carried on for enough stages, the probabilities will converge on to constant values. Furthermore, these same probabilities will result whatever the starting point.

The equilibrium probability $(P_i(e))$ is determined by stating that at equilibrium, the probabilities the following year remain at the equilibrium value, i.e.

$$P_j(e) = \sum_{i=1}^{N} P_i(e)\, P_{ij} \qquad (2.29)$$

We also know that

$$\sum_{i=1}^{N} P_i(e) = 1 \qquad (2.30)$$

Hence we have sufficient equations to determine $P_i(e)$, the equilibrium state for the Transition Matrix. In our example, the equilibrium is 0.2124, 0.2857, 0.2857, 0.2124 for $P_1(e)$ to $P_4(e)$ respectively.

Different Transition Matrices produce different properties of the Markov chain. Many Matrices produce chains which move to a stable state at the limit (called fully ergodic), but Matrices can result in chains with peculiar properties. For instance, when any level or set of levels have particular combinations of zero probability of transferring in or out then it is possible to have <u>disjointed</u> or <u>periodic</u> chains, or have transitory states (15).

2.8 HANDLING THE CORRELATIONS

As has already been explained, the most important single factor in a Risk Analysis study is the determination of correlations between the variables.

If one has adequate historical data, this information appears directly from a Regression Analysis. However, most data we are dealing with is "uncertain," that is there is no historical data, and the few data items with recorded past data available may well not be sufficient to enable the correlation to be determined. For instance, one might have considerable Price/Demand data on which a regression can obtain a correlation between price and demand. However, unless we have very complete data, that is:

| High price/ | High price/ | Low Price/ | Low Price/ |
| Low Demand | High Demand | Low Demand | High Demand |

then the correlation provided by the regression is indicating the "possibility of a correlation between the variables cannot be excluded, based on the available data." We require to know "the data shows the following correlation exists." The chance of having such complete data is small, since price changes are determined to maximize business interests, and not to obtain orthogonalized data for forecasting.

This aside, there is no theoretical method of treating correlation and the following methods have been reported as possible approaches to the problem (21).

2.8a Limiting Disaggregation

Although it has been recommended that the variables should be disaggregated in order to reduce the problem of estimating the probabilities, this disaggregation increases the chance of correlation because after the disaggregation one must determine the degree of correlation between the new variables created by the disaggregation. For example, to consider only annual sales realisations in place of sales volume and price overcomes the problem of defining the amount of correlation between sales volume and price. It however introduces the problem of defining a probability distribution for sales realisations without thinking in terms of sales volume and price.

2.8b Identifying the Source of the Uncertainties

Once one understands why the uncertainties have arisen, one is in a position to make a subjective judgement as to the degree of correlation to be expected between the variables.

Consider a new product which may find a number of different applications displacing established products: The yearly sales volume of such a product will rise and the steepness of this rise will depend upon the success of this product in the different applications. If it proves only to displace one product its sales growth will be low. If it finds a wide application its sales will be high. If its sales are high in the early years, due to successful application in many areas, one can expect its sales to be high in later years too. One can expect a high correlation between sales volume in any year and sales volume in the previous year. An understanding of the reason for the uncertainty in the sales forecast has enabled one to estimate subjectively the correlation to be expected.

Consider the forecasted sales volume of an existing product in a competitive situation: the demand is therefore elastic, an increase in price lowering the sales volume. However, the sales volume may also be dependent on the GNP which could have a Normal distribution. Hence, there would be partial correlation between price and volume; it will not be complete. One can define this by the following equation:

$$\text{Sales volume} = a + b\,(\text{Price}) + e$$

where a is a stable base sales volume, b introduces the effect of price on volume, and e is a random value with a known standard deviation which

introduces the effect of GNP. The understanding of the situation has enabled
us to write a specific model, equation 2.24, which overcomes the correlation
problem. All that is now required is data, or subjective judgements to
define a, b, and e. This approach is briefly mentioned as being used to describe the sales price for copper by Reichard (22).

2.8c Testing Extremes

Because of the dearth of methods available for dealing with correlation,
it is sensible to revert to the simple method of testing extremes. First,
carry out an analysis assuming correlation is present, and then repeat it
assuming it is not present. This at least indicates the cases where correlation is important and gives some idea of the magnitude involved. Any
very serious cases can then be treated very carefully by the few methods
that are available.

2.8d Quantitative Treatment with a Correlation Coefficient

In statistical work and regression analysis, given sufficient experimental data it is possible to obtain a quantitative description of the dependence of error distribution between variables in the form of a correlation coefficient, r, introduced in section 2.5c. This correlation coefficient can be defined as:

$$r = \frac{Covariance\ (xy)}{Variance\ (x) \times Variance\ (y)} \qquad (2.31)$$

where

$$Covariance\ (xy) = \sum[(x_i - E(x)) \times (y_i - E(y))] \qquad (2.32)$$

The correlation coefficient can therefore be estimated if there are
sufficient results available, covering the whole experimental surface.

This approach however cannot easily be introduced into forecasting
because adequate experimental results are never available and the relation-
ships between the distribution must be estimated subjectively.

By defining another correlation coefficient, which lends itself to
being more easily subjectively judged, then the following method for describing and treating the correlation between variables can be employed (20).

Let a correlation coefficient relating the two variables described by
distributions (C_{xy}) be defined as:

"the fraction of the variation of variable y, directly caused by
the variation in variable x, with the remaining variation due to
random effects."

Now, when C_{xy} = 1, then once the value of x is determined, the value
of y is immediately fixed (complete correlation). If C_{xy} = 0, then whatever the value of x, the value of y is determined entirely on a random
basis--i.e. zero correlation. Intermediate degrees of correlation can be
obtained by linear interpolation.

$$y = C_{xy} y_x + (1-C_{xy}) y_0 \qquad (2.33)$$

where y_x represents the value of y if completely correlated with x

y_0 represents the value of y if zero correlation with x

The subjective estimation of C_{xy}, though not easy, should be generally
possible because it has a fairly simple physical meaning.

2.8e Inflation

Inflation produces a special form of correlation in investment calculations because any distribution that is introduced into any cost data to
cover the uncertainty of future inflation is correlated with similar distributions in all the other data that are in money units. This produces an
impossibly difficult matrix of partial correlations. To avoid this problem,
Risk Analysis should be done in Constant Money Terms. The most appropriate
constant money is the worth of the money unit at the beginning of the project. This means that the only data that can see inflation as an uncertainty are the capital costs which are only quoted in money units about two
years before the beginning of a project. The inflation term will also
enter into the tax calculation since in Constant Money Terms future tax
allowances are progressively worth less under inflationary conditions, as
already described in Section 1.7.

2.9 THE RELEVANCE OF UNCERTAINTY TO DATA FOR CAPITAL INVESTMENT

Having reviewed the sources of uncertainty in the data required for
capital investment decisions, and having discussed the various ways of

recording, handling and obtaining the uncertainty data, we are now in a
position to return to the data items required for capital investment
analysis, and to discuss specifically how the data for each should be
handled.

2.9a Sales Volume

    As described in Section 2.2a, the uncertainty in Sales Volume is the
greatest problem in an investment analysis. This discussion on forecasting
techniques showed that the future sales volume is partly the result of a
Time-Series Extrapolation and partly the result of discrete future events.
Extrapolations produce mean forecasts with Normal distributions and discrete
events, such as the product being acceptable in that particular application
or not, produce a discrete distribution. An accurate distribution for a
sales forecast would therefore be expected to be a very complex one, taking
the form of a multi-peaked continuous distribution, as shown by Figure 2.8
which shows four discrete future developments.

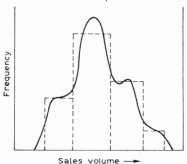

Figure 2.8 Expected Distribution for a Sales Volume Forecast

    In working with such complex distributions, Discrete Distributions are
to be preferred. The advantages are: firstly, that in medium term forecasting the discrete events are probably more significant than the random
scatter, and secondly, the discrete characteristic is more suitable than
the continuous distribution for the discussion of plans and strategies.
This will be dealt with more fully in the following chapters.

    Two correlations are important with sales volume: the yearly correl-
ation--volume last year/volume this year; and the price correlation--
volume this year/price this year relationships. These correlations will

usually be strong but not necessarily 100%. There will be reasons for low
sales without it being due to high prices, and high sales in the first year
will not guarantee high sales in the last year, although one would not
expect very high sales in one year and very low sales the next.

2.9b Sales Price

Second to sales volume, sales price has been described as the most
important data item, which again has to be forecast year by year and a distribution attached to each yearly value.

Price forecasting is almost entirely subjective. How much and when
will competition depress the price; whether future management policy will
be high turnover/low price or the reverse. The possibility of future technical improvements leading to price reductions or the disproportionate
inflation of raw material costs are all subjective judgements better estimated as a set of discrete possibilities than represented by a continuous
function. The distribution would again look like Figure 2.8 which is best
represented by a Discrete Distribution.

The correlation with Sales Volume is simplified by making the same
number of sections to the distribution as the sales volume distribution.
Figure 2.9 shows a Sales Volume-Sales Price pair, both as frequency and

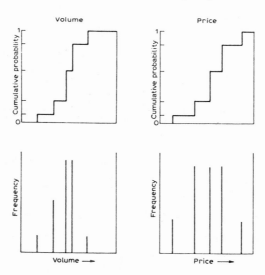

Figure 2.9   Matched Sales Volume/Price Forecast Distributions

cumulative distributions. There will obviously be some stability of price
from year to year. This can be accounted for by a Price/year correlation
factor. However, care must be taken not to overspecify the system with
correlation factors. Of the three possible correlations,

                    Sales/Year, Price/Year and Sales/Price
if any two are 1.0 the third must be zero, otherwise the system is over-
specified.

2.9c Plant Capacity

A manufacturing plant consists of a series of units. For instance a
chemical plant could consist of reaction, distillation, blending and packing.
An engineering plant could consist of forging, drilling, grinding, polish-
ing and assembling. Each process has only second order effects on the next.
Figure 2.10 represents such a series of operations.

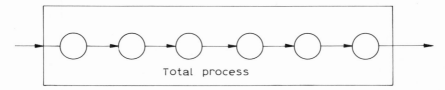

Total process

Figure 2.10
The Manufacturing Process as a Series of Operations

For the total process to have a capacity of C units/unit time each
operation must have at least this capacity. Now assume that the capacity
of each unit is not known with certainty, but only as a distribution with
a mean value and standard deviation. Assume also that the designer has
used a 10% safety margin, that is, he has sized all his equipment for a
capacity of 1.1C per unit time. If the standard deviation for the design
is 0.08C, Appendix II shows, from the Normal distribution, that the probability that each unit alone will meet the required production level is
0.90 (0.9 is the probability that a variable of mean μ and standard deviation
σ will be greater than (μ - 1.25σ). The probability that a 6-stage process
will meet the required capacity will be 0.9^6 which equals 0.53. The total
cumulative probability distribution for the individual items and also for
the plant capacity are given by Figure 2.11. Notice that the distribution

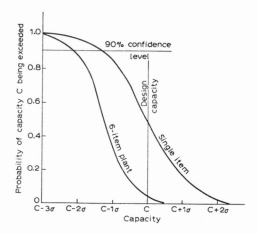

<u>Figure 2.11   Single Item and Total Plant Capacity Distributions</u>

for the total plant is sharper than the individual item distributions with
a mean 1.3 standard deviation less than the mean of the individual process
units. Even more important is that the total plant has a 7% less (0.8
standard deviations) capacity than the individual items at a 90% probability.
The plant items must therefore be 17% oversized to maintain a 90% confidence
limit on the total plant capacity.

    This effect, caused by <u>concatonation</u> of the individual distributions
is important in determining design <u>Safety Factors</u>, and the subject is treated in far more detail in Chapter 7.

    A probability distribution of plant capacity is a combination of the
distributions of the component parts and this provides one way of obtaining
probability distribution data for plant capacity in plant investment
studies. A second method would be to use data from earlier engineering projects to determine the overall probability distribution if distribution
data for individual plant items are not available. The distribution is
likely to be approximated by a Normal distribution since it is composed of
a number of distributions. Alternatively, the Trapezoidal distribution
could be used since it can adequately represent the Normal distribution and
is easily estimated using credibilities.

2.9d <u>Capital Costs</u>

The accuracy of capital cost estimates depends on the level of detail put into the estimate by the Engineering Department. Accurate cost esti-mates where every detail has been specified, have very small standard deviations and are probably near-normal distributions, since they are aggregates of hundreds of distributions of the components of the estimate. Such distributions can be obtained from a comparison of estimated and achieved costs of historical data.

However, the rougher estimates always have a tendency to be low. This is easily understood because the two steps in producing an estimate are: first to decide what is necessary, and then to estimate the cost of it. The greatest source of error in an estimate arises in the first step, in not knowing all that the process needs in the early stages. Since one is more likely to omit an item or part of a process than include too much, the tendency is to produce low estimates. This shows itself in the distri-bution of the final achieved capital cost around the single point capital cost estimate, which cannot be normal but must be asymmetrical or "skew" with a greater probability of the cost being high than low (Figure 2.12).

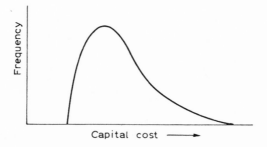

Figure 2.12   Skew Distribution for Capital Cost

This skew distribution must be obtained from past data from the engin-eering department. Data on estimated cost (at the particular estimation level) and final achieved costs will show an occasional overestimate and frequent underestimates, sometimes considerably in error.

The expert from the engineering department can then give a distribu-tion with his estimate--most suitably using credibilities and a Trapezoidal distribution. However, this tendency to underestimate is so predominant, that even after an engineering department has added its contingencies there

is still such a high probability of an underestimate that there is some
excuse for going to the unprofessional extreme of "calibrating the expert"
(24). If an expert distribution has been shown in the past to have a bias,
it is an acceptable technique to modify this distribution before using it
in a Risk Analysis.

2.9e and f Variable Cost and Fixed Costs

Both these items are composed of very many parts which will narrow the
distribution and also make it tend to Normal. If there is any single
component with a dominant magnitude or distribution, this will dominate the
composite distribution.

In the event of any doubt as to a reasonable distribution for these
overall costs, they can be determined by aggregating the individual distributions which should be easier to estimate. This aggregation can be done
by combining the Expected Values and Variances as described in Section 2.5c.

Generally, uncertainties in these costs do not constitute areas of
major importance in investment analysis. Probabilistic distributions can
be adequately estimated using Creditibilities and the Trapezoidal distribution.

2.9g Plant Scrap Value

When the plant is scrapped at the end of the project, the uncertainty
in its scrap value is very high--say between zero and 20% of the original
capital value. So vague is the distribution that a simple uniform distribution could be used. However, the matter will have no noticeable effect
on the project because of the affects of discounting.

2.10 SOME NOTES ON THE TRAPEZOIDAL DISTRIBUTION

The contents of this chapter frequently refer to the suitability of
the Trapezoidal form of distribution. It is very flexible: able to represent symmetrical or skew situations; it does not have long tails, which
are normally impossible in physical situations, and it is comparatively
easy to evaluate the four necessary parameters subjectively in terms of
Credibilities.

Since this distribution has these advantages it has been chosen to
generalize most of the distributions considered in this book. It is useful
to know how to specify other distributions as a Trapezoidal, so that this
generalization does not produce problems when data expressed as other distributions is available. For example, the Normal distribution is particularly useful, but it is not obvious how to define an equivalent Trapezoidal distribution from the mean and standard deviation from a Normal distribution. Table 2.6 summarizes the way in which the Trapezoidal distributions can be used to describe the other forms. Figure 2.13 compares the
standard normal cumulative distributions with the Trapezoidal distribution
determined from the formula in Table 2.6. The fit is acceptable--in particular, the median and mean values are identical, the 95% confidence limits
are very close and the 99.9% Normal distribution point corresponds to 100%
on the trapezoidal distribution.

The mean of a Trapedoidal distribution is given by:

$$\bar{x} = \frac{(x_4^2 + x_4 x_3 + x_3^2) - (x_1^2 + x_1 x_2 + x_2^2)}{3 \left[(x_4 + x_3) - (x_1 + x_2)\right]} \qquad (2.34)$$

TABLE 2.6

Conversion of Other Distributions to the Trapezoidal

	Trapezoidal Representation	
Triangular; 3 values; a,b,c	$x_1 = a$ $x_2 = b$ $x_3 = b$ $x_4 = c$	Exact fit
Rectangular; 2 values; a,b	$x_1 = a$ $x_2 = a$ $x_3 = b$ $x_4 = b$	Exact fit
Normal; mean and standard deviations μ and σ	$x_1 = \mu - 2.6\sigma$ $x_2 = \mu - 0.4\sigma$ $x_3 = \mu + 0.4\sigma$ $x_4 = \mu + 2.6\sigma$	Fit shown by Figure 2.13

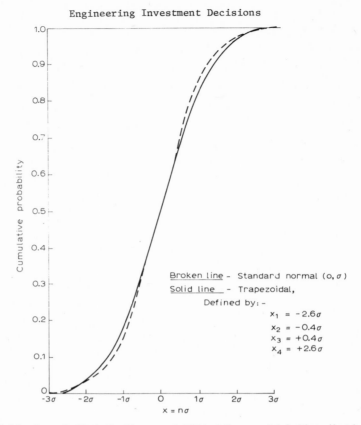

Figure 2.13   Normal Distribution   and Fitted Trapezoidal Distribution

In these first two chapters we have described the data required for a capital investment analysis, the uncertainties associated with this data, and how these uncertainties can be evaluated and recorded.

The following chapters are devoted to methods of decision-taking using this data, taking into consideration the uncertainties expressed in them.

2.11   BIBLIOGRAPHY FOR CHAPTER II

1.   Abernathy, W.J., and Wayne, C.K., Limits to the Learning Curve, Harvard Business Review, 109, (Sept.-Oct., 1974).

2.   Ahmed, S.B., Special Issue on Economical Models, Systems, Man and Cybernetics, SMC-3, 6, (1973).

3.   Allen, D.H., The Concept of Credibilities and its use in Forecasting for Novel Projects, Chem. Engr., 267, (1969).

4. Baumann, H.C., Fundamentals of Cost Engineering in the Chemical Industry, Reinhold Book Co., New York, 1964.

5. Beattie, C.J. and Reader, R.D., Quantitative Management in R and D, Chapman and Hall Ltd., London, 1971.

6. Boot, J.C.G. and Cox, E.B., Statistical Analysis for Managerial Decisions, McGraw-Hill, New York, 1970.

7. Box, G.E.P.and Jenkins, G.M., Time Series Analysis, Forecasting and Control, Holden Day, San Francisco, 1970.

8. Daniel, C., and Wood, F.S., Fitting Equations to Data, Wiley, New York, 1971.

9. Dixon, W.J. and Massey, F.J., Introduction to Statistical Analysis, McGraw-Hill, New York, 1969.

10. Fontela, E., and Gabus, A., Events and Economic Forecasting Models, Futures, $\underline{6}$, 329, (1974).

11. Forecasting Sales, Studies in Business Policy, No. 106, National Industrial Conference Board Inc., New York, 1964.

12. Forrester, J.W., World Dynamics, Wright-Allen Press, Massachusetts, 1971.

13. Giragosian, N.H., Chemical Marketing Research, Reinhold Publishing Co., New York, 1967.

14. ICI., Short-Term Forecasting--ICI Monograph No.2, Oliver and Boyd, London, 1964.

15. Kaufman, A., and Faure, R., Introduction to Operation Research, Academic Press, New York, 1968.

16. Malloy, J.B., Projecting Chemical Product Prices, Chem. Eng. Proc., $\underline{70}$, 9, 77, (1974).

17. Martino, J.P., Technological Forecasting for Decision Making, American Elsevier, New York, 1972.

18. Naylor, T.H., Computer Simulation Experiments with Models of Economic Systems, John Wiley, New York, 1971.

19. Newman, J.W., Management Application of Decision Theory, Harper and Row, New York, 1971.

20. Phipps, A.J., and Gershefski, G., A High Order Computer Language for Data Analysis, in Decisions and Risk Analysis--Powerful New Tools for Management, Engineering Economist Symp. Series VI, Hoboken, New Jersey, 1972.

21. Pouliquen, L.Y., Project Analysis in Project Appraisal, World Bank Occasional Paper, No.11, John Hopkins Press, Baltimore, 1970.

22. Reichard, R.S., Practical Techniques of Sales Forecasting, McGraw-Hill, New York, 1966.

23. Reutlinger, S., Techniques for Project Appraisal and Uncertainty, World Bank Occasional Paper, No.10, John Hopkins Press, Baltimore, 1970.

24. Schlaifer, R., Analysis of Decisions under Uncertainty, McGraw-Hill,
     New York, 1969.

25. Siddall, J.N., Analytical Decision-Making in Engineering Design,
     Prentice-Hall, Englewood Cliffs, New Jersey, 1972.

26. Spetzler, C.S. and Zamora R.M., Decision Analysis of a Facilities
     Investment and Expansion Problem, in Decision and Risk Analysis--
     Powerful New Tools for Management, Engineering Economist Symp. Series
     VI, Hoboken, New Jersey, 1972.

CHAPTER III

DECISION THEORY

Aazelle, Bölle schelle,
d Chatz gaht uf Walliselle,
chunnt si wider hei,
hät si chrummi bei,
piff, paff, puff,
und du bisch ehr und redli
druss.

3.1 INTRODUCTION

Decision Theory is concerned with the understanding and formulisation
of Decision Making. Decisions have been made for centuries, but now at
last we have a theory for it. There is a tendency for some to be cynical
about this subject, as there is about Operations Research in general--a
feeling that some otherwise unemployed mathematicians are attempting to put
a subject, once generally understandable, on a level where only mathematicians will be able to follow it.

Even taking a neutral stand in such a debate one has to admit that,
by creating the subject, the component parts have been identified and
analysed which is of some benefit, particularly when it comes to discussing the subject. Also a number of useful techniques, particularly Decision
Trees, have been put forward.

The subject of decision taking is discussed in general terms in this
chapter, using the framework of Decision Theory to develop the description.
The chapter is not a complete survey of Decision Theory but covers only
those areas relevant to engineering investment decision-making. The following chapters apply these generalities to the investment decision.

3.2 THE ELEMENTS OF A DECISION SITUATION

Any system involving decision-making consists of five elements:

  a) The Decision
  b) The Alternatives

c)   The Criteria

d)   The Constraints

e)   The Events

The Decision is the act of choosing, and it is normally assumed to be within the control of the decision-maker. He decides the course of action and this will be followed. He can decide "install a plant" or "do not install a plant". The out-come is clear, there will either be a product to sell, or there will not be; there is no uncertainty and no question of a probability.

The Alternatives are the possibilities open to the decision-maker as means of achieving his goal. In theory, every conceivable way of achieving the goal is an alternative, but in practice the number of proposals must be limited to "Sensible Alternatives"--that is, Alternatives that cannot be discounted as being unattractive by inspection. The data which defines the Alternatives may be certain or uncertain.

The Criteria are the desired results that the decisions are trying to maximise. They do not have to be economic, though they more often than not are. They can be aesthetic or functional, or simply pleasurable. They are sometimes called the Utilities but in our case we will not use this term because it is more commonly used with a specific money scale, which will be discussed later.

As will be seen in the next section, the criteria need not be singular. It is possible to consider a criterion composed of money plus appearance, for instance.

The Constraints are conditions that must not be violated. For example, the laws that restrict the quantity of effluent a plant is allowed to pro-duce are constraints. Sometimes the availability of labour or capital is limited and these become constraints. Also under constraints comes the restrictions of the field of treatment. Company management can decide that they are to compete in a particular market. This policy is transferred to the planner as a constraint--"Maximise profit (criteria) by manufacturing sulphuric acid (constraint)". The constraints constrain the decisions. Any proposed decision that would violate a constraint must be rejected.

Finally there are the Events. These are the factors outside the control of the decision-maker. They are produced by outside influences such as competition, governments, consumer tastes and so on. The events can be known with certainty, but it is more usual that each has a possible number of outcomes, each with a probability. In discussing this subject we will begin with the simple case of systems containing events with only one certain outcome, for example, "the product sells for ten years." After this we will introduce the more usual "uncertain event," that is, "the product may sell for either seven, or ten, or twelve years."

3.3 THE SINGLE DECISION

Decision Theory is concerned particularly with making a choice from a number of possible alternatives. Making a decision implies identifying some preference amongst the alternatives which makes the choice possible. The quality which gives the preference is called the criterion. Preferences and criteria need not be quantitative or quantifiable.

Probably the most basic decision process is required when you are choosing a dress for your wife. There are many alternatives--if you are un- lucky there may be ten--and one must be selected. A systematic approach suitable in such cases is to make a comparison in pairs. Consider the problem of choosing a dress out of four: A,B,C and D. Let us assume that no one dress is outstanding, but you are able to make a comparison between any two and say whether you prefer the one or the other. We now have a method of making a choice: in the shop we could enact the whole algorithm:

Compare A with B	-	take preferred dress and
compare with C	-	take preferred dress and
compare with D	-	take the preferred dress as the preference out of A,B,C and D.

Let us assume that the following resulted:

$$A/B \longrightarrow A$$
$$A/C \longrightarrow C$$
$$C/D \longrightarrow \underline{D}$$

Notice that in this example the criterion is as vague as the personal attractiveness of the dress. This is a particularly simple example, with the comment "indifferent to the two under comparison" not allowed. Also, no allowance was made for the decision-maker to be inconsistent. In a direct comparison he may prefer B to D, but our algorithm did not make this

check. Finally, the decision-maker must carry out the excercise sequentially.

These three points must all be catered for in a satisfactory binary
comparison method. Indifference must be an allowed reply, unsequenced
comparisons must be allowed and inconsistent results accepted.

A simple technique has been developed for recording the results of
binary comparisons in a preference graph (12). A lattice is drawn as shown
in Figure 3.1; the results of a comparison can now be recorded. If A is
preferred to B then a point is marked on the A successful/B unsuccessful
intersection. If the result of the comparison was indifference, a point

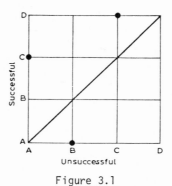

Figure 3.1

Preference Graph for the dress selection example

would be put on the A successful/B unsuccession intersection also. This
enables one to see quickly whether all comparisons have been made; whether
there are any inconsistencies; and which is the preferred.

This example is being taken no further because its application to
investment decision analysis is remote. It does however have application in
other engineering fields such as detailed design or research work where
quantitative models are unsuitable. Interested readers should refer to a
more detailed text (12). It has been described here particularly to show
many of the characteristics of decision theory: it attempts to describe
real situations, it uses mathematical precision in the approach, it employs
useful techniques to aid manipulation, and it tackles systems without needing a quantifiable criterion.

The criterion need not be a single characteristic. Often one is faced
with a cost plus appearance--back to the dress again--or cost plus reliability, or public acceptance plus completion date plus cost. The combinations

are endless--any characteristic of importance to the decision can be included.

In investment decisions, every attempt is made to convert non-monetary qualities into monetary terms by determining what sum of money the quality is equal to. In public investment this tendency is not so strong, although the technique of cost/benefit analysis described in Chapter I, is attempting to convert "Qualities of Life" criteria into money equivalents, as well as identifying who benefits. However, many public sector engineering decisions are made without using a criterion based on money.

A technique for dealing with non-quantifiable multiple criteria is as follows:

Each characteristic is given a number of grades, say 1 to 3, or more if one feels capable of dividing the classification more finally. Then each alternative is judged against each characteristic and appropriate grades are awarded. The alternative with the maximum points can then be chosen. For example:

The dress we are still trying to buy should: suit your wife, preferably be blue to match existing accessories, be easily cleaned, and should not be too expensive. Say we find two dresses: both suit your wife, one is blue but very expensive, the other red and not so expensive, but of course not cheap. The red dress is non-iron, the blue is not. A selection can be made as follows:

	Red Dress	Blue Dress
Suits wife	3	3
Colour	1	3
Easy clean	3	1
Cost	2	1
Total Points:	9	8

The red dress is slightly preferred to the blue. This method is very rough, but it does provide a way of assessing alternatives which is systematic, and it enables a choice to be made when the preference is not obvious.

Though the method is rough, it has been used for decisions on large projects. For instance, it has been used with 10 grades, not just 3, to choose one from a number of proposed schemes for a motorway link into the town of Zurich.

The fact that it may be a rough method is not so serious as it first seems. Often a number of alternatives are equally good, and this produces

an impasse when a decision is required. In such a case a rough formalized
method can be a help in obtaining progress and when either alternative
would be equally good there is no point in applying particularly accurate
methods. If we think back to our dress--red is just preferred; if a more
accurate analysis showed blue to be just ahead, your wife would have probably been equally pleased. The rough method made up your mind and finished
the shopping.

But what if you really feel you should take the blue dress--this may
be because your wife needs a blue dress and you could predict a second
shopping expedition next month if you bought the red dress now.

The selection method can be extended by weighting the various characteristics according to their importance (12). As an initial assessment one
might subjectively set the following weights for the various characteristics.

	Weighting Factor
Suits wife	3
Colour	4
Easy clean	1
Cost	2

The choice of a weighting factor is very similar to the selection of
probabilities in Chapter 2. Having made an initial estimate one should
look at the various combinations to see that the weightings are reasonable
--   that is, one should "Test for Consistency".

Question: Is the combination "suiting wife plus easy clean" (weighting
           3+1) equal to the importance of colour (weighting 4)?

Answer : No--still a slight preference for the blue; increase the colour
           weighting to 5.

	Weighting Factor
Suits wife	3
Colour	5
Easy clean	1
Cost	2

Question: Is the combination "Cost plus suiting wife" equal to "Colour"?

Answer : No--colour would still be preferred, but cost is more important
           than easy clean; set cost weighting down from 2 to 1.5.

The revised weighting would then look like:

	Weighting Factor
Suits wife	3
Colour	5
Easy clean	1
Cost	1.5

Multiplying the points for each class by the weighting for that class, our choice of dress would now be as follows:

	Weighting	Red Dress	Blue Dress
Suits wife	3	3 x 3 = 9	3 x 3 = 9
Colour	5	1 x 5 = 5	3 x 5 =15
Easy clean	1	3 x 1 = 3	1 x 1 = 1
Cost	1.5	2 x1.5= 3	1 x1.5= 1.5
Total points:		20	26.1

With this analysis the blue dress is preferred and you are spared a second shopping expedition.

These methods are very useful in engineering design. Often details of a project must be decided where performance, reliability or "technological grounds" are more important than could be shown by any "normal" cost estimate. It is less useful in decisions concerning total capital projects, although as already mentioned, it finds use in some community projects, where service, rather than profit is the primary motive.

In these examples it is easy to see the relation between criteria and constraints. In the dress example, is colour a criterion or constraint? As the weighting increases, the characteristic changes from criterion towards constraint. The preference is almost constrained by "the dress must be blue".

The problem of determining the preference becomes much simpler when the criterion can be properly quantified. If, for instance, the cost can be used to summarize each alternative, the choice is the simple matter of finding the one with the lowest cost.

The problem is, of course, to summarize all the factors involved in terms of cost. Discussions, however, can now centre around the weighting and cost equivalent of the non-cost factors, which often enables more progress to be made than trying to compare the qualitative factors themselves.

3.4   DECISIONS-EVENT SEQUENCES

3.4a   Ṯhe Single Decision and Certain Event

    In real life there is often interaction between decisions and events.
The decisions are made to maximize criteria which are related to future
events. A simple example is the decision "Build" or "Do not build" a plant
for a product. The associated following event is the market outcome. The
criterion on which the decision is made is the profitability of the project
over its whole life. In other words, the event--the market outcome--must
be known to evaluate the profitability.

    Let us assume that, after due consideration given to possible criteria,
it has been agreed that NPW is the most appropriate and the NPW of a particular project under consideration is £100,000. The total situation, to
build or not, together with the resulting NPW's can be summarized as a
Decision Tree, shown by Figure 3.2.

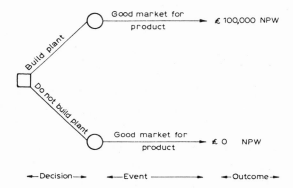

Figure 3.2 Decision Tree: Single Decision and Certain Event

    Notice the convention of such a diagram: the square represents a
decision point, the paths leaving it describe the various possible decision
outcomes, and the circles represent event points. Since the events in this
example are certain, each event point has only one outlet: the "good market
for product".

    On the right hand side of the diagram is the value of the criterion
associated with the various paths from the left hand side of the diagram.
The diagram is used to locate the best decision by finding the most profitable outcome on the right hand side and then retracing the path to this

from the first decision. Figure 3:2 shows this to be the top path,
£100,000 involving "build plant"--which, in this simple case is self-evident.

3.4b The Single Decision and Uncertain Event

We have just discussed the situation where the future event was certain
--there will be a good market for the product. This is rarely the real-
life situation, since future events are generally associated with uncertainty. There may be a good market for the product, or the product may not
sell at all. In real life, therefore, events have more than one possibility.
We might build a plant, and the market may be good, or it may be bad. This
information can be summarized diagrammatically as a <u>Conditional Decision
Tree</u> shown by Figure 3.3, as a development of Figure 3.2.

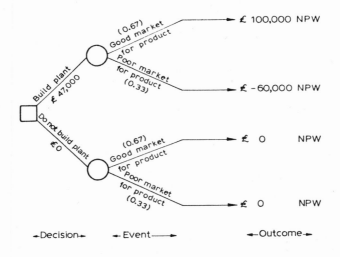

Figure 3.3
Decision Tree: Single Decision and Uncertain Event

This diagram has branching both at decision points and event points.
However, there is an important difference between them. Following the
definition of decision and event, a decision is entirely determined by the
decision-maker, but he has no control over the outcome of an event. That
is, he can decide which path of a decision fork the project will take, but
he cannot do this for the event forks. The project in reality may take

any event path. Notice we are allowing only specific events to occur. We
are not considering that intermediate situations can arise--the market may
be good or it may be poor, but we are excluding the possibility of the
medium market. If it is a possibility then we must have three or more forks
from the event point. The concept of Decision Trees requires that all
events have discrete outcomes. If event data is available as a continuous
function, it must first be converted into a series of discrete outcomes.

All the decision-maker knows about the event fork is its subjective
probability. He can estimate for instance that the market is twice as
likely to be good as to be poor. Hence the probability of the top event
path is 0.67, and the bottom is 0.33. These probabilites can be entered on
a diagram by each fork as shown in Figure 3.3.

Since we have no better information regarding the event fork, we must
work with Expected Values. We are faced with the alternative "build plant"
or "do not build plant" and we can determine the Expected Value of each
path as:

"Build plant" Expected Value = 0.67x100,000 + 0.33 x (-60,000)
                             = £ +47,000   NPW

"Do Not Build plant" Expected Value = 0.67x0 + 0.67x0
                                    = £ 0   NPW

These values can be entered on our Conditional Decision Tree against
the appropriate decision path, as again shown on Figure 3.3.

The decision-maker has full control on the path taken from a decision
point. In this example, if he chooses the top path--"build plant"--he will
have a higher Expected Value than if he does not build a plant.

Notice that we now have to work with Expected Values in Decision Trees
because of the uncertainty of events. Note that the Expected Value is not
the most likely--in our example the most likely is £100,000. In fact, the
Expected Value will never occur--the result will be either £100,000 or
£-60,000 if the plant is built.

The justification for using the Expected Value as the criterion in
Decision Analysis is as follows:

If the Expected Value is systematically used in all decisions taken,
then, if these decisions are independent, one has the greatest probability
of maximizing the total value from the whole set of decisions.

Hence, if a firm made tens of decisions a year, all of approximately the same magnitude, the Expected Value would be an appropriate criterion. However, if a firm made fewer than ten such decisions a year, or if a few of the decisions involved greater sums than the others, the Expected Value may not be a good criterion. This point is discussed at more length later in section 3.7a of this chapter.

It is normal for Conditional Decision Trees to work with Expected Values, and we will therefore do so here. However, one must bear in mind that if the Expected Value is not appropriate the Decision Tree must be re-examined with a more suitable criterion.

Schlaifer (20) in fact presents Decision Diagrams, which are basically Conditional Decision Trees except that instead of using probabilities to obtain single values from event forks, he asks the decision-maker to name a sum which he considers indifferent to the gamble represented by the uncer-tain event fork. This, in effect, is combining probability and utility which will be discussed later. In this way, the decision-maker can intro-duce his personal judgement at each event fork of the suitability of the criterion. However, it does require the personal intervention of a decision-maker who is capable of expressing life as a gamble--the method comes from America. Schlaifer is careful to point out that Expected Values can only be used when Utilities are linear.

3.4c A Sequence of Decisions and Certain Events

Let us move now to a time sequence of decisions, with events being taken as certain. Consider our earlier example of the manufacture of a product with a certain "good market". However, this time let us consider that the market can be satisfied in more than one step. The decision to build a plant is now followed by a decision to build again in three years' time. The decision now is the choice between three alternatives "build large plant," "build medium plant, " or "build no plant." The decision in three years is "build medium plant" or "build no plant." The market is always good, the Decision Tree is given as Figure 3.4, with appropriate outcomes. Each outcome is calculated by an economic analysis of the total path from the first decision to the end of the last event. We can now take the highest NPW, and choose this as the best path or plan out of the 6 paths shown in Figure 3.4. We can state that our best plan, which has a

NPW of £110,000 is to build a medium sized plant now, and a medium sized plant again in 3 years' time. Since the events are certain we can define all future decisions. A series of defined future decisions is called a plan. This is a very important concept for most of the discussions that follow in this book. The best plan for Figure 3.4 is shown in heavy type.

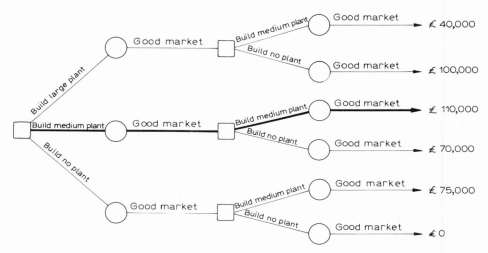

Figure 3.4   Decision Tree: Multiple Decisions and Certain Events

3.4d A Sequence of Decisions and Uncertain Events

The above case, which assumed that the events are known with certainty, does not normally apply to real situations because there is generally un-certainty prevailing about future events. This last case we will now consider involves sequences of decisions and uncertain events. This can represent real situations and is generally the form in which Decision Trees are used.

Consider a two-decision, two-event sequence as the previous example, but now we have uncertain events in that the market could be good or poor and in the second period the market grows or does not grow. Figure 3.5 shows the Decision Tree for the situation.

With two decision points, one with three and the second with two branches, and two event points, each with two branches, we have a total of

$$2 \times 3 \times 2 \times 2 = 24 \text{ different outcomes.}$$

Each of these outcomes has to be determined by an economic analysis

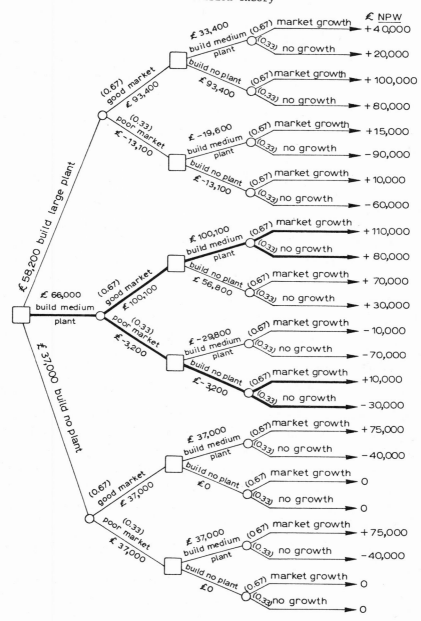

Figure 3.5

Conditional Decision Tree for a

Sequence of Decisions and Uncertain Events

for the path leading to that outcome.

Figure 3.5 shows the highest outcome is £110,000, which is obtained by the plan "build two medium sized plants" when the market turns out to be good for both periods. However, Figure 3.5 shows that this plan could give values of:

£110,000

£ 80,000

£-10,000

£-70,000

depending on the outcome of the uncertain events.

In order to deal with this problem of uncertainty, the Conditional Decision Tree can again be used to determine the best decision to make, based on Expected values. Let us assume again that the probabilities are 0.67 and 0.33 for both events. This data is written in Figure 3.5--again in brackets against the appropriate event fork. It is now possible to work back from the outcome, to determine the Expected Value before the second event. For example, the top value is:

$$0.67 \times 40,000 + 0.33 \times 20,000 = 33,400$$

Similarly, for the second:

$$0.67 \times 100,000 + 0.33 \times 80,000 = 93,400$$

Now consider the second decision on the top part of the Decision Tree: Faced with a choice between an Expected outcome of either 33,400 or 93,400, one would take the 93,400. Since the decision is completely within the control of the decision-maker, we can assume he will make this choice and so the Expected value of the plan, as seen before the decision, is £93,400. Hence, we can write £93,400 down as the value of the plan before this second decision on Figure 3.5.

This can be repeated for the second to the top branch, giving a value of £-13,100 before that second decision.

Since we now have the two Expected values after the first event and also the probabilities of these events occurring, we can again calculate the Expected value as seen before the event. This is:

$$93,400 \times 0.67 + -13,100 \times 0.33 = 58,200$$

This then is the Expected value the decision-maker sees as the outcome for building a large plant as his first "decision". If the whole exercise

were repeated twice more, we can obtain the Expected value which the
decision-maker sees as the outcome for building a medium plant, and for
building no plant as his first decision. These values are written on the
Decision Tree as:

    £66,000 for building a medium sized plant

    £37,000 for building no plant

    £58,200 for building a large plant.

  The decision-maker is now in a position to make a decision; if he
accepts Expected value as his criterion, he should build a medium sized
plant as this will give the greatest Expected NPW. Note that this decision
has taken into consideration all the possible outcomes by using Expected
values.

  Notice that we are interested in making only the first decision. We
must not "make" the later decisions but we have to give consideration as to
what they might be before we are able to take the first decision. Because
the events are uncertain, we have to wait for the outcome of the past event
before making the following decision. We have said the first decision is
"build a medium sized plant." Now, if the market turns out to be good,
from the Decision Tree one can see the best second decision is again "build
a medium sized plant" which has the highest Expected Net Present Worth of
£100,100. However, if the sales realized are poor, our second decision
should be: "do not build" since this produces an Expected NPW of £-3,200
compared with £-29,800 for building a second plant. In fact we have a
"plan" for every future event branch, each plan being conditional on the
outcome of each event. A set of conditional plans is called a strategy.
A strategy is defined by (20):

    'A strategy is a rule which prescribes exactly what decision
    should be made in every situation in which a choice may have to
    be made before the evaluation date is reached.'

  The best strategy for our example is therefore:

     Condition   Action

                Build a medium sized plant initially

and, if the market is good: $\Longrightarrow$ build a second, medium sized plant in
                the 3rd year

but, if the sales are poor: $\Longrightarrow$ build nothing in the 3rd year.

This strategy is shown in heavy type on Figure 3.5. Note that the strategy has a range of outcomes to which probabilities can be assigned and an Expected value calculated.

The probability of each outcome and the Expected value of this particular strategy can be calculated as follows:

Outcome	Probability	
110,000	0.67 x 0.67 = 0.449	49,390
80,000	0.33 x 0.67 = 0.221	17,680
10,000	0.67 x 0.33 = 0.221	2,210
-30,000	0.33 x 0.33 = 0.109	-3,370
	Expected value:	66,010

The strategy, like the plan, is an important concept that will occupy much of the following discussion in this book. A strategy is alway involved when future experience is going to provide information which affects future decisions. The plan is used when future experience cannot provide better information, either because the future is certain, or because the distant future is so independent of the near future, that there will never be any better information available than at present. Under such circumstances one must use a plan even in situations involving uncertain events.

3.4e Correlation between Events

The problem of correlation introduced in Chapter 2 was shown to be very important to the combination of probability distributions. Correlations of probabilities of uncertain events also occur in Conditional Decision Tree Analyses. This is taken into account by using different probabilities for the same event on different forks of the Tree.

Our last example gave equal probabilities to the same event wherever it appeared on the diagram. This means that the chance of market growth (the second event) is equally likely if the market is good, or if the market is poor. Usually, a greater probability of growth can be expected if the market has previously been good. That is, the first and second events are related--there is some correlation between the events.

What should be used in the Decision Tree is one probability for the growth where the market is good which might be 0.8, and a different probability of growth when the market has been poor, possibly 0.1.

The probabilities required are the Conditional probabilities. Firstly, market growth, A, given the market has been good, B, or P (A | B) and the probability of market growth given that the market has been poor, C, P (A | C).

The probability used in Decision Trees should be the Conditional probability given that all the prior events on the Decision Tree have occurred in the particular way required to get to that position on the Tree.

This presents no problem when the events are independent, but, when there is a degree of correlation, conditional probability must be subjectively estimated and the allocation of probabilities to the event forks must be done with considerable care.

In arriving at reasonable probabilities for Decision Trees, one needs to use some rules for determining the probability of combinations of events. The following rules should be adequate for most decision theory problems (17).

The probability that A or B occur (written as P(A ∪ B)) is given by

$$P(A \cup B) \;=\; P(A) + P(B) \tag{3.1}$$

where P(A) and P(B) are the independent probabilities of the two events where A and B are not mutually exclusive.

Where A and B are mutually exclusive, i.e. it is impossible for B to happen if A has already happened and vice-versa, then

$$P(A \cup B) \;=\; P(A) + P(B) - P(A \cap B) \tag{3.2}$$

where P(A ∩ B) means the probability that A and B both happen. "Mutually Exclusive" means alternative ways of doing the same thing. For example, A and B could be different causes of a single deficiency in a product which results in poor sales.

The probability that A and B will occur is given by

$$P(A \cap B) \;=\; P(A) \times P(B) \tag{3.3}$$

if A and B are independent events, i.e. with no correlation.

However, if there is correlation between A and B, the probability of B occurring depends on whether A has occurred. It is no longer possible to use a simple probability of B occurring P(B), but the conditional probability of B occurring given that A has occurred--written as P(B | A).

Hence, for dependent events, the probability of A and B occurring is given by:

$$P(A \cap B) = P(A) \times P(B|A) \tag{3.4}$$

The dependence means that $P(B|A)$ would be different from $P(B)$. If A and B were 100% positively correlated, then $P(B|A)$ would be 1.0; a 100% negative correlation would give $P(B|A) = 0.0$. Intermediate degrees of correlation will produce intermediate values.

3.4f Markov Chains and Decision Theory

There are types of problems in which the uncertain future can be described by a Markov chain. That is to say, the situation tomorrow, or next month, or next year, is uncertain, but it can be related to the situation now by a Table of Transitional probabilities as described in Chapter 2 section (2.7). If these problems involve making decisions, and these decisions affect the Transitional probabilities, then we can apply Markov chain theory to choose the best decision.

Consider a marketing situation which can be described as a 3 month stage-wise problem, in which the marketing manager wants to know whether the best policy when sales fall is to

(a) Run an advertising campaign (policy 1) or

(b) Reduce the price by 10% for the period (policy 2)

Suppose he can give Transitional probabilities for the market next quarter for each of his policies as:

Policy	Statement of policy	Probability of outcome in next quarter
Policy 1	This quarter sales successful-- "do nothing" for next quarter.	0.80 probability of successful sales. 0.20 probability of poor sales.
	This quarter sales poor-- policy "Advertising Campaign" for next quarter.	0.60 probability of successful sales. 0.40 probability of poor sales.

Policy	Statement of policy	Probability of outcome in next quarter
Policy 2	This quarter sales successful-- "do nothing" for next quarter.	0.80 probability of successful sales 0.20 probability of poor sales
	This quarter sales poor-- "reduce price for the period" for next quarter	0.70 probability of successful sales 0.30 probability of poor sales.

The Transitional Matrix for policy 1 is:

Policy 1

This quarter	Next quarter	
	success	poor
success	0.8	0.2
poor	0.6	0.4

The Transition Matrix for policy 2 is:

Policy 2

This quarter	Next quarter	
	sucess	poor
success	0.8	0.2
poor	0.7	0.3

The different policies give different Transitional Matrices, and these will give different probability for successful and poor sales at equilibrium. Markov chain theory enables these equilibrium probabilities to be calculated, from which Expected Values can be determined and compared with the costs of carrying out the various policies. In this way an economic basis for deciding on the policy to follow can be obtained.

Markov chains can also be used in conjunction with integer searching techniques to determine the optimum decisions sequences to use in stocastic situations (13).

Markov decision theory finds more application in operating policy decisions than investment decisions. Although it appears to be a promising technique for determining investment policies in a stocastic market, no

particular applications have appeared in the literature. The disadvantage
may be that investment problems have comparatively short chains with equilibrium not being achieved, whereas operational problems have much longer
chains, with correspondingly more importance being put on the equilibrium
condition.

3.5 THE VALUE OF INFORMATION ON EVENT PROBABILITIES

As described in Chapter 2, forecasting methods can be either quick and
approximate, or lengthy and sometimes more accurate. The question whether
to improve the forecast by an expensive study, such as a full-scale market
research programme is a question of comparing the cost of such a study with
the savings made by having the more accurate forecast. There are many
occasions where the decision remains the same over a very wide range of
forecasts for a future event. For instance, one might be going to de-bottle-
neck a plant as long as there was a forecasted sales increase. Whether
the increase was 2 or 20% may be irrelevant, the decision would remain the
same. In such a case there is no advantage in having accurate forecast
data. This approach is contrary to the normal training of engineers and
scientists who normally deal with continuous functions. Decisions are
integer functions, which mean a whole range of data can result in the same
decision being taken.

The Decision Tree offers a method of assessing the value of perfect
information--that is, complete removal of uncertainty (20, 30). This is
best described by means of an example.

Referring back to Figure 3.5 and the first uncertain event, the immediate future market, let us assume we could carry out a market research programme to establish with certainty whether the market will be good or poor.
How much would this information be worth to us?

If we know that it was to be a good market, we would neglect the poor
sales branch of the first decision on Figure 3.5 and choose to build a
medium sized plant with an Expected value of £100,100. Since this is
exactly the same decision we would have taken without the extra information,
the information saves us £0.

If, on the other hand, the information showed that sales would be poor,
Figure 3.5 shows we would decide to build no plant (Expected value £37,000)
and not build a medium plant with an Expected value of £-3,200 for poor

sales. Hence, in this case, we would change our decision and there would
be an Expected saving of £37,000 - £3,200 = £33,800.

Since our a priori probability estimate for this event is 0.67 good
market, 0.33 poor market, this information gives an Expected saving of:

$$0.67 \times 0 + 0.33 \times 33,800 = £11,154$$

This then is the value of the information, assuming that the study
would produce certain information. Since no market survey can produce
complete certainty, the market survey should cost less than £11,154 for it
to be justifiable.

3.6 DECISIONS INVOLVING THE COLLECTION OF INFORMATION

The previous paragraph showed how the Decision Tree could determine a
value of perfect information. However, information is rarely perfect,
either because events will occur beyond any power of prediction, or because
the cost of obtaining perfect information is too high.

The application of Decision Trees to the decision to collect imperfect
information receives considerable attention in decision theory literature
(17, 23). The question as to whether money should be spent or the project
delayed in order to get better forecast information must frequently be
asked. Decision Trees provide a clear method of answering these questions.

Let us re-consider the expansion decision given on Figure 3.5. Suppose
we had the opportunity of having a sales forecast carried out to predict
the outcome of the first event. From previous experience of this forecasting method we believe the results have a 80% probability of being correct
and it will cost £1,000. We want to know whether to go ahead with the forecast or not.

Reverting to the Decision Tree, our very first decision should now be
"conduct a sales forecast" or "do not conduct a sales forecast". The Tree
on the "do not conduct a forecast" branch is the same tree as Figure 3.5,
but the "conduct a forecast" branch will be different.

Since the analysis must be done before we know the result of the
survey, the survey itself must be considered as an event with the two outcomes "survey forecasts good market" and "survey forecasts poor market,"
each with probabilities attached to them. This second half of the Decision
Tree, with the extra event point included is shown as Figure 3.6.

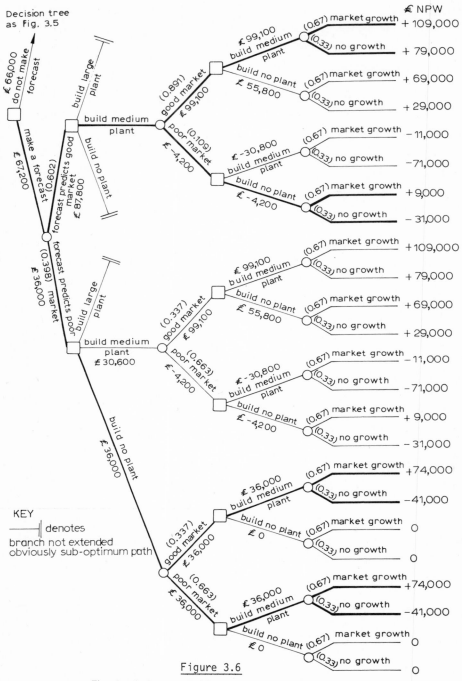

Figure 3.6

The Decision to Collect more Information

We now have the problem of assigning probabilities to these branches, knowing only that there is a 0.67 probability that the market will be good and the forecasts have a 0.8 probability of being correct.

To do this, let us look at the possible combinations of their being a good market or a poor one, and the probability of the forecast indicating a good or a poor market. The set of possibilities are conveniently summarized on Figure 3.7. On this figure we can enter our known data concerning the probabilities of the market and the forecast accuracy on the event branches as this enables us to calculate the probabilities of the four outcomes. These are entered on Figure 3.7

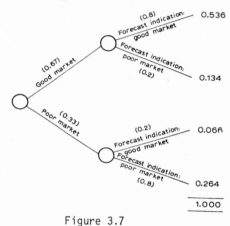

Figure 3.7

The Combination of an Uncertain Event

and Imperfect Information

Now, for our Decision Tree we must assign probabilities to the forecast event forks. The probability of a forecast good market from Figure 3.7 is

$$0.536 + 0.066 = 0.602$$

and the probability of a forecast poor market is

$$0.134 + 0.264 = 0.398$$

To complete the Decision Tree we also need the probability that the market will be good, given that it has been forecast good and the other corresponding three-conditional probabilities to go with the event forks concerning the market. Figure 3.7 shows the probability of a good market, when the forecast is "good market," is 0.536, and we have just calculated that the probability of a forecast good market is 0.602.

Therefore,

the probability of a good market, <u>given</u> a predicted good market is 0.536 / 0.602 = 0.890

Similarly,

the probability of a poor market, <u>given</u> a good forecast is 0.066 / 0.602 = 0.110

the probability of a good market, <u>given</u> a poor forecast is 0.134 / 0.398 = 0.337

the probability of a poor market, <u>given</u> a poor forecast is 0.264 / 0.398 = 0.663.

We now have all the probabilities required on the event forks on Figure 3.6. The Tree has now sufficient data for it to be evaluated in the normal way.

This is an important concept in decision analysis--the modification of prior knowledge as a result of imperfect posterior information, and it is used whenever there is a decision required on the value of obtaining more information, be it forecasting market trials, research development work or prototype tests.

These manipulations with probabilities can alternatively be done concisely using Bayesian Statistics, as follows:

Using the notation M for market outcome, and subscript 1 or 2 for good or poor, and F for forecast, subscripted 1 or 2 for forecasted good or poor market, one can write the problem down as:

Given $P(M_1)$, $P(M_2)$ and $P(F_1 \mid M_1)$ and $P(F_2 \mid M_2)$, we require to know

$P(M_1 \mid F_1)$, $P(M_2 \mid F_1)$ $P(M_1 \mid F_2)$, $P(M_2 \mid F_2)$ and also $P(F_1)$ and $P(F_2)$ for entry on the Decision Tree.

Using the probability rules given in Section 3.4e, we can summarize the necessary manipulations as follows:

$$P(M_i \cap F_j) \;=\; P(M_i) \times P(F_j \mid M_i) \tag{3.5}$$

$$P(F_j) \;=\; P(M_j \cap F_j) + P(M_i \cap F_j) \tag{3.6}$$

and

$$P(M_i \mid F_j) \;=\; \frac{P(M_i \cap F_j)}{P(F_j)} \tag{3.7}$$

where i, j = 1,1 1,2 2,1 and 2,2.

We can now finish the Decision Tree using the calculation procedure as for the previous Decision Trees, writing down intermediate values on Figure 3.6. Note that all final outcomes are £1,000 less than Figure 3.5 because of the cost of the forecast. Not the whole analysis is shown because Decision Tree Diagrams become complex and it is an accepted technique to ignore branches that can be shown by inspection not to lie on an optimum path (20). This can be done for three branches on Figure 3.6.

The important point to note is that if the forecast is a poor market, the decision with the highest Expected value is "do not build"--£36,000 compared with the earlier decision to build a medium plant which now has a value of £30,600. This change in strategy causes the decision to "make the forecast" to be better than "do not make the forecast" (£67,200 cf £66,000).

The best strategy would then be:

Condition		Action
		Obtain a forecast
if the forecast is good market	$\Longrightarrow$	build a medium sized plant in the first year
and if the market is good	$\Longrightarrow$	build a second medium sized plant in year 3
or if the market is poor	$\Longrightarrow$	build nothing in year 3
But if the forecast is a poor market	$\longrightarrow$	build no plant in the 1st year
and if the market is good	$\Longrightarrow$	build a medium plant in year 3
or if the market is poor	$\Longrightarrow$	build no plant in year 3

This strategy is marked in heavy type on Figure 3.6

The treatment in this Section is applicable to Decision Trees which involve the collection of information upon which future decisions are based. The technique is known as Preposterior Analysis (17, 20).

Reduction of the Tree Size

From the examples already given it is clear that Decision Trees very quickly become large and complex. We have been considering up to 3 decisions and 3 events, mostly with only two branches each, and already we have had to discuss only partial diagrams. A reasonable problem may have four decisions and four events with four branches on each--a total of 4^8 or 65,536 paths.

Furthermore, we have been omitting the major work involved which is the separate economic evaluation of each path. The above case would require 65,536 separate project evaluations to be carried out before the Decision Tree analysis could begin.

It is therefore necessary to reduce the number of paths by, for instance excluding those paths which are primâ facia not the optional path.

In the example given on Figure 3.5 the following path is redundant and need never be evaluated:

large plant, poor sales — build medium plant — leading to 2 paths.
In Figure 3.6 the following paths would not be worth evaluating:
good forecast — do not build — leading to 8 paths
and
poor forecast — build large plant — leading to 8 paths.

When the arithmetic becomes too much, we can always fall back on the computer (17, 20), although it is surprising how little computerization of Decision Trees seems to have been done. This will be discussed at length in later chapters.

3.7 THE UNCERTAIN EVENT

Chapter 2 discussed the uncertainties associated with data used in economic evaluations. This chapter has, so far, only discussed uncertainty in events, which usually only implies market volume. Market volume is, however, only one of the seven data inputs, all of which are subject to uncertainty.

Theoretically, one could consider each uncertain data item as an event, and if it has the continuous distribution, employ the concept of event fans instead of branches (20), and treat the whole as one enormous Decision Tree. However, this would involve an impossible amount of computation and so it will not be considered further here.

If we use Decision Trees for some uncertain events, and exclude others, the effect of this uncertain data would be to give every outcome on the Decision Tree a distribution instead of it being single valued. When the distribution of each outcome is small compared with the total distribution of all the outcomes, it is reasonable to separate the problem into two stages:

1. Use the Decision Tree to locate the best plan and strategy by including, in the Tree, only gross uncertainties--e.g. market volumes or prices.

2. Investigate the effect of the more accurately known data on the economic analysis in a later separate stage, using a method much more suited to the aggregation of larger quantities of probabilistic data.

If, however, the uncertainty in the data is so high as to make alternative paths and strategies of a Decision Tree statistically insignificantly different from each other, it is impossible to choose a best plan, or strategy. The benefits of a Decision Tree study are largely lost and one must be satisfied with a number of strategies being approximately equal.

3.7a Criteria for Use with a Range of Outcomes

So far, we have worked entirely with Expected values when faced with probabilistic sets of data. This was defended by noting that, should the Expected value be used a sufficient number of times, the sum total will be maximised.

However, many decisions, particularly investment decisions, are not taken a sufficient number of times within one firm for the high and low fluctuations to cancel each other out. It is no good telling the Bankruptcy Court that it was just a down period and because you are using Expected Value as the criterion you will be all right in the end!

The maximum possible loss that any decision can incur should not be greater than the company can stand, however small the probability. Nor are decisions involving losses of somewhat smaller magnitude acceptable to a firm. A firm has to budget its future finances, and an unexpected loss would cause a change in plan or an embarrassing money-raising action to be necessary. One can also look within a firm at the various teams and managers controlling the projects. At this level, individuals are never going to handle enough projects to allow the Expected value to even out their records. These individuals do not want to be associated with any project if it has a probability of making an embarassingly high loss.

The decision-maker is never indifferent to two projects with identical Expected values but with differing distributions; usually the narrower distribution is preferred. Taking the argument one step further we can say that a decision-maker is normally willing to trade off some of the Expected value for a narrower distribution.

We again, therefore, have a problem of choosing a criterion by which to make the decision. When we are faced with a distribution of values, what rule or "criterion" do we employ to make the choice? Notice that now, in any evaluation, we have two "criteria". Firstly, the quality being judged--attractiveness of a dress or a semi-quantitative points scheme, or NPW, or DCF for an economic analysis; and secondly this new criterion--the treatment of the distribution of the values of the first criterion. To differentiate, one will be called "evaluation criterion" and the second, "decision criterion". Unfortunately there is no standard nomenclature yet developed in the literature.

To discuss various decision criteria, let us consider two alternatives which each have three outcomes. A choice from these two must be made. The system contains alternative actions which are the actions we can take, the states of nature over which we have no control, and the outcomes. Every alternative action and state of nature will produce a corresponding out-come. The results can be conveniently presented in tabular form as Table 3.1. This is called a Decision Matrix and is particularly useful in decision work.

TABLE 3.1

Decision Matrix

		Outcomes		
		States of Nature		
		(1)	(2)	(3)
Alternative Actions	Alternative A	102	50	-2
	Alternative B	105	35	10

Some decision criteria are based on the assumption that no probability can be attached to the various outcomes. For instance, subjective probability estimates may be completely impossible to obtain, or one might not accept the philosophy that subjective probabilities can be worked with. The more important ones in this class are: MINIMAX, Equal Likelihood, and Minimal Regret.

If one accepts the Bayesian Statistics approach--that meaningful subjective probabilities can be allocated to the various outcomes--this leads to the Expected value and Certainty Equivalent methods.

The choice of a decision criterion is important to the outcome of a decision. This will be shown by making a choice for the problem put forward in Table 3.1 by each of the above criteria in turn.

The MINIMAX or Wald Criterion

This criterion chooses the alternative that has a Minimum maximum loss. Applying this criterion to our example we would choose alternative B, because its maximum lowest outcome is 10, compared with -2 for alternative A.

This criterion is unduly pessimistic, and is unsatisfactory in that it rules out the choice of projects with high gains if they are associated with any chance, however slight, of a high loss.

The Equal Likelihood or Laplace Criterion

The Equal Likelihood criterion attempts to make use of all the information in the absence of probabilities by assuming that each outcome is equally likely. It therefore determines the average outcome of each alternative, and chooses the one with the highest average. In our example, this criterion would rank both alternatives equal since they both have average values of 50.

This method can be criticized in that there are no grounds for assuming the chance of the outcomes will be equal, and there is no restraint imposed when very high losses are involved. A chance of a high gain balances a high loss, even though the loss may be sufficient to bankrupt the organisation concerned.

The Minimum Regret Criterion

This criterion requires the construction of a Regret Decision Matrix.
This is similar to the Decision Matrix, but the entries are the losses
involved through not choosing the best alternative for the state of nature
in question. Table 3.2 gives the Regret Matrix for our example. The
criterion now chooses that alternative with the smallest maximum regret.
This is A in Table 3.2 because the maximum regret is 12, compared with 15
for alternative B.

TABLE 3.2

Regret Decision Matrix

		Regret Outcome States of Nature		
		(1)	(2)	(3)
Alternative Actions	Alternative A	3	0	12
	Alternative B	0	15	0

This method is attractive in that it concentrates on the opportunity
loss. In a competitive situation this is probably more important than the
calculated actual loss, since, if all the competitors are in the same
situation, in the event of an unfavourable state of nature developing,
changes will be made--probably price adjustments--to enable the industry to
survive. This theme was discussed in Chapter 2 under Price and Uncertainty.
The man who is really in trouble is the one who falls behind his competitors,
whether the "State of Nature" is favourable or unfavourable.

Hence the Maximum Regret criterion is attractive when competiton is
significant. However, like the MINIMAX criterion, it concentrates entirely
on the pessimistic side and the chance of high gains may be lost because of
a slightly higher opportunity loss.

There are other similar criteria: the optimistic MAXIMIN, maximum
payoff, or the Optimistic/Pessimistic criterion of Hurwicz (21) which
weights the maximum and minimum values differently, and chooses the project
with the highest sum of weighted maximum and minimum. However, since this
book assumes that one can assign probabilities to all the data items in
capital investment, we will proceed no further with these methods which
presume complete uncertainty, and move on to methods which assume probabilities can be allocated to each State of Nature.

Let us assume the following probabilities can be allocated to the example given in Table 3.1

State of Nature	(1)	(2)	(3)
Probability	0.30	0.60	0.10

Expected Value

This criterion has already been introduced, both in Chapter 2, Section 2.5a and during the development of Decision Trees in Section 3.4b.

The Expected value, calculated as

$$E(x) = \sum_{i=1}^{n} P(x_i)x_i \qquad (3.8)$$

where x_i is the value of the component i with probability $P(x_i)$ for a total of m outcomes.

Or, for a continuous distribution, with a frequency distribution $f(x)$:

$$E(x) = \int_{-\infty}^{+\infty} x \ f(x) \ dx \qquad (3.9)$$

The advantages of this criterion are that it weighs all the outcomes correctly, and if it were to be used with many decisions it would produce the maximum total payoff.

Its disadvantage is that it allows losses of serious magnitude to be hidden by less important chances of high gains. Before a sufficient number of decisions can be made to even out fluctuations, a single high loss may have financially damaged the company.

The Certainty Equivalent Criterion

The Certainty Equivalent method has been developed specifically to account for the degree of risk and yet still employ probability distributions in the same was as the Expected Value method. This is done by weighting each value according to the magnitude of its loss or gain and then finding the Expected value of the weighted values.

These weighted values are called Utilities and the method is the Cardinal Utility Theory, developed by Von Neumann and Morgenstern (8,28,29). The theory assumes that there is something other than the normal value

that one tries to optimize in a gambling situation. This "something" is a
combination of value and risk, and it is a personal characteristic of the
decision-maker. It is the utility. Hence, given the person's own relation
between value with risk and utility, one can choose the decision which gives
the greatest Expected Utility. This would be the decision that that individual would make as long as he had a consistent attitude to risk. By
using the same value/risk utility once more, it is possible to convert the
Expected Utility back into an equivalent risk-free money value. This value
is the Certainty Equivalent of the decision, and is effectively the
Expected value, after weighting the various outcomes with the risk involved.
According to the Certainty Equivalent criterion, one chooses the alternative
with the highest Certainty Equivalent.

This method requires the calibration of the individual decision-maker's
attitude to risk. This has to be done by interview, in which the decision-
maker is asked to make a number of choices in risk situations. With the
interview questions carefully designed, it is possible to construct a
calibration curve.

A series of gambles has to be presented to the decision-maker, and he
has to give the equivalent single sum that he would be willing to give,
or receive, in exchange for the gamble.

A series of such questions and the decision-maker's answers could be
given in Table 3.3:

TABLE 3.3

	Gamble Presented (Question)	Decision-Maker's Equivalent Single Value (Answer)
1.	0.5 probability of + £1,000,000 and 0.5 probability of £0	£200,000
2.	0.5 probability of £200,000 and 0.5 probability of £0	£ 80,000
3.	0.6 probability of £-80,000 and 0.4 probability of £200,000	£ 0
4.	0.2 probability of £-200,000 and 0.8 probability of £1,000,000	£150,000

Now the utility of the value is that property which linearizes the desirability of the value, so that the mean desirability can be obtained by addition. It has an arbitrary scale with an arbitrary zero.

Let us define the Utility of £0 (U(£0)) as 0 "Utiles" and the Utility of £1,000,000 (U(£1,000,000)) as 1.0 "Utiles," then by simple substitution for each line of Table 3.3:

From question 1:

$$0.5 \ U(+£1m) + 0.5 \ U(£0) = 1.0 \ U \ (£200,000)$$

therefore U (£200,000) = 0.5 x 1.0 + 0.5 x 0 = 0.5 Utiles

From question 2:

$$0.5 \ U(+£200,000) + 0.5 \ U(£0) = 1.0 \ U \ (+ £80,000)$$

therefore U (£80,000) = 0.5 x 0.5 + 0.5 x 0 = 0.25 Utiles

From question 3:

$$0.6 \ U(£-80,000) + 0.4 \ U(+£200,000) = 1.0 \ U(£0)$$

therefore U(£-80,000) = (1.0 x 0 - 0.4 x 0.5)/0.6 = -0.33 Utiles

From question 4:

$$0.2 \ U(-£200,000) + 0.8 \ U(+£1,000,000) = 1.0 \ U(+£150,000)$$

Now U(£150,000) can be obtained by graphical interpolation from the earlier values for U(£200,000) and U(£80,000), as 0.4 Utiles,

therefore U(-£200,000) = (1.0 x 0.4 - 0.8 x 1.0)/0.2

= -2.0 Utiles

These 6 points are plotted on Figure 3.8 to give the Utility Function for the individual who was interviewed.

Figure 3.8 shows the general characteristics of all Utility Functions. They are monotonically increasing functions which diverge at one end to minus infinity and asymptote at the other to a maximum utility value. These characteristics are equivalent to saying that there is a certain loss beyond which no individual will go, and that when high gains are involved one becomes indifferent to even higher gains. The establishing of the Utility Function can be obtained from a series of questions, as already described. The questions however, must be carefully tailored to ensure that only one unknown is involved in each question. Interpolation can be used, as seen in the 4th question of Table 3.3 to reduce this restriction, but it is a very elementary method of curve fitting, particularly as in

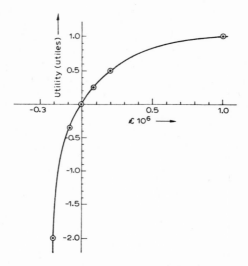

Figure 3.8   A Utility Function

such a subjective analysis one would expect inconsistencies which upset
single interpolations. Normal regression analysis can be used to fit a
curve through a set of data points from an interview once an equation for
the Utility Function is known.

A suitable equation has been reported as (25):

$$U(x) \;=\; A + B \; \ln(x+C) \qquad\qquad (3.10)$$

Interview data can be fitted to this equation by any standard non-
linear regression package. Any serious inconsistencies in the data brought
to light by fitting can be clarified by having a second interview. A
discussion on methods of interview is given in a paper by Spetzler (25).

Having now discussed Utility somewhat more mathematically, we can
define the certainty equivalent (CE) as: the monetary value with the same
Utility as the Expected Utility (E(U)), that is:

$$CE \;=\; \exp[(E(U)-A)/B] - C \qquad\qquad (3.11)$$

where

$$E(U) \;=\; \sum_{i=1}^{n} P_i \; U(x_i) \qquad\qquad (3.12)$$

for an alternative with n possible outcomes of x_i, each of probability p_i.

A further interesting figure that can be obtained from a Utility study is the <u>Risk Premium</u>. This is effectively the premium that one is willing to pay to insure against the risks.

    Risk Premium = Expected Value - Certainty Equivalent

A Corporate Utility Function

Cardinal Utility Theory is based on finding the Utility Function for the individual decision-maker. Strictly speaking, this individual must also be the originator of the subjective probabilities associated with the distribution of outcomes.

However, in practice, decisions are not made by one individual but by a body of individuals, usually the top executives of a firm.

Hence, what is required is a "corporate" Utility Function and not an individual one. Similarly, the estimated probabilities are "corporate estimates," probably by the company expert whose judgement the decision-makers are willing to accept.

Such a Corporate Utility Function enables consistent decisions to be made throughout the firm, and it would also define to middle management the company policy towards risk.

An attempt at obtaining a Corporate Utility Function for a large company is described in a very interesting paper by Spetzler (25). He interviewed 36 managers from one company and discovered a very wide range of attitudes to risk. He describes the development of an effective interview technique which was then used by the seven top executives of the company. This led to discussions on risk policy from which a company policy to risk could be evolved. The work however does not seem to have gone so far as to produce a single Utility Function that the executives agreed would be "company policy."

This same work also compared the Utility Function for small and large projects. The results of the interview confirmed what one might expect if utility theory is to be at all valuable, namely, that the same utility function could represent both small and large projects. A person's attitude to risk is dependent only on the magnitude of the risk and it is not pre-conditioned by the magnitude of the project.

Advantages of the Certainty Equivalent Criterion

Assuming the probabilities of various possible outcomes are available, there is little doubt that the criteria that use probability--Expected Value and Certainty Equivalent--are better than the earlier described criteria.

The lack of protection against the possibility of a large loss associated with the Expected Value criterion is also well accepted, and a common approach is to use a Conditional Expected Value (17), that is to:

"Choose among the alternative courses of action on the basis of the combination of (a) Expected Value and (b) exposure to gain and loss that is most consistent with the decision-maker's objectives and attitudes towards risk."

This definition is rather vague to be considered as a "criterion" since it is leaving reference to risk to some ill-defined individual judgement.

We could make this definition less vague by saying, "The choice should be based on Expected value, subject to the condition that there is no chance of an outcome with a loss of more than £x, where x is defined by the company with reference to its financial situation."

This criterion still has a serious disadvantage: the stepwise nature of the condition. A project with a chance of a loss of £$(x-1)$ would be equal to one involving the same Expected value but no risk.

The Certainty Equivalent Criterion is doing no more than making this discontinuous step into a smooth effect. It is therefore preferred to the Expected Value criterion, as well as to the other criteria that do not use probabilities of the outcomes.

The use of the Certainty Equivalent has also the following secondary advantages:

a) The determination of the Utility Function has an educational effect on the decision-maker who has to concentrate on his own attitude to risk, which he may never have done before. In this way, it makes him analyse more carefully his attitude to risk, which may result in more balanced decisions. It is often reported that individuals take over-conservative decisions in situations involving risk (27).

b) The attitude to risk can be defined as a company policy and so decision-making can be delegated to middle management, and still consistent decisions can be taken.

c) Certainty Equivalent is a quantitative continuous function and so it
provides a means of making "optimum" decisions in risk situations.
This applies only when "other factors" are equal. Chapter 1 emphasized that non-quantitative factors should, and must, be taken into
account before a decision is made, and earlier in this chapter
"multi-characteristic" criteria are discussed. However, should
there be no such "other factors" to include, the Certainty Equivalent method can be used to make the best decision.

Disadvantages of the Certainty Equivalent Method

Though Utility Theory was presented in 1947 (29) it is still rare to
find it being applied. One reason for this is that it is one further stage
of development beyond the use of Risk Analysis and the presentation of outcomes as distributions. Since few companies employ Risk Analysis for the
selection of alternatives, there is little interest yet in various forms of
decision criteria.

The method was first reported as being used in decision-making by
Grayson in his book on Oil-Drilling Decisions (6). Its use seems to have
been restricted to research and consultancy applications, for instance, by
the Standard Research Institute and the Harvard Business School. All, of
course, report favourably on the results of its application!

There is a particular disadvantage that restricts its free introduction
within companies: the need to interview the management. Many techniques
are introduced into companies quietly, at a low level, and if they are
successful they are then sold to higher management. Any method which, as
its first step, involves interviewing the senior management, asking them a
lot of silly hypothetical questions about characterizing their intuitive
attitude to decision taking is bound to be met with a very icy response.

If a simpler method of determining the Utility Function could be
developed, trial schemes could be made by middle management in the planning
department. If such trials were successful, a firm proposal could be made
to senior management.

3.7b The Use of Decision Criteria with Decision Trees

Conditional Decision Trees select outcomes with known probability distributions. This selection must be done using a decision criterion capable of incorporating the probabilities, that is, either Expected Value or Certainty Equivalent.

The description of Decision Trees has assumed that Expected Value is used for the selection of optimum plans and strategies. However, if Utility and Certainty Equivalent are accepted and preferred by the planning group, the choice of path on the Decision Tree should be based on Certainty Equivalents and not Expected Values. Compared with Expected value, a Certainty Equivalent criterion has the effect of penalizing those paths which involve some high losses, and reducing the credit given to paths with above average outcomes. These changes can result in different paths being chosen, and so can give a different optimum strategy. It is therefore important to use the same Decision Criterion during the evaluation of a Decision Tree, as used later in discussing results given by the Decision Tree.

Schlaifer considers this point so important that his description of Decision Trees starts by treating event branches as gambles to show that Utility rather than Expected Value is the significant criterion (20).

3.8   OTHER TECHNIQUES USEFUL IN DECISION THEORY

Probably the most useful technique developed in decision theory is the Conditional Decision Tree which has been the subject of most of this chapter. The second most useful is probably the Simulation of probabilistic events using the Monte-Carlo random sampling method to obtain distributions of outcomes from probabilistic input data. Sometimes the simulation models can be used for a Sensitivity Analysis rather than with a Monte-Carlo treatment to isolate the relative importance of individual components. Optimisation methods also deserve a mention as they are often treated as part of decision theory. They have application in many fields but always in association with decisions.

3.8a The Monte-Carlo Simulation

Chapter 2 explained that the Monte-Carlo simulation is a method of determining the outcome resulting in the combination of a number of components, each having a distribution, by taking only a fraction of the total number of combinations that actually occur. This is a very useful method in evaluation work where much of the data is probabilistic and it is important to know the resulting outcome in the form of a distribution to assess the magnitude of the risks involved.

The Monte-Carlo analysis is the basis of the normal approach to Risk Analysis in project evaluation. The main reason for carrying out a Monte-Carlo simulation is that it correctly shows the combined effect of all the uncertainties. The presence of a large number of uncertainties does have a stabilizing effect which results in a narrower distribution than might originally be expected from an inspection of the individual distributions. A further advantage over the deterministic analysis is the removal of the step of selecting deterministic data to represent uncertain information. Such a selection often has to be illogical and revised when the resulting implications are not agreeable. The removal of this step therefore results in a quicker and more systematic overall analysis (18).

As an example of the problem, consider a sales volume that can only be given probabilistically. The plant capacity should be chosen to be higher than average sales volume to make use of possible favourable market development, but a deterministic economic analysis should use mean sales volume. If the predicted profitability is low, it can be improved by sizing the plant for the mean sales volume, although this would probably be a wrong decision. Hence, we have some difficulty in choosing the deterministic value by which to represent the sales volume and this choice can be justifiably revised to alter the results of the economic analysis.

The Monte-Carlo technique is now being used by many companies and there are frequent reports of its application within companies for investment evaluation (3,4,14,18).

Monte-Carlo simulations also find their use in simulating real-time stocastic events to enable design decisions to be made. For instance, the size of a transport facility--a dock, or a loading wharf--can be checked by running a simulation model with the arrival of goods and transport being chosen randomly. The simulation then runs for as many days or years as are necessary to give confidence that overloading will not occur.

Hence, there are two forms of simulation: the repeated evaluation of a function representing a single outcome, or the progressive, rather than the repetitive simulation of a continuous operation of an activity. It is sometimes useful to use both forms and to have progressive simulation for a limited time, which is then repeated a sufficient number of times to obtain a distribution with the required degree of confidence. This technique is directly applicable to project evaluations where the project is simulated year by year over the project life, and the process is then repeated until the required accuracy is reached.

3.8b The Sensitivity Analysis

A sensitivity analysis consists of repeating a calculation with only one input change by a pre-determined amount--a capital cost increased by 10% for instance. The result of the calculation then shows the importance of the capital cost in the evaluation. It does not show the variation there will be due to the capital cost uncertainty because the change, 10% in our example, was not chosen specifically to represent the estimated capital cost distribution. It does not give any information on the distribution of the outcome, even if the 10% happens to represent the capital cost distribution, unless all the other data are known with certainty. The effect on the outcome of a 10% change in variable cost and a 10% change in capital cost cannot simply be added together to give the combined effect of them both. As described in Section 2.5 distributions cannot be aggregated in this way. However, the Sensitivity Analysis is useful because it pin-points the important areas where future work should concentrate. This future work could be the more careful analysis of the data and a more accurate risk analysis, or it could be more concrete in the form of Research and Development, or information gathering to reduce the risk in the sensitive areas.

3.8c Optimization Techniques

Optimization techniques are methods of finding the value of a set of variables--the independent variables--that maximize any required function of these variables. This function is generally known as the response or objective function.

Optimization techniques can find either the maximum or the minimum of a response since changing the sign of the response converts a maximization problem into a minimization problem, and vice-versa.

Optimization techniques can be divided into methods for <u>continuous functions</u> and for <u>integer functions</u>.

Optimization of Continuous Functions

The continuous function methods can be further sub-divided into linear and non-linear methods. The technique for linear optimization is known as <u>linear programming</u>. This is a powerful technique for locating the optimum value for hundreds of variables as long as all the functions involved are linear. Non-linear techniques are much less powerful than linear programming, handling generally around 10 as a maximum number of variables. The non-linear nature of the function exclude the systematic search that is possible in the linear case, and the much more slower and less efficient step-wise hill-climbing methods must be used.

No more will be said about the optimization of continuous functions because they are not normally applied directly in the area of the investment decision. They are however, very important in the stage before the investment analysis. Each alternative under comparison should be at its optimum within the definition of that alternative. Operating conditions should have been optimized because one can only make proper comparisons with optimized alternatives. These continuous function optimizations are therefore often very useful in locating this optimum for each alternative (19). There is much literature available on continuous function optimization (5, 9, 16).

Integer Optimization Techniques

Integer optimization techniques have to be based on very different principles from the continuous function methods because the independent variables can only have integer values. All functions are discontinuous and this leads to searching techniques having to be based on systematically trying all combinations to locate the optimum rather than any form of progressive improvement such as hill-climbing.

Since integer optimization techniques can play an important role in investment planning, as will be seen in later chapters, the principles of

integer optimization, and Dynamic Programming, in particular, will be
described in some length.

Dynamic Programming

    Consider that we have to build a road to serve five regions but the
route this road should take is undecided. In each region there are a
number of possible towns it could go through. For the five regions, A, B,
C, D and E, let there be 1,3,2,3,3 towns respectively, named A_1, B_1-B_3,
C_1-C_2, D_1-D_3 and E_1-E_3 through which the road could go. These can be shown
diagrammatically on Figure 3.9 which can be presented more systematically
on Figure 3.10.

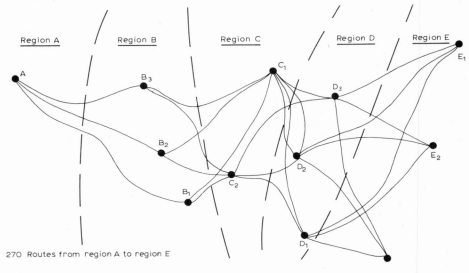

Figure 3.9

Dynamic Programming Example: Choice of Road Route

    From the geography of the situation it is clear that the road must go
through A to E alphabetically. However, it is not clear which towns to
choose. This choice has to be based on lowest cost. We therefore have to
find the lowest-cost complete route.

    The total cost is simply the addition of the costs of the individual
sections of the route. The cost of going between pairs of towns have been
estimated and are given in Table 3.4.

TABLE 3.4

Available Data for the Dynamic Programming Example.

Cost of each Section of Road   (£mm)

Section A-B		Section B-C				Section C-D			Section D-E			
From To	A	From To	B_1	B_2	B_3	From To	C_1	C_2	From To	D_1	D_2	D_3
B_1	10	C_1	5	7	15	D_1	1	13	E_1	1	4	7
B_2	5	C_2	4	11	14	D_2	3	7	E_2	5	5	9
B_3	1					D_3	11	2	E_3	7	5	8

If we consider the first section of the route, A-B, we find from Table 3.4 A-B_3 is by far the cheapest. However, we cannot discard the possibility of A-B_1 or -B_2 being on the best complete route, without considering the route B-C, because all routes from B_3 may be very expensive, leaving the cheapest route A-C to be the one going through B_1 or B_2. This argument repeats itself to the last stage E. Only then can one choose which B town is on the best route. This means fully evaluating all the possibilities, that is: 1 x 3 x 2 x 3 x 3 = 54 routes. Since each route has five stages this requires 5 x 54 = 270 evaluations to be made.

On a large problem the number of evaluations can easily run into millions in which case it is impossible to consider this as a realistic technique because of the enormous number of calculations involved.

However, although one cannot cut out any of the towns (the "states")

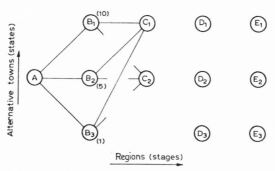

Figure 3.10   DYNAMIC PROGRAMMING EXAMPLE:
Choice of Road Route--Systematic Presentation

from the problem before the end of the calculation, the number of complete
routes evaluated can be drastically reduced by a simple process known as
Dynamic Programming (2).

We can determine the cost of getting from $A-B_1$, $-B_2$ and $-B_3$ from Table
3.4 and this can be written on Figure 3.10 beside each state. These are
10, 5 and 1.

We can now move on to stage C. There are three routes to C_1 and one
can calculate the total road cost to this point by adding the cost at B
written in Figure 3.10 to the cost of going from B to C from Table 3.4.

$$A - B_1 - C_1 \quad = \quad 10 + 5 = 15$$

$$A - B_2 - C_1 \quad = \quad 5 + 7 = 12$$

$$A - B_3 - C_1 \quad = \quad 1 + 15 = 16$$

Obviously the cheapest way of getting from $A - C_1$ is through B_2. The
cheapest way of getting from C_1 to E is unknown, but it will be the same
whether the route to C_1 is through B_1, B_2 or B_3. Therefore, since there is
independence between the later and earlier stages, we are justified in
saying that if the optimum path goes through C_1, it will come from B_2.
Hence, all the routes involving $A-B_1-C_1$ and $A-B_3-C_1$ can be discarded from
further study.

This same argument goes for C_2. Hence only one route per state needs
to be remembered.

So the problem can be resolved and the optimum route found with only
$3 + 3 \times 2 + 2 \times 3 + 3 \times 3 = 24$ sections to be evaluated, compared with 270
if every possible complete route has to be considered. Readers with
enthusiasm could use Table 3.4 to verify this procedure and determine the
optimum route for this problem which should be: $A-B_2-C_1-D_1-E_1$ with a
cost of £14mm.

This procedure of discarding routes as the calculation progresses is
called Dynamic Programming and its primary requirements are that: the problem can be divided into a states and stages, the criterion by which the
optimum is judged must result from moving from stage to stage, and that
this criterion is independent of later stages. Without this independence,
one cannot apply Dynamic Programming because routes cannot be discarded,
since the later stages would affect the attractiveness of the partial
routes.

This analysis could have equally well been carried out from E to A.
It is fashionable to work in the backwards direction in Dynamic Programming,
but the reasons for this are obscure. There are some cases which have to
be carried out in a certain direction to maintain the requirement that the
later stages must be independent of the earlier ones. However, in many
cases, where all stages are independent--as our example here--it is equally
good to use a forward or backward direction. Dynamic Programming is a very
simple procedure, the difficulties are in manipulating real problems into
the format required for dynamic programming. That is, properly defining
the stage and the state, and the criterion for the problem in hand.

Branch and Bound

     A second integer programming technique--Branch and Bound--is also a
useful technique in some decision-making situations. Basically, if one
has a situation which can be represented by a tree structure, as shown by
Figure 3.11, a criterion monatonically increasing as it progresses from the
trunk out along the branches, Branch and Bound enables a selection to be
made without the complete evaluation of every branch.

     Consider, as an example, a number of plant designs which arise out of
three decisions. Decision 1 could be to install one of three types of

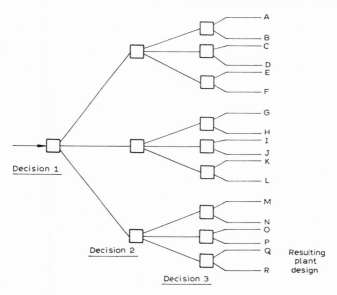

Figure 3.11   Branch and Bound Plant Design Example

reactor. Decision 2, three types of distillation column, and Decision 3,
two types of absorption equipment. This results in 18 different plant
designs. The sledge-hammer approach is to cost these 18 designs in order
to find the one with the lowest total capital cost.

The Branch and Bound method consists of evaluating each design
separately by evaluating one branch completely before moving on to the next.
Now, if the partial cost of a branch becomes greater than the present best
total branch cost, then that combination is discarded, together will all
later branches on that route. Hence, a computational saving is obtained
through not fully evaluating every possible outcome.

For our example given by Fig. 3.11, the systematic search may evaluate
outcome A, then B. Outcome B may be cheaper than A, so A is forgotten, and
only B is remembered. Outcomes C and D may be more expensive than B and
so they are discarded. On calculating the E and F branch, it may be that
the third branch of the second decision already results in a higher cost
than B. Hence there is no need to complete the evaluation of E and F. It
may also be that the second branch of the first decision is already more
expensive than the complete plant B. Hence this branch can be forgotten,
so that the evaluation of G, H, I, J, K and L can be omitted. The third
branch of the first decision may be relatively cheap and so the calculation
may have to proceed to M. M may be cheaper than B, in which case, B is
forgotten and M remembered. The routes N to R are similarly considered,
resulting in the optimum being found without the need for every alternative
to be completely evaluated.

The technique has been successfully applied to the selection of equipment for batch plant design (24).

3.9 GAME THEORY

A part of Decision Theory is a subject called Game Theory. Game
Theory is concerned with analysing and determining optimum strategies when
there is competition between one person and nature, or between two or more
people (10, 12). A Game can be defined as a recurring situation where a
player has to make a choice from a number of alternatives placed before
him. The following choice is made either randomly (by Nature) or by the
second player and a combination of these two choices results in the player
getting a prize or paying a forfeit, the magnitude of which depends upon

the combination of the choices.

Since some choices produce gains, and others losses, it is possible to use probability theory to determine the optimum proportion of each of the different choices to make to obtain the best Expected value from the game. This proportion of choices is the strategy of the game.

Games can be one player against Nature--that is when the second choice is determined by some truly random process--or against a second player, where the choice can be aggressive rather than random--or against any number of players. The important factor in Game Theory is the quantity of information available. Against Nature one has considerable information, at least in a statistical sense, since the "opponent" is behaving truly randomly. Against other players there is uncertainty, because their behaviour is not random, but aggressive; with two or more players the possibility of coalitions between a number of players against other players or other coalitions is present. Coalitions infer that information exchange and collaboration over decisions occur.

Each of these types of game have been studied and theories produced concerning the optimum strategies. Game Theory has been applied to military problems, but there is only little reported of its application in civil situations (11, 22). At present, the theory can only handle very simple idealized situations that do not approach a complexity of real life situations. However, it is an attempt at a science of competition and so eventually must find some use in industrial situations. For example, when an extension is proposed to cover an increase in market, Game Theory may give the optimum strategy to use in a competitive situation. Obviously, one should not assume your firm will capture the whole market, but the other extreme--a too modest estimate--may never produce a profitable plant.

Business game simulations are another branch of Game Theory. In such games a number of players--or teams, representing companies--compete for a market. Each team makes decisions concerning production levels, price, advertising, sales effort and research, and a simulation model uses these decisions as input to provide the simulated sales and resulting profit. The game is played over a number of periods, each period representing a quarter, or a year. The results are generally arranged in the form of a company balance sheet and each team endeavours to produce a "satisfactory performance" for his firm. Many such business games have been produced, some have models of considerable complexity. Usually, but not always, they are used in conjunction with a computer to reduce the efforts required for the calculation, and to produce the company balance sheets (15).

The main use of such games is educational. All the normal management decisions regarding financing, investment and resource allocations must be taken, and it gives the players a feel for business management as a whole.

Game Theory is an interesting subject but it seems to be still far from being applicable to industrial problems. It will not be referred to again in this book.

3.10 BIBLIOGRAPHY FOR CHAPTER III

1. Allen, D.H., The Concept of Creditability and its Use in Forecasting for Novel Projects, Chem. Engr. 47, 267, (1969).

2. Bellman, R., and Dreyfus, S., Applied Dynamic Programming, Princeton University Press, New Jersey, 1962.

3. Cameron, D.A., Risk Analysis and Plant Design, Long Range Planning, 6, 30, (1973).

4. Ezzati, A., A Probabilistic Cost Analysis of Two-Stack Flue Gas-Desulphurisation Systems, Engineering Economist 19, 2, 63, (1974).

5. Fox, R.L., Optimisation Methods for Engineering Design, Addison-Wesley, London, 1971.

6. Grayson, C.J., Decisions Under Uncertainty: Drilling Decisions by Oil and Gas Operators, Harvard University, Boston, 1960.

7. Hillier, F.S., and Liebermann, G.J., Introduction to Operational Research, Holden-Day Inc., San Francisco, 1967.

8. Hillier, F.S. The Evaluation of Risky Inter-related Investments, North-Holland Publishing, Amsterdam, 1969.

9. Himmelblau, D.M., Applied Non Linear Programming, McGraw-Hill, New York, 1972.

10. Horowitz, I., Decision Making and the Theory of the Firm, Holt,Rinehart, and Winston, New York, 1970.

11. Hughes, R.R., and Ornea, J.C., Decision Making in Competitive Situations, Proceedings of the 7th World Petroleum Congress, 537, 1972.

12. Kaufmann, A., The Science of Decision Making, Weidenfeld & Nicholson Ltd., London, 1968.

13. Kaufman, A. and Faure, R., Introduction to Operations Research, Academic Press, New York, 1968.

14. Kryanowski, L., Lusztig, P., and Schwab, B., Monte-Carlo Simulation and Capital Expenditure Decisions, Engineering Economist 18, 1, 31, (1972).

15. McKenny, J.L., Simulation Gaming for Management Development, Harvard Business Management School, Boston, 1967.

16. Murray, W. (Ed.), Numerical Methods for Unconstrained Optimisation, Academic Press, London, 1972.

17. Newmann, J.A., Management Applications of Decision Theory, Harper and Row, New York, 1971.

18. Pouliquen, L.Y., Risk Analysis in Project Appraisal, World Bank Occassional Paper No. 11, John Hopkins Press, Baltimore, 1970.

19. Rose, L.M., The Application of Mathematical Modelling to Process Development and Design, Applied Science, London, 1974.

20. Schlaifer, R., Analysis of Decisions Under Uncertainty, McGraw-Hill, New York, 1969.

21. Siddall, J.N., Analytical Decision Making in Engineering Design, Prentice Hall, Englewood Cliffs, New Jersey, 1972.

22. Singvi, S.S., Game Theory in Investment Planning, Long Range Planning, $\underline{7}$, 4, 59, (1974).

23. Smith, B.E., Introduction to Decision Theory, in Decision and Risk Analysis--Powerful New Tools for Managment, Proceedings of the 6th Symposium, The Engineering Economist, Hoboken, N.J., 1972.

24. Sparrow, R.E., Forder, G.F., Rippin, D.W.T., The Choice of Equipment Sizes for Multiproduct Batch Plants--Heuristics vs. Branch and Bound, IEC Process Design and Development, $\underline{14}$, 197, (1975).

25. Spetzler, C.S., The Development of a Corporate Risk Policy for Capital Investment Decisions, IEEE Transactions on Systems Science and Cybernetics, vol. SSC-4, 3, 279, (1968).

26. Spetzler, C.S., and Zamora, R.A., Decision Analysis of a Facilities Investment and Expansion Problem, in Decision and Risk Analysis-- Powerful New Tools for Management, Proceedings of the 6th Symposium, The Engineering Economist, Hoboken, N.J., 1972.

27. Stulberg, D., Calculating the Calculated Risk, Chem.Eng. 152, (1968).

28. Swalm, R.D., Introduction to Utility Theory, in Decision and Risk Analysis--Powerful New Tools for Management, Proceedings of the 6th Symposium, The Engineering Economist, Hoboken, N.J., 1972.

29. Von Neumann, J., and Morgenstern, Theory of Games and Economic Behavior, Princeton University Press, Princeton, 1947.

30. Winkler, L.R., Introduction to Bayesian Inference and Decisions, Holt, Rinehart and Winston, New York, 1972.

CHAPTER **IV**

THE DECISION TO INVEST IN PLANT

> Wer nüd chunnt zur rechte Zyt,
> der mues ha, was übrig blybt.

In this Chapter we shall consider the requirements for making a decision concerning the installation of plant capacity--either a production increase for an existing product, or establishing production for a new product.

Company management knows what it wants from a decision without having to refer to probability theory, Decision Theory and Risk Analysis. We should therefore also be able similarly to define what is required in a good decision without being over-technical. This definition forms the contents of this chapter.

4.1 THE CHARACTERISTICS OF A "GOOD DECISION"

The best way of judging a decision is probably ten years after it has been taken. One is then in a position to say "that was a good decision" or the reverse. Such a judgement will show the best decisions as those where the future was correctly predicted. This may be no more than coincidence, but the fact remains, they would have been good decisions.

A second class of decisions is that where, as luck would have it, the future was not properly predicted. For instance, sales might be 50% of those forecasted. In this class we will still have some profitable projects and some unprofitable. The profitable projects are more likely to have been those that have considered the probability of low sales, and installed a small plant, or used the plant for additional product, or achieved compensation in another way. The unprofitable plants are probably those that were too big and are now running at half-capacity. A similar situation could be described where the forecast underestimated the market. One decision could have foreseen the possibility of a good market and allowed a cheap expansion to be possible, resulting in extra economies by a higher scale of production. Another decision may necessitate a duplicate plant

being built with no cost advantage being realized for the larger scale
production.

A third class of decisions involve projects in which everything conceivable went wrong: the market failed, competition appeared, the capital cost
estimate was low, and raw material costs rose. Though this situation may
represent incredibly bad luck, for the decision to be classed as "good at
the time" the company should not have been seriously damaged by the failure
of the project.

The retrospective judgement of whether a decision has been good, has
been summarized by asking: "has the economic performance of the project
been good, and was there scope for corrective action when the unexpected
arose?"

The measure of "economic performance" should be that measure used
throughout the company for profitability. Hopefully, the company standard
will be some form of discounted method, because, particularly for the case
of expansions, the income grows over a period of years which makes it important to involve the time value of money.

This means, for a good decision, the profitability should be good for
the most likely set of future conditions, but before the decision is made,
the less likely outcomes should have been investigated, the corresponding
corrective actions determined and the resulting less likely profitabilities
calculated. Then, at the time the decision is taken, the decision-maker
is confident that his decision will handle all possible outcomes reasonably
well. In ten years' time it should be seen to have been a good decision,
whatever the future is likely to bring.

In Decision Theory terminology, the decision should have an attractive
profitability for the most likely outcome, and be the initial decision in a
strategy which has acceptable profitabilities over a wide range of possible
outcomes. And finally, there should be zero risk of involving a loss
greater than an amount that would seriously damage the company.

These criteria of a "good decision" seem to be universally held regardless of whether a company uses formal Risk Analysis or not. It is normal
practice to carry out a Sensitivity Analysis before coming to a decision.
This is really a crude check that the profitability is still acceptable
when there are unexpected changes in the input data.

It is also normal practice to have a fall-back plan, particularly on
large projects. Whenever there is any uncertainty, a fall-back plan must

be defined before the decision is taken. This is in fact a crude form of
strategy.

        In the last stages before a decision taken, it is discussed by the
Board of Directors. These discussions are, in part, to assess the possibilities of loss and the resilience to unexpected changes. This is a further
informal attempt at Risk Analysis and strategy creation.

4.2 THE THREE DECISION STAGES

        Projects normally go through three decision stages. If one is compar-
ing two processes, the operation and equipment requirements of each process
must firstly be near-optimum to present each alternative in its best light.
It is pointless comparing a process designed optimally with another one
using highly unsuitable operating conditions. The first decision stage
therefore selects the operating conditions for each alternative. This stage
demands engineering competence and is too large a subject to be summarized
here. Readers are recommended to refer to--or even to buy--other books in
this field (1, 2).

        This book is concerned only with the next two decision stages. The
first of these is where optimized alternatives have to be compared and one
alternative chosen. This stage can be difficult, because a good decision
cannot be defined by a single criterion. It is not the one with the highest profit, or the lowest risk, but generally some combination of the two.

        Companies are often involved with insurance and so the concept of
insurance in decision taking is second nature. The idea of accepting a project with a lower Expected profit because it has a better potential if the
future turns unfavourable is a normal insurance against an unfavourable
future, and the difference in Expected profits is the insurance premium.
These alternatives are rarely evaluated but are usually considered at a
very subjective level. One alternative might "obviously" be more versatile
than another--it could make another product, or be more easily extendable,
or it may be less technologically uncertain--the technology may not be so
new, or it may be more comparable with the company's expertise. But the
surer alternative may have a lower Expected profit. However, if this were
still acceptable as an investment, and the difference did not represent an
enormous loss of opportunity, the choice would probably go to the surer
alternative. Such discussion shows no conflict between the manager's

concept and those of Utility Theory, except that the manager has to work
without numerical data.

Having chosen the alternative, the third decision stage is to present
more carefully this alternative as a capital investment. This stage is to
show that it is a sensible use of the company's capital. This is the stage
where the project receives permission to go ahead from a theoretical study
to reality. The project is usually compared with a profitability measure
that the company considers acceptable.

At this stage, besides presenting the profitability, it is usually
necessary to provide a Sensitivity Analysis and also to give some description of the expected accuracy of the data. For instance, one has to describe the market, and also to give some indication of the technical reliability of the process. Some companies provide an official scale ranging
from "Class 1--repeat of an existing operating process" down to "Class 5--
new process with new technology." It is then necessary to state which
class the project belongs to with the final submission. All this information is to enable a subjective Risk Analysis to be performed in the mind of
the decision-maker.

In some cases it may be that the company is very short of capital, or
the project is very big. The problem then facing the management is the
choice of a limited number of projects out of a longer list of possibles
that have been submitted for judgement. This opens up the whole subject of
Portfolio Analysis, a branch of operations research which deals with choosing a balanced set of projects to give the company stability and profitability. This subject is beyond the intent of this book. Portfolio
Analysis and Corporate Planning belong to high levels of company and financial planning that are normally not met by engineers until they are no
longer performing an engineering function.

4.3 WHAT DECISION IS MADE AND WHEN IS IT MADE?

Before a plant is operating and producing goods, there must have been
a series of decisions. One does not suddenly decide to build and operate
process X.

In reality, one firstly decides to: spend money and effort on research
                                         into process X

and later   spend money and effort on <u>development</u> of
                                 process X

and possibly, after review,   <u>continue</u> to spend money and effort on develop-
                                 ment of process X

then,   proceed to <u>design</u> and cost process X

followed by,   <u>complete</u> detailed <u>design</u> of process X

then,   <u>engage</u> a contractor for <u>building</u> and <u>erecting</u>
                                 process X

and sometimes, after   allow contractor to <u>finish erection</u> and start
appraisal of the latest   <u>operation</u>.
market figures

Decisions are required whenever the process changes stages, or whenever new market or technical data become available, and often after a yearly review, just to exercise good control on long-term projects. The alternative decision at each stage is "drop the project" though the decision may
be more positively worded as "move our resources on to a different project."

However, each of these decisions is made using the same criterion--the
predicted economics if a plant using process X was erected and run for a
fixed number of years. Although the evaluation assumes the investment in
the plant is made, the evaluation is not being carried out to make the
decision to invest the plant but only to make the decision to proceed one
stage further in the project.

In principle, all decisions should be as late as possible because they
involve the use of uncertain forecast information and the greatest accuracy
for this data is obtained by using the most recent information. Hence, any
analysis only commits one to a decision <u>now</u>. The fact that you decide to
carry out research on process X now, because the analysis shows that a
plant using process X built in five years' time would be economically
attractive, does not bind you to any decision concerning the plant in five
years' time. This is a particularly important principle that must always
be remembered: that no matter what plans or strategies are taken in any
economic analysis, they are not binding conditions to the immediate decision;
they do not have to be followed. The analysis is a <u>simulation</u> of what
could happen in the future if a particular decision were made now, as a
means of evaluating the profitability of the immediate decision.

Because each decision is so limited--deciding only the immediate
future--it is possible to reverse the decision with no great difficulty. A
reversed decision will normally involve some loss, the magnitude of which

depends on the stage the project is at.

One can identify two losses to a reversal of a decision. One is the instantaneous and irrecoverable loss associated with the decision. To decide "give to contractor to build" to be reversed some time later to "tell the contractor not to build" involves a legal cost that the contractor will levy--the cancellation charges--which may be 10% of the project's estimated capital cost.

The second loss is the continuously increasing investment of all the work that has gone into the project. This is very small at the research stage, but by the time the plant has been built this is the cost of the total research, plus the development, plus the capital cost of the plant.

This second loss is not a cost incurred by cancellation since this money has already been spent. However, it enters into any economic assessment by reducing the investment the project needs from that point on in time. Hence, as a project progresses, justifications for completing the project become easier, all other things being equal, because there is no point in including money already spent and now irrecoverable as part of the required investment for the project.

In industrial expansion projects there are very few of the first type of loss--the immediate irrecoverable commitment--virtually all are the continuously increasing type. This gives decisions concerning expansion projects the particular characteristic of being easily modified or reversed. This enables management to review past decisions in the light of new information and, if necessary, produce counter-decisions. The time scale between such decisions can be relatively short. In fact, it is better to regard management as controlling the project in a continuous fashion rather than having to steer the project by a limited number of discrete irreversible "decisions."

The more usual concept of a decision is that it is irreversible. Probably the most extreme being the decision of the U.S. President to press the red button and to unleash atomic war. Within seconds of his decision it is impossible for him to reverse it. One could say the decision immediately incurs an infinite cost. Contrast this with the decision to design a new plant: One could--and does--stop the design, re-consider, re-start, at some small and perfectly bearable cost, in an attempt to ensure the right outcome in the face of changing data.

This flexibility should be looked on as an advantage associated with expansion-type decisions, although it must be used intelligently by management, as too much interference will result in no progress at all. In particular, if a reversed decision is again reversed, there is an easily identifiable loss due to the delay caused by the indecisiveness. The cost of delays can be considerable, particularly if it results in a company entering a market too late. There is, of course, a limit to the reversability of decisions. As time progresses and later decisions are made, the older decisions have so much built round them that they become harder and harder to reverse.

4.4 THE ALTERNATIVES

It is possible to divide expansion decisions into a limited number of different classes. The following list covers the possible methods of expanding production:

A. A new plant using the same process as the existing plant could be built in addition to the existing plants to give the required increase.

B. A new plant, using the same process could be built to replace an existing plant which also gives the required production increase. This may be sensible if sufficient economies of scale or use of modern equipment justifies it.

C. A new plant using a new process could be built in addition to the existing plant to give the required increase.

D. A new plant using a new process could be built to replace the existing plants which also gives the required production increase.

E. Any plant could be expanded once it is existing. "Expansion" infers the addition of some equipment items to a plant to increase its output. Since outputs of existing plants can often be increased by adding only a few items, (de-bottlenecking), expansions are often cheap in both capital and fixed costs.

F. The difference between production capacity and the market demand (the short-fall) could be covered by buying product from a competitor. This is a common technique, particularly on a temporary basis.

To each of these cases must be added the manufacturing capacity before it can be considered a complete definition of an alternative:

"Add a new plant for 2,000 ton/year using the same process" or

"Buy in 200 ton/year"

are concise definitions of alternatives.

In theory, the optimum capacity for any alternative can be calculated, and this exact figure used as the capacity of the alternative. However, in practice this is generally not done and round-figure capacities are more usual. 1500 ton/year is used rather than 1474 ton/year; 500 ton/day and not 471 ton/day, and so on. This is because uncertainties in the forecast are so large and optima are so flat that any accurate optimization is pointless. It is adequate therefore to talk in terms of a limited number of capacity levels for the above list of alternatives.

Capacity expansions are generally made in discrete steps, some years apart. If sales go well, the next step may be sooner, or if poor, later. Arguments against more frequent, smaller steps are that there is no advantage taken of "economies-of-scale" and also that a company's engineering and management resources are inefficiently tied up considering the next expansion before the old one is finished.

One exception to this is the expansion or de-bottlenecking of existing facilities. A relatively short-term de-bottlenecking project can, for a relatively low cost, delay a major capital expense for say 12 months. This is a highly profitable exercise and so extension proposals with capacities equivalent to only one or two years sales growth are common.

Expansions are extremely attractive alternatives. The first expansion of a plant has the double advantage of benefiting from the over design in all the items of equipment except for those forming the bottlenecks as well as the advantage that all expansion projects enjoy not having to supply a total plant. Items such as buildings, control rooms, services and storage are often adequate for twice the original design output.

Because of the economic advantage of expansions, plants become repeatedly expanded until there is a shortage of space, even though the original design usually leaves plenty of room "for later expansion." The engineers then show just that bit more initiative and propose the removal and rebuilding of sections of the old plant to leave room for further expansions of other sections. The expansion gives way to a revamp--the major revision of an existing plant. It is amazing how much modification is possible to an existing plant leaving it still more economic than a new plant. Frequently it is not the capital cost that makes further expansion of a plant uneconomic, but the difficulty in maintaining the production--exactly when it has to be maximum production--and at the same time make the major modifications necessary to create the space for the expansion.

Because plant expansions are so important they must always be given special consideration. There may be alternative ways of expanding which are mutually exclusive, that is, either A or B but not both. Expansions may be consecutive, for instance, Stage I must firstly be installed, followed optionally by Stage II.

The comparison of alternatives consisting of rather low capacity expansions and longer term new plants is difficult, because they do not have the same project life span. It may be an extension can satisfy the further two years demand, after which a new plant has to be built. This has to be compared with neglecting the expansion and building a new plant immediately. The best size of new plant may be equivalent to 6 years demand increase. This means the two projects are still not comparable since one satisfies the market for two years, or with a new plant, 8 years, and the second proposal 6 years. Once could artificially reduce the size of the plant if it followed the expansion--but in reality this would not happen. One could devise some form of credit for the better 6th year position of the expansion plus plant, but this is also rather artificial. This situation can sometimes result in serious consequences. A cheap expansion may "kill" completely the development of an improved process discovered by a research department by completely removing all incentive to work on it.

After the extension is fully utilized, the only remaining option may be to build a small additional plant of the old process. In a competitive situation this could be disastrous compared with the alternative choice of ignoring the expansion and developing the better process. Such situations can be properly assessed by carrying out the economic evaluation well into the future, which means looking much further than the first investment.

A further problem that arises in comparing various alternatives is the dates by which they can be operating. A simple de-bottlenecking extension may take only 6 months from design to start-up. A more complicated expansion may take 12 months, a single new plant with a known process

2-3 years, and a new process--still only a gleam in the research chemist's eye--five years or more. These figures apply to chemical plants. They are industry dependent, consumer goods usually taking less time, and highly technological industries, such as the aircraft industry taking considerably longer.

This causes further restrictions on possible expansion plans. For instance, a new process may have to be excluded as a possibility because there is insufficient time for its development unless there is a possibility of a small extension which could delay the big investment and give time for the new process to be worked up.

Clearly, the choice of alternatives is not straightforward, and it generally requires the proposal of both medium and long-term decisions--an investment plan in fact--to enable the next best investment to be properly evaluated.

4.5 FROM FORECAST TO REALITY

The overriding characteristic of investment decisions is the uncertainty of the data. This was discussed at length in Chapter 2 and referred to again at the beginning of this chapter.

The major uncertainty is the sales volume, and the second most important are prices and costs. Capital cost estimation can also be a serious source of error. In this section we will look more closely at the characteristics of these uncertainties and try to relate the probability forecasts to reality.

The uncertainty in the sales estimate arises from not being sure how the product will be used, from not knowing the future performance of the competitors, and from not knowing the future economic climate.

Let us take as an example a proposed investment for a plant to manufacture a new synthetic rubber. The way in which sales will increase will depend on the uses to which the rubber can be put. It may have obviously good physical properties for rubber mountings in cars, and these properties may be considered so outstanding that the company is confident that it can take over the whole market within the next few years. This may be a fairly safe assumption, but one must also reckon with the discovery of some feature that prevents its development. It may have too short a life-time, and perish after 8 years. It may produce poisonous fumes on burning and so be

legislated against by some governments. One normally neglects the remote
chance of any such discovery being overlooked by the company's intensive
development trials, but there always remains a small possibility of such
points affecting the original market expectation. This new rubber may also
have some physical properties suitable for tyres or inner tubes or flexible
cabling or tubing. Each of these applications requires such a collection
of physical properties that any new rubber is likely to excel in some, and
hardly satisfy in others. The difficulty is to assess, balancing the
advantages and disadvantages, whether a new product will find a major use,
or a specialized use, or no use in each possible field.

Now look at the competitors in this field. First of all, we have the
present supplier of rubber for mountings in cars. He will not sit back and
do nothing. In the presence of the new competition he will probably reduce
his price rather than lose his entire business. This alone may be effective. A car producer is usually more interested in a lower price than in an
improved mounting material. This competitor may counter the new material
by making technical improvements himself. He may have already done that
but not yet announced it. Lastly, the new competitor may enter the field
with an equal--or better--development. These same arguments apply to all
fields in which the proposed new product enters.

Beside the problem of end usage and competition there is the effect of
the economic climate. When a country is experiencing economic problems--
balance of payments, or inflation for instance--the government takes steps
such as altering the taxation or the ease with which loans can be obtained,
which directly affect many markets. For instance, car sales might drop
25%. Hence the demand for rubber will drop 25%. Besides government influences on sales, the sales are also affected by customers' behaviour--a
threatened oil shortage makes customers think twice before buying a new car.
The result is that product sales, other than sales for essentials such as
foodstuffs, undergo cyclic behaviours which can be partly correlated with
the Gross National Product (GNP) cycles.

These three sales influences, individually, have different effects on a
15-year sales forecast. The first influence--product end uses--results in
a consistent forecast. It may be consistently good or poor but it would
be consistent.

Figure 4.1 shows a probabilistic forecast for the influence of product use alone, and also a possible actual sales curve that could result. If the product finds many applications in year 3 it will still have many applications in year 5, 10 and 15. The curve will therefore stay on its original path throughout the project's life.

If we look at the influence of competition, shown in Figure 4.2, we get a slightly different picture. There is either no competition, some competition or strong competition initially, but one can imagine a limited amount of interchange, particularly in the direction towards more competitive influence. Figure 4.2 shows a probabilistic forecast and a possible actual sales curve for a case with initially no competition, but later a competitor entering.

The influence of the economic situation however has a very different effect. Figure 4.3 shows possible forecast sales due to economic conditions. The curves represent possible 2%, 4% or 6% mean annual growth -- a normal range between politicians' promises and actuality. The mean growth could be one of these curves, but imposed is a GNP cycle with, for instance, an amplitude of 10% on a cycle of 8 years. In addition, we have the effect of government-controls and customer confidence which affects sales by a further random ±10%. Compounding these gives an actual sales growth by Figure 4.3 This differs considerably from Figures 4.1 and 4.2 in that the performance in any one year gives only limited information as to the performance in future years.

4.5a Correlation

The differences between these three influences is the degree of correlation between sales and year. There is complete correlation in Figure 4.1, a good correlation in Figure 4.2, but a low correlation in Figure 4.3. As already discussed in chapter II, the identification of any correlations associated with uncertain variables is of prime importance in determining the significance of the uncertainty.

A real sales forecast will be a mixture of these influences, and the degree of sales/year correlation will depend upon the individual mixture. New products are likely to have a high sales/year correlation, since the influence of product use will be the most important. Existing and well established products will have a fairly constant sales with only a small sales/year correlation, because the use and competitive influences will have stabilised, leaving the economic influence as the major one.

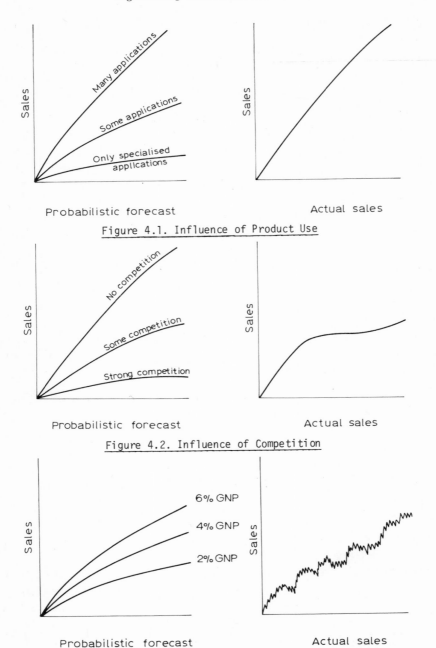

Probabilistic forecast   Actual sales

Figure 4.1. Influence of Product Use

Probabilistic forecast   Actual sales

Figure 4.2. Influence of Competition

Probabilistic forecast   Actual sales

Figure 4.3. Influence of Economic Situation

A further correlation, the sales/price correlation becomes a problem
for products with both use and competitive influences. For a new product
with a number of potential uses, the extent to which the uses can be taken
from existing products depends strongly on the price of the new product
compared with that of the old. Should the technical properties be better
and the price lower, the market is then assured, but if the price is higher,
only partial or even no market share may be obtained.

The price of the new product is so set that the user pays a little
extra for the improved property. This price is then steadily reduced to
maintain a steady rate of conversion of users from the old to new product.
There is a relationship between sales volume and price, and forecasts for
this situation have a high volume/price correlation.

If probabilistic volume/price forecasts represent alternative and as
yet undecided management pricing policies, such as high volume/low price
or low volume/high price, then this infers complete correlation between
volume and price. If probabilistic forecasts represent the genuine uncertainty as to the selling price, regardless of the sales level, there is no
correlation between the price and sales forecasts. Hence the degree of
correlation is dependent on the source of the uncertainty and the basis the
forecaster assumes in making his forecast.

An exact analogous situation exists for the influence of competition.
One could have different undecided pricing policies which infers complete
correlation, or one may have no policy but just a general uncertainty level
as to the future price which infers no correlation.

4.5b Inflation

To discuss the correlation of price and the economic situation is to
be discussing inflation. Inflation affects more than the sales price; all
costs are similarly affected. Without inflation one estimates costs and
sales income, and one estimates profitability, using a reasonable cost of
capital -- say 7-10% on any capital borrwed for the project. Given a little
inflation -- say 1-3% per year, the costs and incomes can be forecast as
rising in step with the inflation rate and the project's profitability
estimated in the normal way, except that the owners of the capital require
to be compensated for inflation. Hence interest rates of 8-13% must be used.

In situations where the inflation is between 5 and 20%, these
corrections to forecast incomes and costs, and the correction to the
interest rate become of major importance and a much clearer basis should be

laid down for the calculation. If one chooses a method using predicted future cost and sales income in the normal monetary terms, one introduces the
inflation uncertainty into every future cash flow. Furthermore, there is
complete correlation between variables whose uncertainties are due entirely
to inflation, and zero correlation for uncertainties from other causes. This
results in an unknown degree of correlation between the uncertainties in the
various cash flows, which renders any attempt at Risk Analysis virtually
meaningless.

There needs to be a more systematic approach for the treatment of uncertain inflation rates than presently in common use by most planners in
engineering companies.

4.5c The Value of Better Forecasts

There is rarely good co-operation between the commerical side of an
organisation and the engineering side when it comes to forecasting. The
engineer feels that any savings he can make by careful engineering can be
dwarfed by an off-hand revision of sales estimates made by the commercial
people. "If only they would realise the work that is built upon their
estimates they would treat the matter more seriously" is commonly heard.
Equally, the comparatively innumerate commercial man thinks future sales
could be "anybody's guess" and it is ridiculous for the engineer to build
so much on to his figures.

Improvements in attitude will only be developed through improvements
in technique. For sales forecasters to deal with probabilities will enable
them to produce sensible forecasts and emphasise from the start that the
data is uncertain. For engineers to be able to inform the sales people of
the cost of the uncertainty will transmit the seriousness of the situation
and produce justification for employing better forecasting techniques.

4.6 THE NON-ECONOMIC FACTORS

After a company's planners have carried out the detailed economic
evaluation, and the chosen alternative has been shown to be economically
attractive, the proposal has to go to the Board of Directors for approval
before the money becomes available. This step is not merely a case of
rubber-stamping -- the Board of Directors does provide a useful function.
There are many factors that can influence a decision besides the economic
or quantifiable ones, as discussed in Chapter I, and these are the reasons

why the Board of Directors sometimes modify the conclusions painstakingly
arrived at using economic criteria.

Consider a proposal which involves working an old process at its quality limit to achieve a saleable quality. There is no sales price or present
competitive advantage in better quality, and so no economic reason to choose
a more expensive newer process with more quality potential. However, there
may be very strong grounds for wanting a process capable of a better quality product -- competition may be becoming tougher, legislation may be
being discussed against lower qualities, the company may want a "high
quality" image. Such reasons would lead to the choice of the more expensive
process. However, the choice should be done by senior company men because
they have the best all-round picture, and they alone should be answerable
for decisions incurring expenses which are not directly economically justifiable. However, before they can make a decision, or reverse any economic
choice, they should know the cost of the more expensive alternative proposal.
They must know the price they are to pay for their policy.

As a second example, consider the situation when a company shows
reluctance to enter into business with another company that may be an integral part economically of the most attractive alternative. This may involve
buying product for two years whilst the new process is developed. The supplier may be a competitor, known for his desire to sell you his product for
re-marketing, but equally well known for backing out of his promises whenever he is a little short of product himself; or the Board may feel, from
previous contact, that this competitor's business practices are rather un-
reliable and it is better not be involved; or there may be delicate nego-
tiations already at stake and the approach for further business may weaken
the company's bargaining position.

A further area where qualitative judgement often outweighs economics
is in meeting customer demands for product. In an increasing market
situation it is never economically attractive to build a new plant immediately there is a shortage of product. Figure 4.4 shows the situation by
indicating the proportion of product sold from the new plant in the first
year. This small quantity cannot be expected to cover the cost of the
capital of the new plant for the first year. It will be economically more
attractive to build a new plant a year or two later, and lose the sales
shown in the black triangle in Figure 4.4

Company policy is often to meet sales demands, even for this uneconomic
period. Should the company feel that, if a customer changed during that
time to a competitor he would not come back, a company policy of meeting

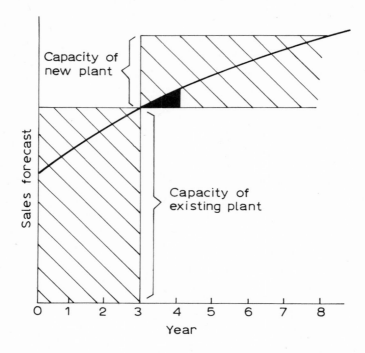

Figure 4.4. New Plant Operation in the First Year

the demand is justified. If a very complete economic analysis, including
the modification of forecasts by losing customers during short-falls could
be carried out, it would anyway be shown to be the most economic.

If, on the other hand, the product was by far the best available or
changes of supplier were frequent, or the market was anyway not very profitable, the company may feel it would lose little on a short-fall, compared
with a gain in the delay in capital outlay.

The year when a new plant should be operating is determined by the
policy towards short-fall. The exact point in time when short-fall in plant
capacity occurs is not accurately predictable, because there is considerable
latitude in the commercial system. Take for example the heavy chemical
industry; the company may hold one month's stock, the customer one month,
leaving some months of safety beyond the time when the demand is greater
than the plant capacity. But in any case the demand month by month is a
fluctuating figure, and so there will be a margin of some months to cover
normal fluctuations. However, a date must be set for the new plant to be

operating some years in advance. To choose between time intervals of one
year apart for this decision is too coarse, whereas to discuss which month
the extension must operate is difficult because of the monthly fluctuations.
Probably a realistic objective is to set the date to the nearest quarter
year. On engineering manpower planning grounds it is necessary to decide
on a fixed date, -- say a month end -- for the completion of the project,
but this fixed data has rarely a commercial basis.

4.7 THE QUICK EVALUATION

This chapter began by describing that investment decisions progress as
a series of small decisions, all of which can be easily reversed if new data
shows this to be necessary and that it is possible to imagine the control
being a continuous process. The detector for this control system is the
economic evaluation, and the control action is the decision. Hence, every
iteration of the control loop requires an economic evaluation to be made,
each time with updated technological and commerical data.

When such a system functions, it is a great support to those working
on the project. Research and Development departments particularly benefit
because in these departments new technological information is continually
being found, and such departments are always needing advice on where to
concentrate their efforts. To have an economic evaluation "on request" helps
them assimilate the worth of new results, directs their efforts into regions
worthwhile to the project considered as a whole, and enables them to stop
work on grossly uneconomic projects at an early date. Design departments
also gain from frequent economic evaluations and planning studies. Changes,
if essential, will be spotted and therefore made earlier; the designs could
centre around the cost-sensitive areas, reducing running costs or capital
cost, whichever is the more significant, and again wasted work on uneconomic
projects could be stopped earlier.

However, such a system rarely functions. It is more usual for the
overloaded planner to work on the most advanced projects first, leaving
projects in their early stages -- particularly the research projects --
to belated studies at 6-month intervals. These analyses often show the
research direction to have been poorly chosen and sometimes show that some
information has rendered the project uneconomic. Often this information
has been known for a long time, but no-one was available to make a proper
evaluation, and though the significance of a series of disappointing dis-
coveries may have been realised by the research team, no one had done the

evaluation, and though the significance of a series of disappointing dis-
coveries may have been realised by the research team, no one had done the
evaluation to show the seriousness of their combination.

There is a need for a method of evaluation and planning that is quick
and that can be updated with the minimum amount of effort by the planner.

Many evaluations are done for comparative purposes. They compare alternative processes, or alternative operating conditions or as a sensitivity
analysis to determine the importance of the individual factors. Such
evaluations need not be thorough, because the project as an investment is
not being questioned, but the relative importance of results must be correct,
since the difference between evaluations is to be examined. It is particularly difficult to carry out quick, incomplete calculations a number of
times in order to compare comparatively small differences. Each evaluation
makes a number of approximations, and these must be identical each time.
This difficulty can be overcome by using a computer program. The program
can be written to incorporate approximations and can run with the different
sets of data. This enables consistent estimates to be produced and hence
valid comparisons can be made. It is a common occurrence to see planners,
when questioned as to why their latest evaluations are so much worse than
their earlier ones, search through pages of calculations and provide no
concise, satisfactory answer. Such sessions usually end with both questioner and planner feeling there must have been a mistake -- in the more optimistic result!

If each evaluation is summarised by the data input and the output of
a computer program, there is a complete summary of the work always available
and confidence in the planner is maintained.

4.8 SOME RULES OF THUMB

Planning and Decision-taking concerning a choice of alternative methods
of production expansion are usually done by applying rules of thumb plus a
few evaluations rather than evaluating the profitability of every possibility. Following the evaluation the possible risks are taken into account in
a subjective manner and from this a decision is made. An example of some
of the "rules of thumb" are:

    a) Alternatives can be ranked for attractiveness based on :
        (i)   Capital Cost/ton/year of product
        (ii)   Variable cost/ton of product
        (iii) Cost/ton based on a simple rate of interest calculation.

Neither (i), (ii) nor (iii) give an unconditionally correct ranking, but alternatives low in all three can be confidently discarded.

b) Low capital projects are more favourable than major investments. The expansion of an existing plant is almost always the most attractive.

c) The capacity of any new plant should be set to satisfy about seven years market growth.

d) (i) If you are presented with probabilistic data try to get the commercial department to define which percentile is to be designed to -- as "company marketing policy"

    (ii) If d) (i) fails, determine the marginal cost and choose the percentile above which the Expected profit breaks even with the marginal cost of supplying the capacity. Design to this percentile as though it were the only data available.

e) Obtain a definition as "Company Policy" that the company will always satisfy the demand. Hence when demand equals existing capacity it is time to expand.

f) If an expansion is not clearly the most attractive alternative, then compose a series of two-level expansions or plans from the better alternative selected by rule (a). Each of these plans has then to be fully evaluated by DCF, NPW or whatever criterion the company uses.

g) Do no embark on major investments in uncertain markets unless there is no alternative. Always define corresponding fall-back plans before making the final choice.

This technique is common practice and is obviously therefore quite workable. It relies heavily on the planner spotting combinations of alternatives in his plans, and on his being able to sort good from bad in his initial comparison of alternatives. It enables non-quantitative factors to be introduced at an early stage in the analysis.

There are however, grounds for looking for improvements to this method:

a) It does not make use of probabilistic data. This data does represent further information and should be fully utilised if at all possible.

b) It does not consider every combination of alternatives. There may be particular combinations that have been overlooked. All planners know the feeling "Have I missed the obvious?"

c) It assumes that rules always hold. There will be
    situations where the rules result in particularly bad
    advice.

d) There is no pretence that the rules will find the
    "optimum" decisions, but it is assumed that the optimum
    is so flat that it gets near enough for all practical
    purposes. This may not be so.

e) It provides no quantitative Risk Analysis. All the
    necessary data may be to hand, yet it is never used.

f) It does not formalise the back-up-plan. This is an
    important part of an evaluation, but it is not normally
    thoroughly evaluated.

g) It is not quick. It relies so much on the planner's
    interaction that it takes him days or weeks to complete
    a study and to revise old studies. An "evaluation on
    request" is out of the question.

h) It is not automated, so there is always the suspicion
    of inconsistency masking the real conclusions, particularly
    in comparative studies.

Because of these shortcomings a development project was started in
the Systems Engineering Group at the Swiss Federal Institute of Technology,
(ETH) Zurich., to produce a method of investment analysis and planning
that was somewhat more systematic than the existing approaches, and which
overcame the weaknesses listed above.

The following chapter describes the development of the method and a
number of application to investment problems of general interest are given
in later chapters.

4.9 BIBLIOGRAPHY FOR CHAPTER IV

1. Rose, L.M. The Application of Mathematical Modelling to Process
    Development and Design, Applied Science, London 1974.

2. Fox, R.L. Optimisation Methods for Engineering Design,
    Addison - Wesley, London, 1971.

CHAPTER V

THE DEVELOPMENT OF A TECHNIQUE
FOR PRODUCTION CAPACITY PLANNING
AND EVALUATION

> Sächs mal sächs isch sächsadryssig,
> isch de Lehrer no so flyssig,
> isch de Schüeler no so dumm,
> macht de Lehrer rumpedibum.

The first three chapters of this book discussed economic evaluation, probability analysis and Decision Theory. These are essential components of any study of proposals concerning investment and manufacturing capacity. Chapter 4 discusses the requirements of such a study, and shows that the methods usually applied have a number of short-comings. This chapter combines the techniques discussed in the first three chapters with the requirements given in Chapter 4 in order to develop a procedure to overcome the short-comings.

The need for speed and consistency emphasised in Chapter 4 clearly means the technique should be programmed, so that the analysis is carried out by computer. This is nothing new; economic evaluations have been carried out by computer by most people for many years.

Let us start by defining the requirements of a good technique.

5.1   REQUIREMENTS OF THE TECHNIQUE

5.1a   The Method Should be Easy to Use

First and foremost, the method should be completely generalized. It should handle all sales patterns and alternatives without the need for modification or reprogramming. The need for speed and consistency has already been stressed. The less the preparation of the data before submitting to the computer-programmed analysis, the easier and more consistent will be the method. This means the programme should be used as early as possible in the planning study. For example, the programme could handle annual

cash flows as input data, or it could handle the inidividual costs -- capital,
fixed and variable cost -- and calculate its own annual cash flow. The
latter is preferred because it gives a quicker result with less opportunity
for error, although the former is more common in investment evaluation
programs.

5.1b Two Levels of Analysis are Required

The technique should recognise two levels of analysis. Firstly, the
selection from alternatives, and secondly the more detailed economic evaluation of the selected alternative.

The first level should develop optimum plans as described in Chapter
III in the discussion on Decision Trees. These should extend far enough
into the future to minimise the problem brought about by different capacity
alternatives which is described in Chapter IV, Section 4.4. It is important that a number of good plans be developed, not only the best one, because
the consideration of non-economic factors and risk requires more than one
alternative to be presented, to cover the eventuality that the optimum has
some non-quantifiable/or risk disadvantage. The planning stage should also
consider back-up plans for the selected alternative.

The second level of analysis should be capable   of the evaluation of
more than a single evaluation criterion. As Chapter I shows, there is no
single criterion that is universally acceptable, and so the project should
be judged in the last stage of analysis against a number of different
criteria to enable a judgement to be made based on a composite of more than
one criterion. This second level should also be able to handle Sensitivity
Analyses.

5.1c It Should Handle Probabilistic Data

Chapter II shows that none of the data needed for the evaluation are
known with certainty, and some -- future sales volume in particular -- are
almost always very uncertain. The combination of these uncertainties to
show the expected distribution in the profitability should be done rigorously. The use of Sensitivity Analysis alone is unsatisfactory because it
does not indicate the combined effect; a technique for the aggregation of
distributions such as the Monte-Carlo method is necessary. The particular
correlation effect between sales volume and year and sales colume and price
must be properly handled since this has a marked effect on the final

distribution. Some method of handling the probabilistic data at the first
level, the planning stage, is also required.

The tendency to choose single values for data input, even when the
data are very uncertain, is inevitable when there are no computational tools
for probabilistic analyses. However, if computational methods are developed
for handling probabilistic data conveniently, it might open the way for the
probabilistic approach to project analysis as the normal procedure.

The resulting distribution obtained from the probabilistic analysis
should have the option of being judged by some decision criterion, such as
Utility, which concisely expresses the distribution as a single figure and
can easily be handled and discussed, in addition to the straightforward
display of the distribution to enable any other form of decision criterion
to be employed.

5.1d It Should Handle Inflation

Inflation should be accounted for in a consistent way so that it is
exactly clear what assumptions are being made. The influence of a change
in the inflation rate should also be shown by the analysis.

5.2 THE DEVELOPMENT OF THE ELEMENTS OF A METHOD

To fulfil these requirements a number of techniques have to be collected together and others developed. The individual techniques are discussed
in the following sections. Having described each method the chapter
continues by describing how these components are built up into an overall
technique.

5.2a The Generalization of the Problem

The system under consideration is the managerial control of development,
selection and building of a manufacturing plant.

This system comprises the following parts:

      a)   The alternatives
      b)   The forecast sales information
      c)   The decisions

Each of these parts must be expressed in some completely general way
if a generalised technique is to be obtained.

5.2a (1) The Alternatives

The Alternatives can be characterised by five data items: The Capital
Cost, Manufacturing Cost, Variable Cost, Fixed Cost and Scrap Value, as described in Chapter I. For example, the new plant may have a high capital and
fixed cost and a fairly low variable cost. An alternative proposing the
purchase of a product from a competitor would have zero capital and fixed
costs, but a high variable cost. An extension may have low capital and
fixed costs, but also low capacity. But the extension is still not completely defined because an extension is always an extension of a particular plant.
Extensions therefore require more information than the 5 data items described
above for their complete definition. Furthermore, extensions are sometimes
built in stages which must be carried out in a fixed order. One could increase a reactor section of a plant which was already a bottleneck in stage
I, so that a later increase to a purification section would give a further
increase in plant capacity, (Stage II). To implement Stage II before Stage
I would produce no capacity increase because the reactor bottleneck would
still limit production. Hence we have a second property for extension: they
may have to be carried out in a specific order. The transfer of all this
information can be achieved by including the identification of the parent
plant for every expansion as a 6th data item. If the parent plant is itself
an extension, we can adequately represent a fixed sequence of extensions.

Besides some extensions being conditional on others being built, it is
possible to have mutually exclusive extensions. A parent plant with limited
clear ground could be expanded by putting an item of equipment on that ground
which could be either large or small, but if there is only room for one item,
when the small has been installed this excludes the possibility of installing the large also. It remains possible to remove the small and install
the large, i.e. "scrap" the first extension, but not to have both. Hence a
further property of extensions is that they may be mutually exclusive.

This "exclusion" characteristic can be summarised by one further data
item, this being a 0 or 1 flag with every expansion to denote whether the
expansion excludes all others or not.

Figure 5.1 summarises this convention and shows the types of extension
allowed by it.

The 5 data items plus the two extensions codes therefore characterise
any alternative for the planning and investment analysis study.

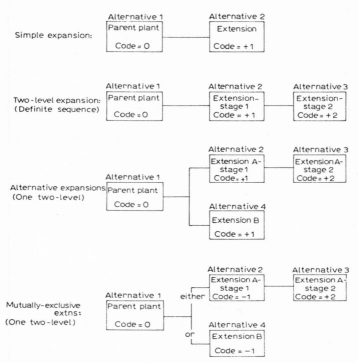

Figure 5.1   Alternative Extension Structures and their Coding

5.2a (2) Forecast Sales Information

A generalised technique should deal with any future sales pattern.
Interesting techniques have been developed to plan investments for definite
sales volume growth patterns (4), but in practice, sales forecasts do not
fit into simple linear or exponential growth curves and so the methods reported in the literature cannot be applied as a general technique. This
means that yearly sales figures should be given as data input. There should
be no restriction on their pattern. It should be possible to represent
steady growth of any form, or growth followed by decline, or decline alone.
This wide specification is necessary because the most usual and most difficult situations to plan for are growth followed by decline. Once this
growth followed by decline can be handled by a planning technique,   total
growth or total decline patterns can also be handled.

Sales price information is also needed, and this again should be entered as yearly data so as to cope with any pattern of pricing policy. The
important sales volume/year and sales volume/price correlations must be

properly handled. This is an extensive subject and will be discussed separately later in this chapter.

Plants frequently produce more than one product. When this is a by-
product over which there is virtually no control it can be treated as a raw
material with a negative price, equivalent to its credit on disposal. However, frequently plants make more than one product intentionally and in
controllable amounts -- either simultaneously or sequentially in campaigns.
A generalised technique should be capable of handling all the data for a
single plant -- a number of products, each with probabilistic forecasts, for
demand and price, and each with its own variable cost -- and not require all
this data to be predigested by hand or be fed to the program as a pseudo-
single-component.

5.2a (3) The Decisions

Before deciding how to generalise the decisions, we should define what
the decisions are. Chapter IV, Section 4.3 argues that most decisions in
manufacturing capacity are easily reversible and are of the type "proceed
one stage further with the development of alternative X" rather than "build
and operate alternative X". However, to choose from the alternatives we
must carry out a simulation well into the future, assuming that alternative
X is built and operated. If this simulation is attractive, then we make
the decision to "proceed one stage further". Hence although the real
decision is only to move forward one step, the decision simulated and represented by the planning and evaluation stages is to "build and operate".
Similarly, although the real decision may be being taken four years before
the operation of the plant, since the planning stage decision is not the real
decision; there is no need to simulate reality by incorporating this time
delay, which simply causes further complication. For planning purposes,
one can decide in year 3 to build and operate plant X to produce in year 3.

The previous section on Sales Forecasts argued that increases and decreases in forecast should be handled. This implies that decisions involving
reduction, as well as increases in capacity should be part of the generalised
technique.

The types of decision assumed with manufacturing capacity are summarised
in Table 5.1. This Table also includes an example of how each of these decisions can be represented by a combination of "scrap" and "build" decisions.
In fact, all decisions can be represented by combinations of "scrap", "build"
or "do nothing" for every alternative. Extensions of plants again are

slightly different in that if a parent plant does not exist an extension
cannot be built, and if a parent plant is scrapped, then the extension must
also be scrapped.

TABLE 5.1
Summary of Decisions involving manufacturing capacity

Plant 1 is existing and operating
Plant 2 is on the drawing board

Proposal	Decision	
	Plant 1	Plant 2
Add a further plant	"do nothing"	"build"
Replace existing by new plant	"scrap"	"build"
Make no changes	"do nothing"	"do nothing"
Reduce manufacturing potential	"scrap"	"do nothing"

5.2a (4) The Timing of the Decisions

Profitability calculation are normally made as a series of discrete
time periods. The Forecast data is considered as discrete data per time
period. The decisions are discrete events in time and normal investment
evaluation is done in discrete time periods. This emphasis on discrete
time period is to simplify the calculation. To allow time to be a continuous variable would considerably increase the complexity of all calculations. The choice of the length of the time periods depends on the significance of the time scale but since the quantity of calculation is directly
proportional to the number of time periods, it is advantageous to keep them
to a minimum. However, since it is reasonable to take decisions only once
per period, the period must not be so big that the analysis becomes too
coarse. In projects typical of the chemical industry, with from two to five
years of development and a fifteen-year project life, time periods of one
year are reasonable, apart from the timing of the first decision.

For the first decision, on which detailed manning and commerical plans
are based, a much finer analysis is required. For projects typical of the
chemical industry, three months is considered suitable, as argued in
Chapter IV.

Having agreed a time period. to be completely general it must be possible to make any decision in any time period.

    Decisions to change production capacity should be made whenever it is economic to do so -- that is at the first point in time when it is more profitable to carry out the decision than to delay it for a further time period. This definition requires an optimisation to be performed which involves further computation to locate the correct time period, but this definition does have the advantage of being completely general. This definition can be overridden by management policy that the sales demands must always be met. With this constraint, and with deterministic sales volume data, this optimisation problem disappears and the time to expand is simply in the first year that sales exceed capacity.

5.2b The Evaluation

    Using as data the 5 data items -- Capital Cost, Capacity, Variable Cost, Fixed Cost and Scrap together with the operating period for every plant in, and foreseen in, a company manufacturing complex, together with the forecast sales volume and price per year, it is possible to calculate annual cash flows and then evaluate the profitability of the complex.

    To do this one must determine whether all plants will be operating fully -- that is, whether sales are greater than total capacity or not. If the latter, then the optimum operating policy for the firm must be determined. The optimum operating policy is to cut down production or shut down plants in the order of their variable costs, -- the more expensive plants should be closed before the more efficient. The evaluation procedure must therefore firstly perform an analysis to determine which plants are operating before an annual cash flow can be obtained. This point is made in the second example of Chapter I, Section 1.9b, for a two-plant system. However, for a generalised technique this must be adequate for any number of plants.

    Evaluations are made to determine the profitability of an investment. It is therefore necessary to obtain the cash flow resulting from the investment, and not the cash flow from the investment plus all previous investments in that product's manufacturing plant. This requires with- and without- investment calculations to be performed to establish the difference. Once more the base-case may consist of more than one plant, and this again may require an operating plant policy to be evaluated before the cash flow can be determined.

    It is important that the with- and without-investment analyses are made within the same computer program. It is a common approach to develop a cash flow difference by hand and evaluate this difference by computer. However, the hand calculation is not a straightforward subtraction since this again requires operating policies to be generated on paper before the annual cash flows are obtained. Though by no means difficult, this stage represents a number of tedious hours of work and the scope for inconsistent operating assumptions or just plain arithmetic errors, -- which a new technique should try to avoid.

The evaluation should include the incurring of a penalty cost for every unit of demand not met. This is a fairly standard technique for representing losses other than normal profit loss. It can represent loss of profit on future sales because the dissatisfied customers may never return, or some fine imposed by some contractual agreement, or just some representation of loss of goodwill.

The evaluation method should also have provision for annual cash flow adjustments to represent the costs incurred by projects that are not directly associated with the alternatives. The intermittent outlay of working capital, and its final recovery would be an example.

Because it is impossible to represent all features of an investment by one economic criterion, a generalised method should evaluate a fairly wide number of criteria so that the projects can be judged from a number of standpoints. A suitable set, that involves   little   extra programming but covers the field fairly well, would be:

NPW	-	a widely accepted criterion based on the total cash earned by the project, having paid for the capital used
EAV	-	an annual cash flow criterion
BCR	-	a criterion related to capital cost
DCF	-	a widely accepted criterion based on considering the project as a money investment
Pay-back Time	-	a widely accepted criterion emphasizing the market forecast risks
Undiscounted Cash Flow	-	To judge the importance of the project on the liquidity of the firm

5.2c The Development of Optimum Plans

Chapter IV introduced the problem of comparing alternatives of different capacities in a steadily increasing market. A direct comparison of the installation of each alternative alone is not possible, because the manufacturing capacities of the alternatives are different at the end of

the project. This may lead to one being in a much more attractive position
than the other, and the best way to evaluate the worth of this better position is to simulate the next expansion stage. Hence, to evaluate the first
decision one is faced with evaluating the second, and possibly later decisions, to include the value of the potential correctly for future expansion.

This situation is clearly shown by the comparison of the installation
of a large plant,   with   a   marginal expansion. No comparison can be
made without including the proposals for further expansions after the extension has been "used up".

If these further proposals do not identically match the capacity of
the large plant proposal, it may be necessary to propose further actions
following the larger plant until the actions are so far in the future that
their effect is made negligible by the discounting.

If we accept that later decisions are essential to the correct choice
of alternatives for the first decision, these later decisions must be
"optimally chosen" since it is always necessary to compare optimal alternatives to obtain valid results from the comparison.

Determination of an optimum plan is difficult. At the present time, no
systematic methods seem to be used and Chapter IV suggests that present
methods rely on the planner "spotting clever combinations of alternatives".
This is not a good basis for a generalised automated method. As one moves
to a method which requires longer term plans to be developed the optimum
combination of year and alternative that makes up these plans becomes
virtually impossible without a systematic method. A truly systematic method
would investigate all possible decision at all possible times at which they
could be made.

We have already argued that it is fair to consider 12 months as a
reasonable time step between decisions. "All possible decisions" means
build everything that has not been built, scrap everything that is existing, in every combination, as well as "do nothing" to provide the range of
alternative plans that then have to be evaluated.

Take as a simple example two alternatives A and B and two time periods.
Figure 5.2 gives a Decision Tree for a two-year period allowing all possible decisions to be made in each time period. The decisions are not all
written in, but it is obvious from the change-in-plants data what the decisions must have been. This simple case, -- two decision stages and two
alternatives -- produces 16 different plans since each path represents a
plan. Each plan will have a different outcome since the monies are spent

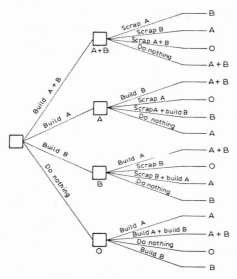

Figure 5.2

A Decision tree for a 2-Alternative, 2-year Project

at different times, even when the end-states are the same. For example, the
top plan on Figure 5.2 O→(A+B)→B, will have a different profitability from
the last plan O→O→B, even though they both have B as the final state.

For this simple example we have four possible states and 16 plans; for
a more reasonable example, say 6 alternatives and a 15-year evaluation,
there is a total of

$$1 + \frac{6}{1} + \frac{6\times5}{1\times2} + \frac{6\times5\times4}{1\times2\times3} + \frac{6\times5\times4\times3}{1\times2\times3\times4} + \frac{6\times5\times4\times3\times2}{1\times2\times3\times4\times5} + 1 = 64$$

different states possible, and all these possible each year. Hence, the
first year will produce 64 paths, the second year will branch into 64×64,
and in the fifteenth year there will be 64^{15}, or 1.24×10^{27} paths. Each
would need evaluating. Obviously such a complete system for generating
plans produces an impossible amount of computation, and the resulting
Decision Tree becomes a decision forest with 1.24×10^{27} branches!

It is obvious from Figure 5.2 and also from the way in which the calculation of the number of paths is made, that there are only a limited
number of different states, and the different paths simply represent
alternative ways of getting to different states. It is, therefore, possible
to summarise the plans more concisely on a figure other than a Decision Tree.
If all the different states are written down each year, the different plans

can be represented by different routes of getting to the end state.
Figure 5.3 displays the information in this way for our first example of 2
plants, A and B. The information contained in Figure 5.3 is the same as in
Figure 5.2.

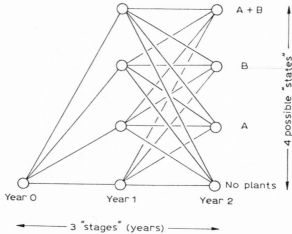

Figure 5.3

An Alternative to the Decision Tree Representation

In looking for the optimum plan one must find:
    a)   the best end-state
    b)   the best way of getting to that end-state

We can divide the search into these two steps. We can take each end-
state, find the best way of getting to it, do this for each end-state in
turn, and then finally select the end-state with the highest worth.

The problem lies in the first step -- finding the best way of getting
to each end-state. In our second example there are 64^{14}, or 1.93×10^{25}
ways of getting to each end state and 64 end-states. To choose the best
from 64 is no problem but the evaluation and choice out of 1.93×10^{25} is.
We have therefore isolated our problem to that of choosing the best route
to a given end-state.

The definition "best" is governed by cash flows. Neglecting, for the
moment, discounting and taxation, we are looking for the plan with the
greatest total positive cash sum at the end state. This cash sum has risen
from yearly cash flows which have been added together to give the end-state
total cash. A yearly cash flow equals <u>sales and income in that year</u>

minus <u>variable cost in that year</u> minus <u>fixed costs for that year</u> minus
<u>capital outlay in new plant commissioned in that year</u> plus <u>scrap value for</u>
<u>plants shut down in that year</u>. Yearly cash flows so defined are independent
of what has happened in previous years, or what will happen in future years.
A selection of a plan going from A to (A+B) in year two has an easily calculated cash flow. It is :

$$(sales\ income) - (fixed + variable\ costs) - (capital\ cost\ of\ B)$$

Adding such partial plans together one can find optimum plans as follows:

On Figure 5.3 there are 4 ways of getting to (A+B) in year 2:

$$
\begin{array}{ccccc}
0 & \to & 0 & \to & (A+B) \\
0 & \to & A & \to & (A+B) \\
0 & \to & B & \to & (A+B) \\
0 & \to & (A+B) & \to & (A+B)
\end{array}
$$

The partial cash flows of these four plans can be evaluated to give
the cost of the four routes for getting to (A+B) in year 2. The best of
these can be located and the corresponding plan remembered; the other
three routes can be forgotton. Now if Figure 5.3 is a part of an evaluation
with more that three stages, all further plans involving passing through
(A+B) in year 2 have only one route from year 1, not the origional 4. The
search is reduced by 75%.

We are describing <u>Dynamic Programming</u> (DP), which has already been
described in Chapter III. The expansion planning problem expressed in this
way gives all the conditions required for Dynamic Programming to be applied
for searching for the optimum plan. (1,5,7)

Since this is the most important single idea being presented in this
book, it is worth one further example to show what DP is doing in this case.

Consider our two-year A and B example extended to a 3-year project.
Fig. 5.4 gives the stages/states matrix and shows all the routes to get
to the end-state (A+B). Four, which pass through A in year 2, are shown in
full, and the remaining 12 only partially. The 16 full plans can easily
be identified without overloading the diagram with lines. Without using
DP these 16 plans would require a total of 16 × 3 = 48 partial plans to be
evaluated before the best plan to (A+B) in year 3 could be chosen.

Applying DP, there would be four partial plans to evaluate in year 1
and 16 in year 2. However, at year 2 we have 4 stages, each with four
partial plans --but only the best partial plan at each stage need be remembered at this point, giving only 4 partial plans for evaluation in the final
stage. Hence, DP has reduced the number of partial plans to be evaluated from
48 to 24.

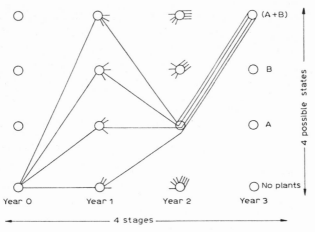

Figure 5.4

Strategy Reduction by Dynamic Programming

The advantage of DP becomes more obvious the larger the problem. For
our earlier example requiring 64^{14} plans to one end-state or 15×64^{14}
partial plans to evaluate,   the application of DP can reduce this to approximately $15 \times 64^2 = 61,440$, or a 4.7×10^{21} fold reduction in computation.

The last step to select the best end-state requires no discussion,
except to point out that the evaluation of all the plans to the other end-
states does not necessitate a new search, but only additional evaluations
in the last year. A single DP search locates the optimal plan to every
end-state in the one search.

Dynamic Programming can only be applied if the partial evaluations are
independent of past or future situations. Hence, any criterion can be used
in the search that obeys this requirement. We have discussed undiscounted
cash flow, before tax. This meets the requirements but is hardly useful as
a criterion for the selection of alternatives. Discounted cash flow can be
used with no problem. At first this seems contrary to the concept of discounting, which is attempting to make the criterion sensitive to the history
of the investment, -- but closer consideration shows that the way the problem is posed does in fact give partial plan discounted cash flows which are
independent of earlier or later decisions. For year N with a partial cash
flow £C, the discounted cash flow is $£C/(1+r)^N$, -- quite independent of past
or future.

The introduction of taxation is rather different. The tax allowance
on capital investment depends on when the capital was spent. Dynamic

Programming cannot therefore be employed if the tax allowance is calculated on a yearly basis. However, if the capital investment figure is credited with the future tax allowances at the time the investment is brought into the cash flow, this dependence on the past is removed and the cash flow, after tax liabilities, can be calculated for partial plans and used with Dynamic Programming. This technique of crediting the capital sum with the tax allowance is a normal accounting technique. It is described in Chapter I Section 1.6b.

Hence, discounting methods after tax can be used as a criterion and so the method can sensibly be used to choose alternatives. NPW can be used without problem. DCF rate of return, on the other hand, cannot be used directly because it is necessary to know the total plan in order to adjust the interest rate to get an overall zero NPW. To use DCF, the search would have to be repeated with different discount rates until the optimum plan had a zero final NPW. Iteration around a Dynamic Program search is enormously expensive in computer time and not to be recommended. Considerable information could be obtained by repeating a Net Present Worth search two or three times at different discount rates. The "optimum" plan with the NPW nearest to zero would, for all intent and purposes, be equivalent to a selection based on DCF.

Although we have made an impressive reduction in the number of evaluations, there is still too much computation required for DP alone to be the basis of a planning technique. Figure 5.4 shows clearly what nonsense most of the plans are. Many propose building more than one plant in a single year. Some involve no plants for part of their duration although the firm is assuming it will be a reliable manufacturer. Many plans involve too much capacity; an example with 10 alternatives will have hundreds of plans that involve all 10 being built! The introduction of a number of sensible constraints will very considerably reduce the number of strategies left for the Dynamic Programming search.

The following constraints were therefore taken as part of the technique. They represent the building of some "common sense" into the procedure:

a) Any state producing more than one pre-set ratio (for instance 1.5) of the maximum demand is immediately discarded

b) Any state producing less than a pre-set ratio (for instance 0.5) of the demand for that year is immediately discarded

c) Only one alternative can be built "in one year"

d) Only one plant can be "scrapped" in one year.

These restrictions reduce considerably the number of partial plan
evaluations necessary. For example, for a 10-- alternative problem with a
15 year life, 1.5×10^7 evaluations would be necessary with an unconstrained
DP search, and with the above constraints it will be reduced typically to
about 10^4 evaluations. This number can be adequately handled by computer
and hence we have now a technique for developing optimum plans. This was
one of the major requirements of an improved technique for planning and
evaluation.

The method so far described is very suitable for finding optimum plans
for "reasonable" problems with 6-8 alternatives. A problem with 10 alternatives would be considered "large".

This method determines the optimum plan for every end-state. This
has a further advantage of providing a series of good plans from which a
decision-maker could exercise judgement, and could allow for non-economic
factors.

Frequently there are timing constraints that any realistic plan must
acknowledge. Some alternatives may not be immediately available, they may
still be at the research stage with a plant being 3 to 5 years away. The
search method cannot recognise this unless the information is introduced
as a further constraint into the search. This is done by describing each
alternative with a further data item -- the first year it can possibly be
available. The search then discards all partial plans that involve any
alternative before this allowed year. The result of the search is then
a series of optimum plans to each end state, none of which involves building any alternative before its first available year. As a bonus, this
extra constraint has reduced the computation required for the search.

Constraints are also sometimes met in the timing of the first
decision. A decision to "go ahead and construct plant" may be involved,
or the market may be such that the plant must be built at a definite time.
In this case the timing of the first decision is fixed; the investigation
is only to determine which alternative should be built. The search method
can easily be constrained further to discard all plans that do not have
their first decision in any specified year. More welcome constraints to
reduce the search!

In the development of the method it was observed that in a 15-year
project, investments after 10 years were heavily penalised because scrapping

them after only 5 years operation was completely uneconomic. Hence, any
plans involving decision in earlier years followed by a decision after year
10,   may   be penalised by the later decision which included the artificially short plant life. This philosophy of choosing the best first decision
by evaluating the whole project plan throws up a new problem of"how to conclude a project"which does not appear in normal evaluation methods.

The problem can be largely overcome by evaluating an "Extended Life"
for the projects whereby projects continue to operate for a certain number
of years at constant conditions beyond their project-life with a yearly
cash flow equal to that of the "last year" of the project.

The need for an Extended Life is illustrated by the following example.
Consider a situation where a product demand could be met for a 15-year
period by the installation of a new plant (Alternative 1) shown on Fig. 5.5,
to give Plan A; or by the installation of a smaller extension (Alternative 2)
to cover seven years, during which time a more efficient process could be
developed to give a new plant (Alternative 3) as a second stage -- Plan B.
This new, cheaper process would run for 8 years, -- until the 15th year
which has been taken as the project life. If, at this time, all the plants
were "scrapped" for the purpose of the evaluation, the scrapping of this

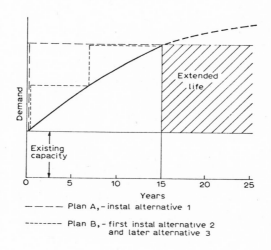

Figure 5.5

Extended Life in Project Evaluation

8-year old improved process plant may give it the appearance of being uneconomic.   This may make the Plan B less economic than Plan A, only

because of the early scrapping of this efficient plant. By allowing all plants to operate beyond the project life such errors are minimised, since this is a closer approximation to reality.

This discussion would be unnecessary if the project life really represented the intended project life. However, the project life is more a statement of a sensible planning period than a time at which manufacture will cease. For instance, it may be impossible to produce forecasts beyond this time. It is, therefore, more reasonable to use some form of simple method of extending the life of the project, than to assume the plants will be scrapped simply because sales forecasting becomes very difficult.

An adequate Extended Life can be calculated by assuming that the performance in the last year is representative of the years beyond the project life. Such a constant cash flow can be discounted by the simple formula described in Section 1.4b.

The application of this formula to extend a Project Life L years by M years, with an overall cash flow in the last year of its life of $£F_L$, gives

$$\text{NPW from Extended Life} \ = \ \frac{F_L}{r} \left(\frac{1}{(1+r)^L} - \frac{1}{(1+r)} (L+M) \right) \qquad (5.1)$$

is therefore a simple way of allowing the plant to operate beyond its Project Life, without requiring extended forecasts to be supplied.

A suitable number of years for the extended life is fairly difficult to determine. Table 5.2 shows the magnitude of the correction for a fairly sensitive project, that is, a new project with initially no sales building up to a maximum at the end of the project.

Table 5.2
Magnitude of Extended Life Correction

Extended Life (years)	Discounted Cash In-Flow 15% Discount Rate	Contribution to Discounted Cash In-Flow from Extended Life. % of 15-year Projects Cash In-Flow.
0	250	0
5	282	13
10	298	19
15	305	22
Infinite	313	25

To take the extension to infinity has certain attractions, but the grounds are not sufficiently strong to justify the correction of 25% of the total project cash in-flow. Ten years is a reasonable extension period for later decisions, and five years is perhaps rather short. A choice between five and ten years would strike a balance between not producing too high a correction (and so to be out of line with accepted 15 year projectlife analysis techniques) and being long enough to overcome the original problem of scrapping plants before their economic lives were over. The actual value chosen, as long as it is more than five years, is probably irrelevant to the outcome of the search for optimum plans, but it may produce differences in profitability estimates.

5.2d The Inclusion of Probabilities

The reason for undertaking a probabilistic analysis is to obtain the distribution of profitability of the project. This information is important to the decision-maker. There may be risks involved that he is unwilling to take,(i.e. he may take the distribution into account in coming to his decision) or distributions may produce justification for obtaining better information.

Again we have to consider the two decision stages -- planning and evaluation -- separately in order to discuss probability analysis. The planning stage requires probabilities to generate conditional plans -- strategies -- whereas the evaluation stage is interested in determining the profitability distribution for the one proposal being analysed in depth. We shall start with the evaluation stage, and return to the planning stage later.

The most important data to consider probabilistically are the sales volumes. Chapter II discusses this at length and proposes that the distribution should be presented as a Discrete Distribution. Each level of the distribution can describe a complete yearly sales forecast pattern for the life of the project, and can be identified by the basis on which the forecast was produced, e.g. "sales are good because ......". Closely connected with volume is price, and so price and volume are best treated in similar ways.

5.2d (1) Probability Analysis at the Evaluation Stage
A limited Treatment

The evaluation of a definite plan, given, for example, four possible
market developments each with associated selling price, can be made by
taking each probability level as a separate project and   combining
the results using the corresponding probabilities, as long as the criterion chosen is additive. As long as NPW, or one of its related crieteria,
is chosen, then one can say:

$$E(NPW) = P(1)NPW_1 + P(2)NPW_2 \ldots + P(n)NPW_n \qquad (5.2)$$

For a 4-level forecast, the Expected Value can be obtained by four
separate evaluations to obtain NPW_1 to NPW_4. The Expected Value should
be calculated in this way rather than by evaluating a single project with
mean sales, because the function is not continuous. The logical branch:

        Product made = demand if capacity > demand
     or Product made = capacity, if capacity < demand
means that one cannot assume

$$E(NPW) = NPW(E(Sales)) \qquad (5.3)$$

Notice that the four evaluations also give the range of NPW
to be expected as well as the Expected Value. We have already achieved
a major advance in risk analysis by providing the range of outcome from
the distribution of the most uncertain variable. What this distribution
is showing is the range of outcomes if the sales fix   on one forecast
level and   stay there for the whole life of the project. In other
words assuming complete correlation between sales volume and year. If
there is less than complete correlation, the distribution of the NPW will
be narrower. This type of analysis cannot say how much narrower.

The inclusion of price variations can be effected in this analysis by
pairing price and sales forecasts -- i.e. probability level 1 has a
certain sales growth and a certain sales price. Should there be two
possible price levels for one sales volume, then there are two probability
levels with the same sales volume curve but different prices, each with
a corresponding probability. The reverse can also hold -- there can be
two probability levels with different sales volume forecasts but identical
prices. We are stipulating a 100% correlation between sales volume and
price. To have zero correlation between sales volume and price on a 4-
level forecast would require 16 probability levels. This is now

becoming rather excessive in terms of data input, and it is best handled
in another way which will be described in the next section.

This first method of including probabilities is therefore limited to
cases where the correlation between both sales volume and year, and sales
volume and prices are 100%. The method however, does have the advantage
that it can be applied with minimum modification to any existing evaluation
program; it requires a minimum of computation and it gives worst case limits
when the correlations are less than 100%. Furthermore there are many
situations where both correlations are in fact 100%.

A Complete Probability Analysis

The above method is not suitable where there are more than a few data
items expressed probabilistically because of the large number of calculations
necessary, and also because of the need for a complicated logic structure
to generate systematically all the combinations for the evaluation. Though
this method is by no means impossible, (10) it is more usual and more flexible to use the Monte-Carlo method described in Sections 2.5b and 3.8a. All
data that is uncertain should be characterised by types of distributions,
and the appropriate parameters for each distribution fed as data to enable
a Monte Carlo analysis to be performed.

The development of a Crude Monte-Carlo program is very straightforward.
Given an evaluation subroutine, which can be the same as that used for the
deterministic analysis, one selects a set of data with the repeated use of
a random number generator, goes into the evaluation subroutine with the
data and exits with the evaluation. In this case, any single-valued
investment criterion is valid, NPW, DCF, Pay-Back for example. This process is repeated an adequate number of times to produce the necessary confidence levels. A few subroutines are needed to produce a tidy output --
cumulative or frequency distributions, -- and a statistical subroutine to
provide the Expected Value and 95% confidence limits, and the program is
finished. Careful attention to computer programming, with a proper
modular subroutine structure makes it possible to produce a Monte-Carlo
analysis of everything that the evaluation subroutine can do deterministically, up to and including a sensitivity analysis. Many such programs
exist, and some are statistically very sophisticated (2).

However Crude Monte-Carlo simulations are generally not very efficient
and they require very many iterations to obtain an accurate estimate of
the Expected Value. By employing Stratified Sampling (3) a very worth-

while reduction (of about 75% of the computer time) was obtained over the
Crude Monte-Carlo run.

This was obtained by choosing each initial market level a number of
times in proportion to the probability associated with that level. For
example, in a 3 level simulation with equal probabilities, one third of
the total number of simulations would start on each demand level. With
high demand/year correlation coefficients, remarkable improvements in the
accuracy of the Expected Value were achieved.

The same random number sequence was also employed for every computer
run. This enables comparisons of Expected Values from different runs to
be made and small differences to be of significance, even though the
Expected Values themselves may be inaccurate.

A point of particular interest in the simulation of economic systems
is the treatment of correlation.

The particular correlations that must be properly treated are sales
volume/year and sales volume/price, and also the overall correlation produced by general inflation. Inflation will be treated in a later section;
we will discuss here only the two correlations of the commercial data.

A generalised Monte-Carlo program should handle any specific degree
of correlation for volume/price/year. This cannot be done by expressing
the degree of correlation by a simple correlation coefficient (C), as
described in Section 2.8d, because such a method can only be applied to
continuous variables, and volume/price data is to be treated as a number of
discrete levels.

The volume/year relationship for instance must represent a market
growth in which discrete changes can occur resulting in a change in demand
level. Furthermore these stochastic changes are more likely to involve
transitions to adjacent levels than jumps over levels. In fact, the market
growth required for the simulation is a form of Markov chain, with definite
probabilities attached to each level. If further information can be given
to characterise the stability of the market in the form of a correlation
coefficient, then we have all the data necessary to carry out a realistic
simulation.

After a number of possible schemes were attempted, the following
approach was developed, and found to give a very satisfactory simulation
of a growing market.

> (a) A Transition Matrix was developed (see Section 2.7) which
>   characterised the stability of the market, and whose
>   equilibrium was in agreement with the forecast probabilities.

(b) Each year of the simulation a random number was generated and this was used to select the new demand level from the old level, using the transition Matrix.

Two details remain -- how to develop the transition Matrix, and how to use it.

The correlation coefficient concept was maintained, in that two transition matrices were involved, the zero correlation and total correlation, and then these were combined to give the partially correlated Transition Matrix.

$$\begin{bmatrix} p_{11} & p_{12} & \cdots\cdots & p_{1N} \\ p_{21} & p_{22} & & \\ \cdot & & & \\ \cdot & & & \\ \cdot & & & \\ p_{N1} & p_{N2} & \cdots\cdots & p_{NN} \end{bmatrix}$$

required Transition Matrix

$$= C_v \begin{bmatrix} p_{11}^t & p_{12}^t & \cdots\cdots & p_{1N}^t \\ p_{21}^t & p_{22}^t & & \\ \cdot & & & \\ \cdot & & & \\ \cdot & & & \\ p_{N1}^t & p_{N2}^t & \cdots\cdots & p_{NN}^t \end{bmatrix} + (1-C_v) \begin{bmatrix} p_{11}^o & p_{12}^o & \cdots\cdots & p_{1N}^o \\ p_{21}^o & p_{22}^o & & \\ \cdot & & & \\ \cdot & & & \\ \cdot & & & \\ p_{N1}^o & p_{N2}^o & \cdots\cdots & p_{NN}^o \end{bmatrix}$$

Total Correlation   Zero Correlation
Transition Matrix   Transition Matrix

$$= \begin{bmatrix} C_v p_{11}^t + (1-C_v)p_{11}^o & C_v p_{12}^t + (1-C_v)p_{12}^o & \cdots\cdots & C_v p_{1N}^t + (1-C_v)p_{1N}^o \\ \cdot & & & \\ \cdot & & & \\ \cdot & & & \\ \cdot & & & \\ \cdot & & & \\ C_v p_{N1}^t + (1-C_v)p_{N1}^o & \cdots\cdots\cdots\cdots\cdots\cdots & & C_v p_{NN}^t + (1-C_v)p_{NN}^o \end{bmatrix} \quad (5.4)$$

The fully correlated Transition Matrix is a simple diagonal matrix, since the probability that the next level is the same as the last level is 1, and all other probabilities are zero:

$$
\begin{bmatrix}
p^t_{11} & p^t_{12} & \cdots & p^t_{1N} \\
p^t_{21} & p^t_{22} & & \\
\cdot & & & \\
\cdot & & & \\
\cdot & & & \\
p^t_{N1} & p^t_{N2} & \cdots & p^t_{NN}
\end{bmatrix}
=
\begin{bmatrix}
1 & 0 & 0 & 0 & 0 \\
0 & 1 & 0 & 0 & 0 \\
0 & 0 & 1 & 0 & 0 \\
0 & 0 & 0 & 1 & 0 \\
0 & 0 & 0 & 0 & 1
\end{bmatrix}
\qquad (5.5)
$$

When N = 5

The zero correlation Transition Matrix was designed to incorporate some characteristics of market changes. It was assumed that the probability of jumps of more than one level in one year was small enough to be neglected. Hence the Transition Matrix (for the five level case) takes the form:-

level next year (j)

P_{11}	P_{12}	0	0	0
P_{21}	P_{22}	P_{23}	0	0
0	P_{32}	P_{33}	P_{34}	0
0	0	P_{43}	P_{44}	P_{45}
0	0	0	P_{54}	P_{55}

level this year (i)

This matrix must satisfy the following constraints:-

(1)   there must always be a level next year

i.e. $\displaystyle\sum_{j=1}^{N} P_{ij} = 1$ (5.6)

(2)   the equilibrium probability for each level must equal the forecast probability for that level ($P_j(f)$) -that is, as derived in section 2.7 :-

$$p_1(f) = p_1(f)\ p_{11} + p_2(f)\ p_{21} \qquad (5.7)$$

$$p_2(f) = p_1(f)\ p_{12} + p_2(f)\ p_{22} + p_3(f)\ p_{32} \qquad (5.8)$$

$$\cdot$$
$$\cdot$$
$$\cdot$$

$$p_5(f) = p_4(f)\ p_{45} + p_5(f)\ p_{55} \qquad (5.9)$$

(3)   the sum of the forecast probabilities must add to unity

$$\text{i.e.} \quad \sum_{i=1}^{N} P_i(f) = 1 \tag{5.10}$$

Since there are more unknowns than equations there still remain many possible Transition Matrices that could be used. To narrow the possibility still further a further characteristic was stipulated -- that the chance of moving up a level or down a level or remaining on the same level should be equal if the forecast probabilities are equal.

The following procedure was found to generate a Transition Matrix which satisfied the above criteria. Notice that this is a modification of the simple $\frac{1}{3} \frac{1}{3} \frac{1}{3}$ matrix, with the modifications altering the value from $\frac{1}{3}$ when the forecast probabilities are not equal.

$$P_{n,n-1} = \frac{1}{3} \left(\frac{2 P_{n-1}(f)}{P_{n-1}(f) + P_n(f)} \right) \tag{5.11}$$

$$P_{n-1,n} = \frac{1}{3} \left(\frac{2 P_n(f)}{P_{n-1}(f) + P_n(f)} \right) \tag{5.12}$$

$$\text{and} \quad P_{n,n} = 1 - P_{n,n+1} - P_{n,n-1} \tag{5.13}$$

except for the first and last levels which have to be:-

$$P_{1,1} = 1.0 - P_{1,2} \tag{5.14}$$

$$\text{and} \quad P_{N,N} = 1.0 - P_{N,N-1} \tag{5.15}$$

What must be guarded against is producing a matrix with any unnecessarily low (or negative!) probabilities. This would impart peculiar properties to the market development. For instance any further zero beyond those already stipulated would produce a disjointed system with no possible interchange between the disjointed sections. This protection against low probabilities is achieved by stating that if adjacent forecast probabilities $p_n(f)$ $p_{n+1}(f)$ differ by more than a factor of 3, then the following modified procedure is used.

if $P_{n-1}(f) \geqq 3\, P_n(f) : -$ $P_{n,n-1} = 0.5$ (5.16)

and $P_{n-1,n} = \dfrac{0.5\, P_n(f)}{P_{n-1}(f)}$ (5.17)

or if $P_n(f) \leqq 3\, P_{n-1}(f) :-$ $P_{n,n-1} = \dfrac{0.5\, P_{n-1}(f)}{P_n(f)}$ (5.18)

By adding this derived zero correlation matrix to the full correlation Transition Matrix according to equation (5.4) a Transition Matrix for any degree of correlation can be obtained.

The author is indebted to Dr. W.R. Johns for the development of this method.

The next year's levels required by the Monte-Carlo simulation were generated from the matrix by taking 0-1 random number (Z), and finding the level at which the cumulative probability becomes greater than the random number:-

Given last years level as i, then the next years level J is such that,

$$\sum_{j=1}^{J-1} P_{ij} \leqq Z \leqq \sum_{j=1}^{J} P_{ij}$$ (5.19)

This technique produced simulated market developments that were very sensible in appearance, and which still maintained the integer nature of the analysis. Examples of generated market developments are shown in Appendix 1, Figure 1.

A similar problem exists for the price/demand correlation since an integer method of handling the correlation is again required. In this case our Transition Matrix required, is:-

Price for this year (new state)

	P_{11} P_{12} ············· P_{1N}
	P_{21} P_{22}
Derived for	.
this year	.
(old state)	.
	P_{N1} ················ P_{NN}

The same technique was therefore employed - - with one exception.

A zero price/volume correlation means the probability of any price occurring is dependent only on the _forecast_ probability of that price occurring and related in absolutely no way to the demand. Contrast this with the demand/year correlation where we assumed that even at zero correlation the level the previous year had some influence because wild fluctuations in demand were unacceptable.

Hence the zero correlation Transition Matrix for price/demand differs from the demand/year matrix in that it is simply a matrix of the forecast price probabilities.

$$\begin{bmatrix} p_1(f) & p_1(f) & p_1(f) & p_1(f) & p_1(f) \\ p_2(f) & p_2(f) & p_2(f) & p_2(f) & p_2(f) \\ p_3(f) & p_3(f) & p_3(f) & p_3(f) & p_3(f) \\ p_4(f) & p_4(f) & p_4(f) & p_4(f) & p_4(f) \\ p_5(f) & p_5(f) & p_5(f) & p_5(f) & p_5(f) \end{bmatrix}$$

This matrix satisfies the necessary constraints summarised by equations 5.6 to 5.10.

There remains the problem of estimating C_v and C_p, but at least this technique does provide the adequate flexibility for handling any degree of correlation. It is likely that only three levels 0, 0.5, and 1.0 can be estimated by most forecasters, with possibly the 0.25 and 0.75 levels being attempted by the more courageous. Methods for obtaining subjective judgements for these correlations coefficients are discussed in detail in Chapter VIII, Section 8.6b.

Figure 1 of Appendix I shows a number of simulated demand and price curves using these Markov techniques, for a number of values for the correlation coefficient. Figures 4.1 to 4.3, Chapter IV, show the types of volume and price fluctuations that these correlation coefficients are trying to simulate.

A final point of interest in the Monte-Carlo simulation concerns the number of simulations required to get an Expected Value within the required confidence limits. Section 2.5b, Chapter II, describes a method of calculating the number of simulations required to obtain a required confidence level on the Expected Value by using estimates of the standard deviation of the distribution. This can easily be made into an automatic procedure by carrying out a few initial simulations to calculate a preliminary estimate for the standard deviation, which then enables the number of

simulations required to reach a pre-defined accuracy to be determined by
Equation 2.16. The Monte-Carlo simulation can then automatically run for
this number of times. The user simply has to state the 95% confidence
limit he requires on the Expected Value and the program can determine when
enough simulations have been done.

5.2d (2)   Probability Analysis at the Planning Stage

The reasons for conducting a probability analysis at the planning
stage are both to devise plans which do not have chances of excessive
losses and also to obtain realistic simulation of the future. The whole of
life is a sequential probabilistic process with later decisions being
taken to maximise criteria in the light of later information. The same
applies to plant expansion capacity decision behaviour. Later decisions
correct for earlier ones as more information becomes available. This has
the effect of raising the profitability and reducing the losses, compared
with an evaluation made at one point in time. To evaluate an investment
decision properly this corrective process should be simulated. To make
a "good decision" there must be scope for these corrective decisions. A
probabilistic planning technique can achieve these ends.

Every uncertain data item can be considered to be an event fork in a
decision tree and if the distribution of the uncertainty is continuous the
event fork becomes an even fan (8). We have already shown the difficulty
of reducing a decision tree with two event forks to tractable proportions.
The thought of carrying out this search with 10 or more event forks or
fans is formidable.

By far the most uncertain and most important data item, as far as
the planning stage is concerned, is sales volume per year. If a planning
technique could be devised to handle this one variable, it would make a
major contribution to a probabilistic analysis at the planning stage. This
therefore formed the limited aim for a generalised technique.

The general problem can be visualised with the help of decision trees.
So far, our planning technique has dealt with a series of decisions and
certain events. This is shown in Figure 5.2 and described in Chapter III
Section 3.4a. Since the events -- the yearly sales figures -- produce no
branches, they can be, and are, omitted from Figure 5.2. If we now include
uncertain yearly sales volumes, -- suppose a 2-probability level forecast
is available, a double fork should be inserted into the decision tree for

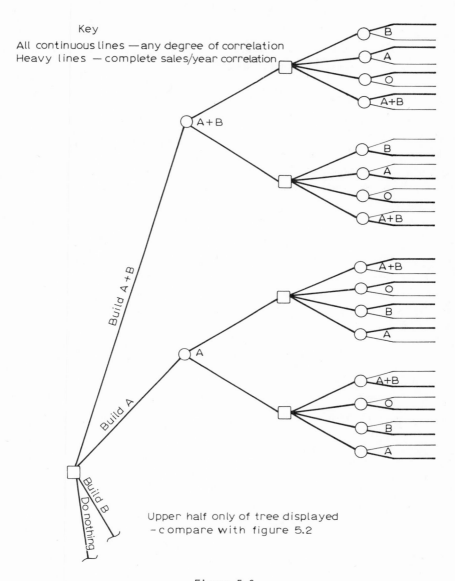

Key

All continuous lines — any degree of correlation
Heavy lines — complete sales/year correlation

B
A
O
A+B

A+B

B
A
O
A+B

A+B
O
B
A

A

A+B
O
B
A

Build A+B

Build A

Build B
Do nothing

Upper half only of tree displayed
– compare with figure 5.2

Figure 5.6

Conditional Decision Tree for Two-Alternative Two-Year Project
Selection with Two Probability Levels of Yearly Sales

for each year. Figure 5.6 shows this for the example equivalent to Figure
5.2. Notice for this simple example we have 64 outcomes, compared with 16

for the deterministic analysis. For a reasonable length project, with 4 or
6 sales probability levels, we once more have an impossibly large search to
handle. In fact, the factor increase is R^L, where R is the number of probability levels and L is the project length.

Complete Sales/Year Correlation

In any case, Chapter II suggests that the most likely and most serious
relation between sales volume and year is the complete correlation. That is,
the future sales are not known, but if they turn out to be good, they are
more likely to be good for the whole project -- and vice versa.

If we consider a conditional decision tree with complete correlation.
the same event fork is taken from every point in the tree, and all other
branches are discarded. The heavy lines on Figure 5.6 show the only allowed
paths in our example. In this case we have reduced the number of possible
plans from 64 to 32. In general, the reduction is $R^{(L-1)}$ times for R
probability levels and L years project life. We have increased our original
deterministic search by a factor of R, -- an acceptable increase.

This method provides us with a way of evaluating the strategies (and
hence of choosing the best strategy)by using the search technique already
described in Section 5.2c R times, repeating the search with each sales
forecast level. The resulting strategy tells us the best set of decisions
to take whatever the sales level, assuming we know before the first decision
is taken what that level would be. This is sometimes useful in research
planning, but more usually, when the decision has to be taken there is no
information available to say which sales level will apply. It is usually
not known until the first decision has been taken, the plant built and
operating, and the development of the market observed.

In this more usual case, the first decision must be made without
information, but it can be assumed that the probability level of the sales
is known before the later decisions are taken.

A really useful method should, therefore, be capable of choosing a
first decision without information, which is part of a good future strategy
what ever the sales outcome.

One obvious method is to find the optimal plan based on the mean sales
forecasts, extract from this the first decision, and the proceed with R
dynamic programming searches to obtain strategies with this fixed first

decision using complete correlation. However, this method is not ideal,
its disadvantage is shown by the arguement that follows.

A primary objective of the study is to determine which first decisions
have the greatest scope for later corrective action, so as to maximise
profit or minimise loss whatever probability level is realised. To
choose the first decision on mean sales grounds will result in a fairly
intelligent choice, and the following searches will determine what corrective actions are possible and give the resulting profitability distribution
but there will be no opportunity actually to choose the first decision
from the risk analysis itself. The first decision has already been fixed
before any risk is investigated. There is no way of choosing a first
decision based on results of a risk analysis without carrying out a risk
analysis for every first decision. There is no short-cut. The total
risk analysis must be done with every alternative that can sensibly be
considered as a first decision.

This in fact presents no difficulty to the dynamic programming
search routine -- in fact such constraints are welcome since they reduce
the search considerably, allowing all plans that do not include the particular alternative as the first decision to be discarded.

This approach is in fact generating all the "back-up plans" automatically, and enabling the first decision to be made having quantitatively
considered the best back-up plans associated with all the alternatives.
This is doing no more than quantifying much of an experienced manager's
non-quantitative judgements.

This risk analysis for the planning stage is valid only for complete
correlation between sales volume and year. For A alternatives and R
probability levels it requires A x R searches to develop A strategies,
one for each alternative as the first decision.

The results can be displayed in a Decision Matrix with the probabilities
being the "states of nature" as the columns, and the alternative first
decisions, the "alternative actions" , as the rows. The Matrix shows the
NPW of the optimum plan for the corresponding probability level having
fixed the corresponding first decision. Table 5.3 shows such a Decision
Matrix for 4 probability levels and 4 alternative first decisions.

How to take decisions with this Decision Matrix will be discussed
later in section 5.2d (3).

Table 5.3

Decision Matrix for a 1st Decision Analysis

First Decision (Alternative Actions)	States of Nature			
	p_1	p_2	p_3	p_4
Small plant A(1)	100	100	50	50
Medium plant A(2)	200	100	0	-200
Large plant A(3)	300	200	-100	-400
Extension to existing plant A(4)	50	50	50	40

Zero Sales/Year Correlation

The other correlation extreme, which may occur occasionally with stable products changing as a result of economic fluctuations, is the zero corre- lation case. This represents a less serious problem in decision analysis because the resulting profit distribution is comparatively narrow. However, it is interesting to study, particularly as a support for cases considered to have an intermediate degree of correlation. The intermediate correlation case's distribution will lie between this and the complete correlation result. In this case the sales in any year are independent of any other year -- before or after. Hence the probabilistic performance of any particular plan is the sum of the discounted cash flows associated with the different sales levels weighted by their probabilities. That is, the Expected Present Worth for year n is given by:

$$E(PW_n) \ = \ \sum_{i=1}^{R} \frac{P_i \left[(F^n_{in})_i - (F^n_{out})_i \right]}{(1+r)^n} \qquad (5.20)$$

where $(F^n_{in})_i$ and $(F^n_{out})_i$ represents the year n in and out cash flows for probability level i.

The concept of strategies when correlation is zero is meaningless, be- cause one has no information on the future and one must always resort to the best average single plan. The conditions for Dynamic Programming hold, and we can find the optimum average plan by using tne same search technique as described earlier, except that the Expected Present Worth is evaluated at

each stage by equation (5.20) in place of the present worth of a single
sales forecast.

There are two points worthy of note: The first is that the Expected
Present Worth should be evaluated from the weighted cash flows and not by
a single evaluation using the Expected Sales. The point has been discussed
in Section 5.2d(1) for the evaluation stage.

The second point concerns the assumption that Expected Present Worth
should be the criterion in the search. In a strict sense this should be
the decision criterion agreed by the planner as being appropriate. It
could well be some form of Certainty Equivalent to weigh against undue
heavy yearly losses instead of the Expected Value. No difficulty is introduced in modifying the search to do this.

Unless there is certainty in the sales/year correlation, it is advi-
sable to provide the data on the zero correlation case in addition to the
Decision Matrix shown as Table 5.3 since this will then give both extreme
situations. Reality is likely to be between the two, usually somewhat
nearer the complete than the zero correlation case. The zero correlation
analysis should therefore be done for every first decision. We now have a
total of $A \times (R+1)$ searches to provide a good risk analysis at the planning
stage.

5.2d(3) The Decision Criterion

In both evaluation and planning stages we can now have probabilistic
evaluations, either as continuous distributions from a Monte-Carlo analysis
or as the Decision Matrix (Table 5.3). How do these help us make decisions?

Clearly, one could simply present this data to the decision-maker.
From this point on he can make a reasonable decision, balancing the non-
quantitative judgements with the distribution of the profitability to choose
an alternative that is profitable, not too risky and with potential for
future modification, that fits in with other non-quantitative company
policies.

One figure that might save him a few moments' work would be the
Expected Value, since this is not presented as such by either the Monte
Carlo distribution or the Decision Matrix.

This then can be the end of the analyst's work.

However, since Decision Theory has studied Decision Criteria in some
depth, it seems a pity to ignore it all and leave it to some subjective
judgement when it is possible to carry the quantitative analysis one stage

further. The decision-maker can make a subjective judgement balancing
Expected profit and maximum loss, or, with a little mental effort to get
the Regret Matrix, he can take into consideration the Maximum Regret criteria.
These criteria can be used directly by the decision-maker from the Decision
Matrix or distribution but the more sophisticated criteria , the Certainty
Equivalent, is not so easily applied.

Chapter III discusses this Certainty Equivalent criterion at some
length and argues that it possess sufficient advantages for Companies to
incorporate it as a standard technique. This then can be the last stage
of both the planning and evaluation stages; the presentation of the results
of the analysis as Certainty Equivalents.

Chapter III, Section 3.7a describes how the Certainty Equivalent of a
probabilistic set of results is calculated. Each data item -- either each
result from the Monte Carlo run, or each entry in the Decision Matrix --is
converted to utiles, and these utiles, weighted by their corresponding probabilities, are summed and re-converted to give the Certainty Equivalent.

The difficulty with applying the Certainty Equivalent criterion, as
explained in Section 3.7a, is generating a suitable Utility Function. If a
simpler method other than interviewing the company's main board could be
devised, it could have a much -wider and freer application.

The Utility Function is amazingly fluid. The zero point is arbitrary
and the scale is also arbitrary. It is only a scaling factor to increase
the weight of losses and reduce the importance of gains. It has been shown
to be adequately represented by an equation of the form (9):

$$U(x) \quad = \quad A + B\ln(x+C)$$
$$\text{but when } x \leq -C, \quad U(x) = -\infty \tag{5.21}$$

which   has   the important property of $U(x)$ becoming infinite at a point
representing the minimum profit tolerated by the company (or the negative
of the maximum tolerable loss). $U(x)$ becomes negative infinite when:

$$\log(x+C) \quad = \quad -\infty \tag{5.22}$$

and this holds when

$$(x + C) = 0 \tag{5.23}$$

therefore
$$x = -C \tag{5.24}$$

Hence the minimum tolerable profit is given by the negative value of
the constant C in the Utility Function.

To define the equation we must obtain values for the constants A and B. These constants effectively set the zero point and the scale for the function, but since these are arbitrary, their actual values chosen are of little importance. This being so, we can devise a very simple method of calculating the Certainty Equivalent as follows:

The Certainty Equivalent is calculated from the Expected Utility, $\bar{U}(x)$, where $\bar{U}(x)$ is obtained from:

$$\bar{U}(x) = \sum_{i=1}^{R} P_i [A + B \ln(x_i + C)] \qquad (5.25)$$

$$\bar{U}(x) = A \sum_{i=1}^{R} P_i + B \sum_{i=1}^{R} P_i \ln(x_i + C) \qquad (2.26)$$

Now by definition of probability levels, $\sum_{i=1}^{R} P_i = 1$, since R is the total number

therefore $\quad \bar{U}(x) = A + B \sum_{i=1}^{R} P_i \ln(x_i + C) \qquad (5.27)$

Now the Certainty Equivalent x_{CE}, is given by the equation:

$$U(x_{CE}) = \bar{U}(x) = A + B \ln(x_{CE} + C) \qquad (5.28)$$

$$\ln(x_{CE} + C) = \frac{\bar{U}(x) - A}{B} \qquad (5.29)$$

substituting (5.27) and (5.29)

$$\ln(x_{CE} + C) = \sum_{i=1}^{R} P_i \ln(x_i + C) \qquad (5.30)$$

or $\quad x_{CE} = (x_1 + C)^{P1} (x_2 + C)^{P2} \ldots \ldots (x_R + C)^{PR} - C \qquad (5.31)$

This equation can be used to obtain the Certainty Equivalent without involving the constants A and B. We need only the one constant C. C can be estimated either from the minimum tolerable return, or if a more accurate Utility Function has been obtained by full scale interview, as the value of C after fitting the Utility Curve with the equation

$$U(x) = A + B \ln(x+C) \tag{5.32}$$

Note that for equation (5.31), the following equation must hold exactly:

$$\sum_{i=1}^{R} P_i = 1.0000000 \tag{5.33}$$

Departure from unity produces considerable errors in x_{CE} because of the form of equations (5.31). For example, an error of 0.01 can result in a 25% error for x_{CE}. The minimum acceptable accuracy for $\sum P_i$ depends on the numerical values of $(x+C)$.

Table 5.4 compares Certainty Equivalents of two, extreme, two-level outcomes calculated by this method, compared with taking utility values directly from an interview-determined graphical Utility Function. The results are, within the scatter of the interviewee's answers, the same. To make this comparison, a range of maximum tolerable losses was determined by extrapolation of the interviewee's points. These results suggest there is little to be gained from developing complicated Utility Functions.

Table 5.4

Comparison of Certainty Equivalents

Distribution	Certainty Equivalent		
	proposed method	interviewee's graph	
Case 1 - $10 mm) 50/50 + $20 mm (chance	$+1.5 to + 2.0 mm (C = 25 to 36×10⁶)	$-4 to + 3 mm	Spetzler (9) Fig. 11
Case 2 - $200 mm) 50/50 + $600 mm(chance	$-42 to + 50 mm (C = 250 to 400×10⁶)	$-200 to - 10 mm	Spetzler (9) Fig. 13

5.2e The Value of More Information

Decision theory also concerns itself with determining the value of information. This is a particularly interesting subject in investment analysis because future forecasts are uncertain, yet by applying more forecasting methods and embarking on expensive surveys, better estimates can be

obtained.   The problem is always to know whether such extra expense is
justified.   If an investment analysis can also estimate the savings in having
better information, this justification could be provided.

The Decision Matrix, can be used to obtain such information in the
following way.

Consider the project given by Table 5.3; assume that the state of
nature is not known and the four outcomes in the Decision Matrix have equal
probabilities.   The Expected Value of each action is our best estimate of
its outcome.

Table 5.5

Expected Outcomes With and Without Information

	Equal Probabilities, no better information	Complete Information: either p_1 or p_2 or p_3 or p_4			
Small Plant A_1	$(100+100+50-50)/4=+$ 50			50	
Medium Plant A_2	$(200+100+ 0-200)/4=+25$				
Large Plant A_3	$(300+200-100-400)/4=0$	300	200		
Extn. to A_4 existing plant	$(50+50+50+40)/4= 47.5$				40
Expected Value	50 without more information	147.5 with complete information			
The value of complete information is 147.5 - 50 = 97.5					

Table 5.5 gives the Expected Values for the four types of plant.   To
maximise income, the plant with the highest Expected value should be chosen
-- Small plant A_1 -- with an Expected value of +50.

Now, if we had complete information : that is, if we carried out a
survey which could tell us which of the 4 probabilities would be certain,
we would make our decision to maximise that probability level's income.
This could be 300, 200, 50 or 40, depending on the result of the survey.
But as we have not yet carried out the survey we do not know which of these
it will be.   Our only information at present is that they are equally likely.
The Expected value - if we considered the survey - is therefore

$$(300 + 200 + 50 + 40)/4 = 147.5$$

The value of the complete information is therefore   147.5 - 50   which
equals 97.5. It would be worth paying up to this amount if the

technique was used a sufficient number of times within the company to accept
Expected values as a criterion, and if the results of the survey were guaranteed to be certain.

In practice, neither of these conditions completely hold but at least
we now have a rough measure by which proposals involving expensive forecasting can be judged.

This rough measure can be improved by modifying the analysis to allow
for using forecasting methods that are not completely certain. Section
3.5 introduced a technique Preposterior Analysis, whereby surveys which do
not produce certain information can be used to guide decision-making.
Given a past history of equivalent surveys, a probability can be given to
the likelihood of the survey being correct. This can be incorporated in a
conditional probability analysis (or a Decision Tree) to determine the value
of this partial information. Section 3.6 adequately covers this method.

Objection to Expected value for the decision criterion is often valid
and it is probably wiser to employ Certainty Equivalents rather than Expected values. There is no difficulty in doing this.

Hence, by presenting a Decision Matrix, adequate information is provided to decide how much effort should be put into obtaining better forecasts.

5.2f The Treatment of Inflation

Any economic analysis should state clearly the treatment it is employing for inflation and then all the input data can rigorously keep to this
basis. Of the two bases possible, constant value money and current money,
the constant value money treatment is the more suitable for probabilistic
analyses because it avoids the correlation problem, as described in section
4.5b.

A constant value money basis means that all data must be converted to
money at year 0 of the project. Hence, capital costs estimated for spending in year -1 or -2 would have to be inflated for inclusion in the analysis.
Likewise the current money tax allowance on depreciation would have to be
deflated back to year 0. This is described by formula in Section 1.7.
Other money flows, where changes are soley the result of inflation, would
appear as constant data. Any changes in cost and price due to other reasons
such as fall in profit margins, or disproportionate rise in price due to
world shortage then show up clearly as changes in the data.

Inflation has an effect on the yearly cash flow of a project. In an inflationary situation the cash flow will increase simply because the value of money has fallen. Even when a constant value money basis is taken for the evaluation, the yearly cash flows should display the current money situation because this particular criterion concerns the liquidity of the firm, which is measured in current money.

5.3 COMBINING THE ELEMENTS TO FORM A TECHNIQUE

Section 5.2 proposes a number of methods of solving individual problems concerned with investment analysis. By themselves they are not useful; they must be built up into a computer program before they pass from the category of being interesting to being useful! This Section describes how the methods of Section 5.2 have been built together to produce a useable investment analysis technique.

5.3a A Technique for Planning

Section 5.2a showed how one could generalise data for the alternatives, data for the sales forecast and the representation of all decisions in a fairly simple way. These recommendations were followed -- the 7-data items to represent alternatives, yearly data to represent the forecasts and the 'scrap', 'build' and 'do nothing' to represent all decisions.

Section 5.2c described how a single best plan to match a deterministic set of sales data can be found by a single dynamic programming search. Section 5.2d(2) showed that by reading in sets of yearly sales data to represent probabilistic sales information, and carrying out

(No. of alternatives as first decisions) × (No. of probability levels) dynamic programming searches, we can obtain the strategies associated with each first decision. These results can be presented in a Decision Matrix, which is useful if there is complete correlation of sales value and year. A further dynamic programming search on each alternative capable of being a first decision gives the corresponding plan for the zero yearly correlation case.

Section 5.2d(3) recommends that the Decision Matrix should then be evaluated, using both Expected value and Certainty Equivalent criteria to assist in the analysis.

The combination of these elements produces a program that provides the planner with an optimum plan, giving both the alternatives and their

timing when the sales data is deterministic. When the sales data is probablistic, the same program produces the Decision Matrix that enables him to
make a choice of first decisions after having examined all the back-up plans
associated with all the possible first decisions. Searches will not only
have located the best back-up plans for every market outcome, they will
also have evaluated them. Any criterion or mixture of criteria can be used
to make the final decision -- Expected value, Certainty Equivalent, MINIMAX,
Maximum Regret, or just plain judgement.

Should the yearly correlation be very small, strategies are inappropriate and the mean plan produced in addition to the Decision Matrix is more
applicable. Partial correlation requires a first decision that has a robust
mean plan.

The timing of the first decision can be given to the nearest 3 months
as described in Section 5.2a(4), the timing of the later anticipated decisions to within 1 year.

The program to carry this out, PLADE, is described in Appendix 1.
Chapter 6 and 7 give examples of its use.

For the program to function properly in a convenient manner, a series
of smaller "elements" were built into the method. Since they form an
integral part of the technique they are listed below:

			See:-
1)	NPW is taken as the evaluation criterion	-	Section 1.4b
2)	When manufacturing capacity is greater than sales volume, the plant with the highest variable cost is assumed to be shut down first	-	Section 5.2b
3)	Extended Life is incorporated to prevent the plans with the late decisions being penalised.	-	Section 5.2c
4)	Processes not immediately available can be introduced into the later years of the search	-	Section 5.2c
5)	A first decision can be made at a fixed time to simulate an irreversible go-ahead situation	-	Section 5.2c
6)	The Certainty Equivalent is determined if the maximum tolerable loss can be stated	-	Section 5.2d(3)
7)	A base-case, non-investment evaluation is done to enable the NPW due to the capital investment to be presented	-	Section 5.2b

8)   When demand exceeds capacity a penalty
     proportional to the short-fall can be
     included in the costs.   - Section 5.2b

5.3b A Technique for Deterministic Economic Evaluation

Section 5.2 explained that no single evaluation criterion is universally
acceptable because the criteria are only guide-lines as to the project's
attractiveness. Therefore, a general evaluation technique should produce a
number of different criteria to enable a subjective assessment to be made.
Section 5.2 goes further by recommending a set of six that should conveniently and adequately cover the field. These six have been programmed as a
set of subroutines for the deterministic economic evaluation. This program
requires, as data,the five data items to define the alternatives, but in
addition, it requires the starting and scrapping years for each alternative.

The program includes an automatic sensitivity routine to enable single
defined + ve or - ve changes in the cost or sales data to be investigated.

The evaluation is done on a constant value money basis, as described
in Section 5.2f. Hence, inflation is used only to modify the initial capital sum, depreciation, the pay-back and year cash flow outputs.

The program must make decisions when capacity is above sales demand --
it again turns plants' outputs down progressively, those with the highest
variable cost first.

A base-case analysis must be automatically incorporated to enable the
differences in profit due to the investment to be properly evaluated.

This program is very straightforward. It is similar to many of the
evaluation programs that are already in existence. It perhaps evaluates
according to more criteria, and it defines inflation, but it alone is no
noteworthy program. However, it is completely compatible with the planning
program. It uses the same data input, and is the essential next step after
planning. Having chosen a plan, it should be evaluated by a number of
criteria, and also a sensitivity analysis should be performed.

This evaluation program has one more feature. It enables probabilistic
sales data to be handled by repeating the calculation with each probability
level in turn. As described in Section 5.2d(1), this provides a set of NPW
results which represents the extreme distribution of using the given plan
in an uncertain sales situation where there is complete yearly correlation.
Assuming no counter-actions are made, and the plan is carried out to the
bitter end, this represents the "worst case" profitability distribution
for the project due to sales uncertainty. This can be useful as an

outside estimate of the magnitude of the uncertainty caused by the sales
information.

5.3c A Technique for Economic Evaluation Under Uncertainty

Chapter II showed that most of the data involved in economic evaluations
is uncertain. The following items are all subject to uncertainty of some
degree:

Sales volume	Fixed cost
Sales price	Scrap value
Working capital	Inflation rate
Capacity	Penalty for inability to supply
Capital cost	demanded product.
Variable cost	

A subroutine has been written which carries out a Monte-Carlo analysis
using distributions for the above data, as described in Section 5.2d(1).
The two sales volume and price data sets are required as discrete, the
other data have to be presented as trapezoidal distributions. The economic
evaluation subroutine is used as a simulation model and the Monte-Carlo
subroutine itself is concerned with data sets and summarising the results.
Monte-Carlo simulations giving either NPW or DCF distributions can be
obtained, and a Monte-Carlo sensitivity analysis can also be performed.

The important feature of any Monte-Carlo analysis is the treatment of
correlation. There are only two correlations allowed for in this program:
the sales volume/year and the sales volume/ sales price, as described in
Section 5.2d(1).

The number of iterations allowed is between 50 and 500. The program
itself calculates the number of iterations required once it is given the
95% confidence limit required for the mean value. Chapter II, Section 2.5b
gives the technique that enables this to be done.

The resulting simulations are displayed as a cumulative distribution
and also the Expected value, standard deviation and Certainty Equivalent
are printed out.

5.3d The Combined Technique

Both planning and evaluation are important stages in decision-making.
They have so far always been treated quite separately. The most complete
planning technique produces a distribution of profitability. The most

complete evaluation procedure produces a different distribution of profitability. However, they both concern the same decisions, somehow they must
be brought together. The project in the end must be represented by only one
distribution.

The distribution from the planning study is the result of carrying out
a strategy -- a set of conditional plans. It excludes uncertainty in all
data other than sales forecasts. Furthermore, it only considers one special
case -- complete year/ sales correlation. It is therefore far from complete
as the representation of the profitability distribution for the projects,
but it does emphasize the planning aspects.

The distribution from the evaluation stage, the Monte-Carlo simulation
includes uncertainties in both sales and the Alternative input data, including any degree of correlation for year/sales price. However, it uses a
fixed plan. There is not provision to alter the plan to minimise losses
or maximise gains where there is some year/sales correlation. As has been
repeatedly emphasized, modification of plans is the most usual action that
a company makes to control its profitability; it has an immediate effect
on the profitability distribution. Monte-Carlo distributions therefore
cannot be considered as a fair representation of a profitability distribution -- only a pessimistic one.

A reasonable attempt at a profitability distribution must include all
that the Monte-Carlo distribution has, and in addition use a strategy, not
just a plan, for the evaluation.

The Monte-Carlo method is a simulation. It is doing no more than
simulating the life of the project many times. Each simulated life chooses
a set of data to describe the Alternatives and then each year a sales
volume and price is chosen. The correlation coefficients ensure that the
next year's sales and price have the right relationship to last year's.
We can therefore take this simulation one stage further and decide, during
the life of the run, what the next decision will be. Supplied with a
strategy instead of a plan, the simulation can determine, from past sales
information, what sales probability level it is on. Having determined this
it can make a second decision in accordance with the best plan for that
sales level.

This now produces the link required between the two levels of planning.
The strategy distribution is included in the distribution produced by the
Monte-Carlo simulation, and this distribution contains all the information

from the total planning study. This is a very close approximation to
reality and it is best looked upon as making a few hundred trial runs at
the project before actually embarking on it.

Planning Decisions During the Simulation

Having roughly described the method of culminating the investment
analysis, let us examine in more detail the problems associated with making
planning decisions during the simulation.

The object is to be able to simulate any degree of year/ sales
correlation and, armed with a strategy assuming complete yearly correlations,
and a plan assuming zero yearly correlations -- both obtained as results
from the planning program -- make reasonable decisions when the correlation
is in fact at neither extreme. In reality, one would note each year's sales
and take some form of running average. If there is much correlation, this
average will tend to indicate the future sales levels. If there is very
little correlation, the average will produce some mean sales level. In
practice, one would expect comparatively little difference between the
best mean sales level plan of the strategy and the mean plan, given by the
zero correlated DP search.

Therefore by taking some weighted average of past sales, weighted
to give emphasis to the most recent information, we have an indicator by
which we can choose the conditional plan of the strategy to follow during
the project.

Following a single life simulation of the project, we start by making
the first decision. There is no choice; all branches of the strategy begin
with this decision because the whole of the analysis is the evaluation of
this decision. As time progresses and our weighted average of past sales
settles towards a particular sales level, the year for the second decision
for that sales level arrives, and this decision is then simulated. As time
goes on the sales may stay on the same level in which all future decisions
for that conditional plan are enacted. However, if the correlation is not
very high and the sales stray on to a different level, then we have a
problem, -- we have no plan from going from the second decision on one level
on to other levels, and we will only obtain one by re-running the planning
program. Re-running such a DP search within a Monte-Carlo simulation is
out of the question and so when this case does arise, there is little
choice but to carry on with the series of decisions associated with the
second decision level. This means that some simulations will be more
pessimistic than would have occurred in reality because in reality one

would have re-planned. However, this deviation from a realistic simulation
is a small price to pay for having a method of combining the two levels of
analysis.

5.4 THE PLADE PLANNING SYSTEM

The total suite of programs, centering on OPLEX for the planning,
INVEST for the evaluation, and MONTE for the simulation have been written
to be mutually compatible and they call on common sub-routines wherever
possible. This has two advantages. Firstly, once the data is collected
and punched it can be used without alternation for each level of analysis.
Secondly, it enables a successful run from the planning stage to move
automatically on to the evaluation and simulation stages without the need
to re-submit the program.

The total program, called PLADE (PLanning And DEcision), is an attempt
at an automated planning system for general use in a planning department.
It was thought that, in the same way that Flowsheet programs are becoming
the corner-stone of process engineering departments of large companies
with a large proportion of their work being handled by one standardised, but
very flexible, computer system (6), so a standardised, flexible planning
program might benefit planning departments.

PLADE has been written as a university Research Project to test this
proposal; the program is non-proprietary, and the success of the research
project will be measured by the number of planning departments that feel
the idea has sufficient merit for them to adopt it.

The next 3 chapters describe various studies in investment decision-
making and show how PLADE helped in them.

More details of the PLADE system are given as Appendix 1, which is
the Program Manual.

5.5 BIBLIOGRAPHY FOR CHAPTER V

1.   Erlenkotter, D., Sequencing Expansion Projects, Op. Res., 21, 2,
      542, (1973).

2.   Gershefski, G., and Phipps, A.J., A Higher Order Computer Language
      for Risk Analysis.
      Decision and Risk Analysis -- Powerful New Tools for Management
      Proceeding of the 6th Symposium. The Engineering Economist,
      Hoboken, N.J. 1972.

3.   Hammersley, J.M. and Handscomb, D.C., Monte-Carlo Methods,
      Methuen, London 1964.

4. Manne, A.S., Investment for Capacity Expansion -- Size Location
    and Timing Phasing, Allen and Unwin, London,1967.

5. Petersen, E.R., A Dynamic Programming Model for the Expansion
    of Electrical Power Systems, Management Science, 20, 4(ii)
    656, (1973).

6. Rose, L.M., The Application of Mathematical Modelling to Process
    Development and Design, Applied Science, London 1974.

7. Rose, L.M., Walter, O.H.D., and Myhre, J., Planning Manufacturing
    Capacity, Chemtech., Aug, 4, 494, (1974)

8. Schlaifer, R., Analysis of Decisions Under Uncertainty,
    McGraw - Hill, New York, 1969.

9. Spetzler, C.S., The Development of a Corporate Risk Policy for
    Capital Investment Decisions, IEEE Trans. on Systems Science and
    Cybernetics, SSC-4, 3, 297, (1968).

10. Spetzler, C.S., and Zamora, R.M., Decision Analysis of a Facilities
    Investment and Expansion Problem.
    Decision and Risk Analysis-Powerful New Tools for Management,
    Proceedings of the 6th Symposium. The Engineering Economist,
    Hoboken, New Jersey, 1972.

CHAPTER   VI

THE SELECTION FROM ALTERNATIVES

Wyssi Chue, schwarzi Chue,
weli wämer usetue?
Glaube immer die.

6.1   INTRODUCTION

Chapter 4 enumerated the three stages of an investment decision:
  (a)   the optimisation of alternative proposals
  (b)   the selection of one proposal
  (c)   the detailed economic analysis of that proposal
This chapter, and the next, are concerned with stage (b)--the selection
of one proposal.

It is worth sub-dividing this selection (b) once more into:
  (a)   the selection of the basic principle--be it expansion, new
        plant, new process, replacement plant, buying-in, etc.
and   (b)   The determination of the optimum size for the chosen
        principle.

In theory, one should not determine the basic principle without optimising each proposal with respect to size. However, as Chapter 3 will show,
optima with respect to size are usually fairly flat and all alternatives
are usually similarly affected by equal changes in capacity. In practice,
therefore, it is acceptable to first make a selection of the basic principle after ensuring that all the proposals have equivalent capacities, and
later, separately determine the optimum capacity.

This chapter is concerned with the selection of the basic principle;
Chapter 7 deals with the fine tuning to get the optimum capacity.

An interesting starting point is to discuss how deep the analysis
should be for this selection. It is common for process selection to be
made from a single, simple economic evaluation. Chapter 5 has involved
Dynamic Programming, conditional plans, probabilities and Decision Matrices.
Is risk analysis justifiable or necessary to select the basic principle?

Following on from this we have the problem of selecting the Decision

Criteria. Is Expected Value an adequate criterion or should some method of
weighing losses--MINIMAX or UTILITY--be employed?

Besides these points concerned with the mere mechanics of the selection
a much deeper subject is the appreciation of the reasons for the selection
of the particular alternative. Planning is a very difficult activity which
consists basically of creating a number of detailed proposals, and then
evaluating these by some criterion to select the best. The art in planning
is to be able to create these alternatives in such a way that the selection
is based on the important differences between the projects and not the
result of random differences introduced by the planner in creating his
alternatives.

The literature contains nothing on this difficult subject and this text
is not going to basically alter the situation. However, though nothing can
be said on the creation of the alternatives, a proper analysis of the
results of a planning study will enable the reasons for the selection to be
established and lead to a deeper understanding of the subject which, in
turn, helps to create a better set of proposals for a second round in the
study.

Following on from the understanding of the results is the question of
the value of more information. This usually implies better forecast information from deeper expensive forecasting studies. Could better forecasts
lead to different decisions, and what are the possible savings? Answers to
these questions can also be drawn out of a planning study.

The aim of this chapter is to discuss all these points and to do so by
means of examples. The first part is concerned with the selection of these
examples--this is particularly important because it is on these examples ·
that the rest of the chapter is based.

6.2 SELECTION OF EXAMPLES

The ideal situation would be to take past real projects for these examples because they would test the validity of all the assumptions concerning market behaviour in addition to the purely mechanical aspects of checking whether the Expected profitability of a strategy is significantly higher than that of an intuitively developed plan. The second advantage of
taking past real projects is that the total behaviour of the intuitive

approach, including later corrective decisions, would be included and this
must be done before we can conclude that there is any benefit from the more
complicated analysis.

The use of real projects implies monitoring projects over their lifetime. Since the problem is stocastic, many hundreds of projects must be
monitored before any conclusions can be drawn. This is obviously impossible.

In such situations it is usual to employ the technique of computer
simulation to replace the stocastic field study. A computer simulation can
be made hundreds of times, and it is possible to include in the simulation
different forms of market development and plan-and-strategy decisions made
throughout the project life. We already have the necessary computer programme for this simulation--MONTE--introduced at the end of Chapter 5.
MONTE, with probabilistic sales data and various values for the year/sales
correlation, can be used to simulate the history of the project with different plans and strategies. When these plans and strategies are taken from
different levels of analysis, we have a method of measuring the value of
the various levels of Risk analysis.

We must now decide which projects should be used in this comparative
study. Again, ideally, one should use real data from past projects. However, the stocastic nature of the problem means that hundreds of old projects must be examined to get a proper cross-section of the various types
of project. Furthermore, it is virtually impossible to obtain any real
planning data. Past data has never been properly recorded and all data,
particularly on current projects, is too confidential to be broadcast.
Again we must resort to synthetic data, and risk the criticism that the
data has been chosen to produce the conclusion we want.

Firstly, the number and type of projects for the comparison must be
decided. They should cover the range of characteristics normally met in
project investment analysis. There is little value in investigating a
number of similar type projects, omitting other important types.

6.2a Choice of Projects

The study was limited to types of projects met in the chemical industry, since this whole work started to fulfil a deficiency met by chemical
engineers. The technique could equally well be applied in any capital-
intensive industry, be it food, heavy engineering, auto or the textile

industry.
     An analysis of types of chemical product identified 4 main types:
         (a)   the stable chemical intermediate--acids, inorganics,
               fertilizers, organic intermediates, e.g. monomers and
               solvents,
         (b)   the newly introduced product--any new product from a research
               department, either replacing an existing product or serving
               a completely new market,
         (c)   the short-lived product with an initially high demand--
               specific agricultural chemicals, e.g. pesticides, fungicides
               and pharmaceutical type products,
         (d)   large-tonnage chemicals, slowly losing their market--this
               could be sodium carbonate, calcium carbide or some of the
               earlier developed plastics or fibres.
     There are also only a limited number of types of capacity expansion
that can occur. These can be listed as:
         (a)   a new plant could be built using the existing process
         (b)   a new plant could be built using a new process
         (c)   existing plant could be replaced by a more efficient plant
         (d)   existing plants could be expanded
         (e)   existing plants could be closed down.
     From these alternative projects and types of capacity expansion, four
separate planning studies were developed. These are summarized in the
following pages.

Case 1   Stable Product

     The product is a basic chemical with a growth rate tied to the
country's GNP. Future sales may rise at a rate of between 2 and 6% per
annum. This is conveniently represented as four separate probability
levels, each with an equal probability. The probabilistic forecast is
given as Figure 6.1. Complete year/sales correlation cannot be expected
because of the characteristic of GNP fluctuations. However, the correlation will be high since there will be virtually no chance of changing from
the highest to the lowest probability level in later years. A reasonable
chance of switching levels is one level change every 3 years. This is

adequately represented by a correlation coefficient of 0.75. It is company policy always to stay in this market, and always to meet demand.

There exists one operating plant, with a capacity of 10,000 tons per year. Possibilities for increasing product are to build:

   a) a small extension of the existing plant
   b) a large extension of the existing plant
   c) an additional plant using the same process
   d) provision of new capacity and possible replacement
       of the existing plant by the installation of a plant
       using an improved process.

To these proposals we must supply plant capacities, since our integer search method chooses alternatives; it does not propose capacity levels. The capacity levels must be chosen intelligently, otherwise a good process can become a poor alternative because of ill-matching capacities. When in doubt extra alternatives must be included with additional capacities.

Table 6.1 shows the chosen Alternatives and their capacities. The capacities of the extension are assumed to have been decided on technical grounds. The capacities of the additional plants have been chosen to produce plants with total manufacturing capacities in the 15-18,000 tons range -- which looks reasonable from inspection of Figure 6.1.

Table 6.1.

Alternatives for Case 1 - Stable Product

Alternative	Description	Capacity tons/year
1	Existing Plant	10,000
2	Small Extension	1,000
3	Large Extension	2,000
4	Additional Plant, existing process	2,500
5	Additional Plant, new process	5,000
6	Replacement Plant, new process	16,000

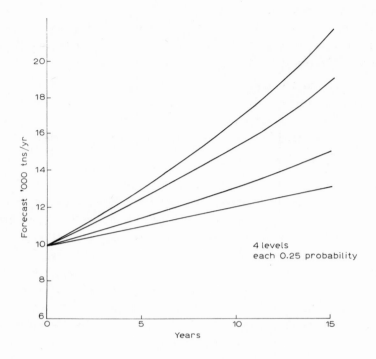

<figure>

Forecast '000 tns/yr

4 levels
each 0.25 probability

Years

</figure>

Figure 6.1.

Case 1 - Stable Product - Sales Volume Forecasts

Intelligent selection is essential unless all possible variations are to be
checked out by permutations of computer runs. In particular, in this
example, before choosing the capacities the following points were considered:

      a)   If the additional plant using the same process is
           attractive, the extension will be even more so.
           Hence the sequence will be : 2 then 4, or 3 then 4.

      b)   If the new process is attractive, it will probably be
           preferred to spending any capital on the old process.

      c)   If the new process is very attractive, it may justify
           complete replacement of all manufacturing facilities.

      d)   A two-stage expansion would probably be sensible in 15 years.
           Hence, 15-18,000 tons/year would be a reasonable, achievable,
           design level.

Case 2 - New Product with Uncertain Potential

The product in this case was taken as a new product from the research
department. The future sales will depend on how well the product is accepted by the market, and whether any disadvantages become apparent after it
has been introduced. There may be market areas where it is certain to take
hold, and others where there are differing levels of probability that it will
be accepted. Such a situation would produce a widely diverging probabilistic sales volume forecast, as given by Figure 6.2. This figure, with a
three-fold difference in market between the worst and best forecasts, was
taken for this example.

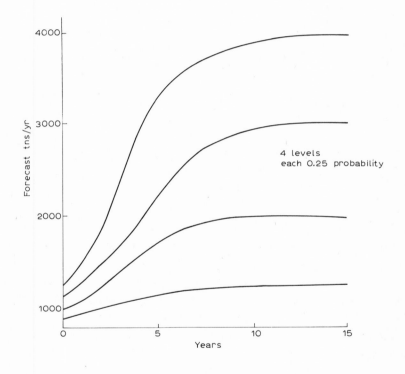

Figure 6.2
Case 2 - New Product - Sales Volume Forecasts

The year/sales correlation is 1.0 since, if the market develops, it
will always result in good sales. The volume/price correlation is also
1.0 because the prices are completely dependent on the sales volume level.

Since this will be the first plant for this product in the company, the Alternatives are very simple. The Research Department has developed one process; the only problem is to determine its size.

The forecast shows clear market saturation levels which give a good basis for providing some capacities with which to begin the study. Also, a frequent approach in this situation is to have a two-stage expansion. Hence, some Alternatives should be presented that together form a set of reasonable two-level plans.

From these arguments a list of Alternatives, given as Table 6.2, was produced for selection and evaluation. Company policy is to install when economic to do so and not to install just to meet demand.

Table 6.2
Alternatives for Case 2 - New Product

Alternative	Description	New Product tons/year
1	Large Plant	3,750
2	Medium Plant	2,750
3	Small Plant	2,000
4	Very Small Plant	1,200
5	Second Small Plant	2,000
6	Second Very Small Plant	1,200

Case 3 - Limited Life Product

This case represents an existing product whose plant is working at capacity and so an expansion is required. The product is not expected to have a long life. It is typical of pharmaceutical and pesticide products, where experiences show that they are generally replaced in ten to twenty years by improved products. However, the present market is good and climbing, but the sales level before the decline is far from certain. Fig. 6.3 gives the sales forecast for this example. The price is forecast as remaining constant for the project life for all the sales levels. The sales/year correlation is taken as 1.0 -- with little chance of interchange between the sales levels, the volume/price correlation is irrelevant as the price is constant.

For Alternatives we have either small extensions of the existing plant, or the installation of additional plants, all using the same process, with economies of scale producing the only cost differences.

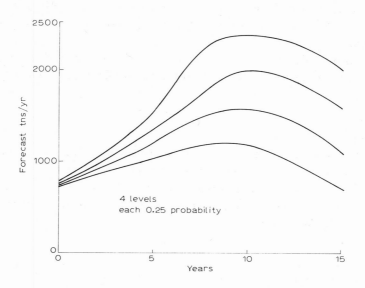

Figure 6.3

Case 3 - Limited Life Product - Sales Volume Forecasts

Table 6.3

Alternatives for Case 3 - Limited Life Product

Alternatives	Description	Capacity tons/year
1	Existing Plant	1,000
2	Extension of existing plant	100
3	Small new Plant	500
4	Second small new Plant	500
5	Medium new Plant	900
6	Large new Plant	1,300

Again the selection of possible capacities generates the range of
Alternatives.   The existing plant has a capacity of 1000 tons/year and can
be extended only to 1,100 tons/year.   There is no question of company
policy being always to meet the market as a decline is expected.   Reasonable

plant capacities should therefore nearly meet the maximum sales levels for
each likelihood, and two-stage expansions to achieve these levels should be
possible. Out of these considerations the Alternatives given in Table 6.3
were developed.

It is assumed that the question has been raised by a planning group
interested in both the best Alternative to choose and the best time to
install it.

Case 4 - Declining Product

This case consists of an established product whose market is expected
to decline, except under very favourable conditions where it may be
expected just to hold its own and remain steady. The decline is expected to
be very slow and so it still represents a reasonable future opportunity.
The forecast is given as Figure 6.4.

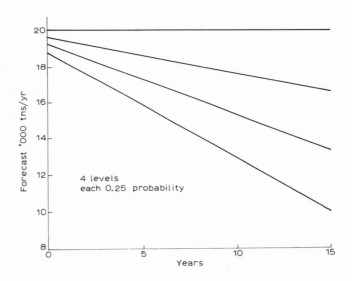

Figure 6.4
Case 4 - Declining Product - Sales Volume Forecasts

The company has already two plants that can adequately satisfy future
demand, but the opportunity has arisen to introduce a new process with
reduced costs. Since there is still a 10,000 to 20,000 tons/year sales

expected for the next 15 years, an improved process would still be attractive, even with the declining market forecast.

Prices are expected to remain constant, and management feels no reason
to operate a plant uneconomically to satisfy demand. There can be some
change between forecast levels depending on fluctuations in business conditions, but full interchange in later years is considered impossible. Hence
the sales/year correlation of 0.75 is considered reasonable to represent
this relationship. The volume/price correlation is again irrelevant as
prices are constant.

The two planning problems are:

      a) when is it economic to close down one of the
          existing plants and which should it be?

and   b) compared with a), is it economic to replace the
          existing plants with a plant using a new process.
          If so, how big should this new plant be?

The Alternatives therefore consist of the two existing plants, and a
set of new plants within the new process, but of different capacities.
The capacities for the new plant must obviously lie between 10,000 and
20,000 tons/year. The optional size is likely to be somewhere between
these two extremes. From these considerations the Alternatives given in
Table 6.4 were evolved.

It is assumed that the study has been raised by the engineering department who must know the effect of completing their design work already begun.
Hence, the study should investigate the economics of operating the new
plant in year 1.

Table 6.4
Alternatives for Case 4 - Declining Product

Alternatives		Capacity tons/year
1	Small existing plant	8,000
2	Large existing plant	12,000
3	Large plant, new process	20,000
4	Medium large plant new process	18,000
5	Medium small plant, new process	12,000
6	Small plant, new process	10,000

6.2b Generation of the Data

These four synthetic projects must now be supplied with data. This
data must be "reasonable"; -- that is, it should be typical of the type of
project, it must result in a reasonably economic proposal and it must present a difficult decision situation. The first point concerns the relative
levels of capital, variable and fixed costs. The second point determines
the price level of the forecast. These points require some imagination but
they present no real difficulty. However, they do not concern factors that
alter the relationship to each other of Alternatives within the same project.
This is covered by the third point which is by far the most difficult to
synthesize; the Alternatives have to present a difficult decision problem.

Consider a project with four Alternatives and four forecast sales
volume levels. There would be a difficult decision problem is each Alternative was the best for a corresponding probability level, say Alternative
1 for level 1, 2 for level 2 and so on. Conversely, Alternative 1 must be
worse than Alternative 2 at level 2, worse than 3 at level 3 and 4 at level
4. Similarly, there are three other sets of conditions for the other 3
alternatives.

The total requirement can be conveniently summarised on Figure 6.5.
If we plot NPW by each Alternative versus sales level, we arrive at Figure
6.5 for the difficult decision problem. This figure satisfies the above
conditions, our synthetic data should therefore produce a similar pattern
if we are to draw any conclusion about risk analysis. There is no point
in working with risk-free data to produce generalisation about risk.

The data was generated in the following way:

a) Reasonable costs were roughly estimated, to be in
    accord with the type of project, and the selling price
    was arranged to give a reasonable return, -- the Rate
    of Interest method (Section 1.4a) was useful in this
    context. Capital and fixed costs were assumed to have an
    Economy of Scale factor of 0.6 (See Chapter VII).

b) Plans centering on each alternative in turn were drawn up.
    Timing was done by the rules of thumb given at the end
    of Chapter VI.

c) Each plan was evaluated at each probability level. This
    was conveniently done by the evaluation mode of the PLADE
    program, since this gives directly the required sets of
    NPW's shown on Figure 6.5.

d) When this data did not appear as in Figure 6.5, that is,
    when some Alternatives were optimum at more than one
    level, then the synthetic data was adjusted by trial

and error until the results appeared as in Figure 6.5.
This adjustment was sometimes difficult since not only
the NPW's, but also the slope of the NPW/ probability
level had to be simultaneously adjusted.

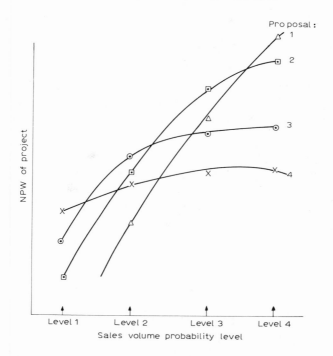

Figure 6.5
The Presentation of a Difficult Decision

The result of this process was data for each case that, based on a
rough analysis, made the alternatives seem equally attractive. This process
was not designed to "manufacture a problem" but to simulate one to two years
of intelligent development in a research or engineering department. The
poor alternatives have already been discarded during the development period
and the planners see only difficult situations.

Case 1 - The Stable Product

The characteristic of this type of product is the low, but relatively
safe profit margin. The Alternatives consist of two exclusive extensions
to the existing process, which have low capital and fixed costs, a new small
plant using the existing process, and two plants using the new process, --

one as an addition to the existing plant, and the second as a replacement
for the old process. The process is characterised by a lower variable
cost than the old process. Its fixed and capital costs are similar to
those of the old process.

It is interesting to note that it took 16 attempts to find data whose
results correspond to Figure 6.5. To achieve this, the difference had to
be so reduced that they were of the order of 1%. That is, when 1) of
Figure 6.5 was better than 2) it was better by only about 1% in the NPW.
To increase this difference always made one Alternative better at all probability levels. There was obviously very little risk associated with this
case; it is not possible to make a really bad decision. Table 6.5 summarises
the synthetic data for this case.

It is important to note that this study is to select between three
alternative principles for future product -- plant extensions, new plant
with the existing process, or a new process. If the study was to select
the best size of plant for a given process, the problem does contain some
risk, and a poor decision could be made. This situation is discussed in
Chapter VII.

Table 6.5
Case 1 - Stable Product - Data Summary

Forecast: as Figure 6.1
Price   : £120 per ton
Policy   : Total market will be satisfied.

	Capacity	Capital	Fixed Cost	Variable Cost
1. Existing Plant	10,000	Existing	200,000	40
2. Small expansion or	1,000	60,000	10,000	40
3. Large expansion	2,000	250,000	15,000	40
4. Small plant, same process	2,500	500,000	60,000	40
5. Medium plant, new process	5,000	850,000	100,000	25
6. Large plant, new process	16,000	2,780,000	160,000	25

Case 2 - New Product

The major characteristic of this class of product is the high potential profit margin to be expected for a new product with an uncertain future. There is only one process assumed to be available and the planning problem is the choice of plant size.

This is clearly a risky situation and it was correspondingly easy to synthesize. Data corresponding to Figure 6.5 was obtained with only four trial-and-error runs. Table 6.6. summarises the synthetic data for this case.

Table 6.6

Case 2 - New Product - Data Summary

Forecast : As Figure 6.2

Price   : £400 per ton

Policy   : Demand satisfied only when economic to do so. Initial
            Investment will be made in year 1

Description	Capacity	Capital	Fixed Cost	Variable Cost
1. Large Plant	3,750	2,000,000	350,000	35
2. Medium Plant	2,750	1,650,000	325,000	35
3. Small Plant	2,000	1,430,000	275,000	35
4. Pilot Plant	1,200	1,100,000	200,000	35
5. 2nd Small Plant	2,000	1,430,000	275,000	35
6. 2nd Pilot Plant	1,200	1,100,000	200,000	35

Case 3 - Limited Life Product

This case represents an existing product, which is making a reasonable profit, whose market is fairly sure to decline some time in the future. The demand in the immediate future is predicted to outstrip capacity and the problem is what capacity increase is reasonable. Only one possible process exists, and the alternatives are created by taking various capacities and using the 0.6 power rule. There is also the possibility of making a small extension to the existing plant.

This case took only five trial-and-error attempts to obtain suitable data. The profitability was fairly sensitive to the decision, but the risk did not seem as high as case 2. Table 6.7 summarises the data for the alternatives.

Table 6.7

Case 3 - Limited Product - Data Summary

Forecast: As Figure 6.3

Price   : £150 per ton

Policy   : Do not meet demand if uneconomic to do so.

Description	Capacity	Capital	Fixed Costs	Variable Costs
1. Existing Plant	1,000	Existing	20,000	35
2. Extension	100	20,000	2,500	35
3. Small Plant	500	110,000	14,000	35
4. Second small Plant	500	110,000	14,000	35
5. Medium Plant	900	150,000	19,000	35
6. Large Plant	1,300	210,000	23,000	35

Case 4 - Declining Product

In this case existing plants were assumed to be earning an average profit but the market is expected to slowly decline. However, the research department has developed a new process which gives a 20% saving in variable costs. To synthesize the data the capital cost had to be such that under average sale conditions it was recovered by the total savings in variable costs. It took nine trial-and-error runs to obtain suitable data. It was easy to devise processes that either economically replaced all the old plants or could not, but to devise a condition where it was economic at the top probability level, but not at the bottom, was very difficult, but eventually possible. Like Case 1, the difference became so small that it appeared not to matter which decision was taken. A difficult risk situation could not be created. The final synthetic data is summarised in Table 6.8.

Table 6.8

Case 4 - Declining Product - Data Summary

Forecast: As Figure 6.4

Price   :   £300 per ton

Policy   :   Assume process replacement will commence in year 1.

Description	Capacity	Capital	Fixed Costs	Variable Costs
1. Small existing plant	8,000	Existing	220,000	200
2. Large existing plant	12,000	Existing	700,000	200
3. Large new process	20,000	6,200,000	450,000	162
4. Medium/Large new process	18,000	5,700,000	420,000	162
5. Medium/Small new process	12,000	4,700,000	340,000	162
6. Small new process	10,000	4,200,000	300,000	162

6.3   THE COMPARISON OF DIFFERENT LEVELS OF RISK ANALYSIS

6.3a   The Alternative Risk Analysis Levels

Tables 6.5 to 6.8 give the data that would be presented to the planner. Using this data he must:

> a)   choose one of the Alternatives presented to him as his recommended first decision
>
> b)   present a predicted profitability of the project with this first decision.

These two points can be considered separately -- in fact, they always remain separate,

> -- if the project is predicted to be profitable
>
> then we will make the first decision

will be the approach taken by the final decision-maker.

Let us list various common levels of analysis used with probabilistic sales data in ascending order of complexity. We assume that all the levels are sophisticated enough to consider future decisions beyond the first, -- we are taking for granted that plans are preferred to single decision evaluations, and we are investigating the depth of the analysis using plans and strategies.

Level 1 : A simple aggressive policy could be adopted, and a plan chosen for the optimistic sales forecast.

Level 2 : A realistic policy could be assumed, and a plan chosen for the mean sales forecast

Level 3 : A simple probabilistic approach could be used, and a plan chosen that produces the highest Expected NPW,
- each year the NPW for each level is separately evaluated and then used to obtain the Expected value.

This can be expressed mathematically as :

$$NPW = \sum_{L=0}^{N} \sum_{j=1}^{J} P_j \ (NPW)_{j,L} \qquad (6.1)$$

where J = No. of sales probability levels

N = Life of Project

p_j = Probability of each level

$(NPW)_{j,L}$ = NPW of the chosen fixed plan for probability level j for year L.

For the rare occurrence when demand/year correlation is zero, the resulting NPW represents the true Expected profitability of the project.

Level 4 : This level makes use of the more usual strong correlation between sales and year and chooses the best first decision as a result of investigating strategies. These strategies have made use of the market development information to guide future decisions which further increase profit or minimise losses in the light of new information.

Mathematically it can be expressed as:

$$NPW = \sum_{j=1}^{J} P_j \ (NPW)_j \qquad (6.2)$$

where, in this case,

$(NPW)_j$ is defined as the NPW over the whole life of the project of the best plan associated with the sales probability level, j.

Here a total of J plans are required for this level, and the set of values $(NPW)_j$ represents the profitability distribution for the project when the demand/year correlation is 1.0. Since this coefficient is usually nearer to 1 than zero, we would expect Level 4 to give a better appreciation of the risk than Level 3.

Levels 1 and 2 are common, rather superficial approaches to the problem. Some companies have a mixture of these levels as their standard procedure, e.g. Level 1 is used to size and cost the equipment, and Level 2 is used to evaluate the income and profitability. This has the advantage that the company is in an agressive position and can quickly take advantage of a favourable market development, but it is not fooling itself as to its Expected profitability. This approach however cannot be used in marginally profitable projects or established products where the spare plant capacity cannot be carried by income from the mean sales forecast.

Level 3 is equivalent to a Monte-Carlo simulation using sales demand as the uncertain variable. As such it is a common approach. Level 4 is only practicable where there is an automated method of developing sets of conditional plans available. Without the aid of a search routine it involves an impracticable amount of work. Since such search routines are not part of standard planning procedure, it is beyond any normally applied level.

6.3b The Comparison of the Various Levels

Since there are two objectives of a planning study, any comparison must cover the two distinct points:

a) What is the cost incurred above the cost of the "optimum first decision" for taking first decisions given by the different levels rather than taking the "optimum first decision"

b) What is the difference in the predicted profitability of the project and the "actual outcome" for each of the different levels?

We have therefore, to define the "Optimum first decision" and the profitability of the "actual outcome" before we can make a comparison.

We must assume that management is free to make, and therefore will make, corrective actions in the future. This means the "optimum first decision" will in fact be the first decision in a strategy. Even though management may not consciously think in terms of strategies, their ability to make corrective decisions will in fact mean that they are using a strategic approach. For simulating the future we can conclude that the "optimum first decision" that can be made is the first decision of the best strategy. In other words, Level 4 has identified the "optimum first decision"

Chapter V argued that a Monte-Carlo evaluation may include simulation of the decision-making for each simulated demand pattern as a realistic attempt at evaluating the "actual outcome" of a project. This method can, therefore, be used to evaluate the "actual outcome" for any first decision.

Combining these two -- the generation of strategies to simulate intelligent management, using the OPLEX search routine and the Monte-Carlo simulation as programmed by MONTE and FLOW; -- we have a technique for generating both the actual outcomes of the "Optimum first decision" and the "actual
outcome" of any other first decision. We are now in a position to compare
various levels of analysis.

The above argument is not strictly accurate in that a manager effect-
ivly re-considers plans on a continuous basis, and need not hold to a set
of plans evolved at one point in time. This can be accurately simulated
by including a DP search each year for each run of the Monte-Carlo simulation. This point has been made in Chapter V, Section 5.3d where it was
argued that the small difference to be expected did not justify the extensive computation involved. Again, in this analysis we do not intend to
make the simulation so realistic by including the DP search within the Monte
carlo analysis. It is assumed that the difference, -- which will show itself
as a slightly lower risk, and slightly improved Expected profitability --
will be small.

The comparison of levels is made by evaluating the two following
expressions:

$$(\text{Cost of using simpler level})_L = (NPW)_{S,4} - (NPW)_{S,L} \qquad (6.3)$$

and

$$(\text{Error in predicted profitability})_L = (NPW)_{P,L} - (NPW)_{S,L} \qquad (6.4)$$

where $(NPW)_{i,L}$ refers to the Expected NPW obtained for level L, determined
by calculation i,

> when i = S, the Monte-Carlo simulation has been used with the
> best strategy corresponding to the first decision given
> by level j.

> when i = P, the NPW has come from the planning calculation, and
> has been extracted directly from the Decision Matrix

6 .3c Presentation of Results

By running the 4 cases with PLADE in the planning phase, a complete
Decision-Matrix is obtained. Taking Case 1 as an example, the following
data can be extracted from the Decision Matrix given as Table 6.9.

1. Employing Level 1, the best plan would be to install alternative
   SM.EXTN and the apparent NPW -- $(NPW)_{P,1}$ -- would be 13450, since
   probability 1 is the optimistic profile and SM.EXTN has the highest
   NPW in this column.

STABLE CASE 1

DECISION MATRIX

INITIAL DECISION	NO YEARLY CORRELATION NPW	----------WITH TOTAL CORRELATION----------					
		MEAN	CERT EQUIV	PROBABILITY 1	2	3	4
2 SM.EXTN	7110	7670	7371.	13450	10590	4970	1680
3 LGE.EXTN	7090	7690	7475.	12880	10060	4580	3240
4 2500TNPLT	2920	4660	4365.	9950	7630	1950	-910
5 5000TNPLT	4630	5520	5124.	11580	8940	3070	-1490
6 NEWPROC	4330	4930	4322.	12150	9280	1850	-3570

BASED ON THE ABOVE FIRST DECISIONS,
THE EXPECTED VALUE OF COMPLETE FORECAST INFORMATION IS   370.

Table 6.9

Decision Matrix for Case 1

PROBABILITY LEVEL   1
FIRST DECISION RESTRICTED TO   2   SM.EXTN

YEAR	DECISION BLD SCRP		CAPY	WORTH	DEMAND 1 LEVEL	STATE	
					INITIAL	10000	0
1	2	0	11000	3563	10200	11000	0
2	0	0	11000	7616	10900	11000	0
3	6	1	16000	-8925	11400	00000	1
4	0	0	16000	-3904	12000	00000	1
5	0	0	16000	924	12600	00000	1
6	0	0	16000	5595	13300	00000	1
7	0	0	16000	10097	14000	00000	1
8	0	0	16000	14456	14800	00000	1
9	0	0	16000	18830	15500	00000	1
10	5	0	21000	18971	16400	00001	1
11	0	0	21000	22582	17200	00001	1
12	0	0	21000	26092	18200	00001	1
13	0	0	21000	29489	19200	00001	1
14	0	0	21000	32765	20200	00001	1
15	0	0	21000	54289	20400	00001	1

Table 6.10a

PROBABILITY LEVEL   2
FIRST DECISION RESTRICTED TO   2   SM.EXTN

YEAR	DECISION BLD SCRP		CAPY	WORTH	DEMAND 1 LEVEL	STATE	
					INITIAL	10000	0
1	2	0	11000	3563	10200	11000	0
2	0	0	11000	7567	10700	11000	0
3	6	1	16000	-9081	11200	00000	1
4	0	0	16000	-4207	11700	00000	1
5	0	0	16000	489	12300	00000	1
6	0	0	16000	4959	12900	00000	1
7	0	0	16000	9242	13400	00000	1
8	0	0	16000	13335	14000	00000	1
9	0	0	16000	17237	14600	00000	1
10	0	0	16000	20949	15200	00000	1
11	0	0	16000	24473	15800	00000	1
12	0	0	16000	27723	16500	00000	1
13	0	0	16000	30677	17200	00000	1
14	4	0	18500	32189	18000	00010	1
15	0	0	18500	51430	18700	00010	1

Table 6.10b

Table 6.10a, b, c, d

PROBABILITY LEVEL   3
FIRST DECISION RESTRICTED TO   2   SM.EXTN

YEAR	DECISION BLD SCRP		CAPY	WORTH	DEMAND 1 LEVEL	STATE	
					INITIAL	10000	0
1	2	0	11000	3563	10200	11000	0
2	0	0	11000	7418	10400	11000	0
3	0	0	11000	11058	10700	11000	0
4	0	0	11000	14490	11000	11000	0
5	5	0	16000	12149	11400	11001	0
6	0	0	16000	15116	11700	11001	0
7	0	0	16000	17906	12000	11001	0
8	0	0	16000	20526	12300	11001	0
9	0	0	16000	23009	12700	11001	0
10	0	0	16000	25359	13100	11001	0
11	0	0	16000	27559	13400	11001	0
12	0	0	16000	29635	13900	11001	0
13	0	0	16000	31592	14200	11001	0
14	0	0	16000	33434	14600	11001	0
15	0	0	16000	45812	15000	11001	0

Table 6.10c

PROBABILITY LEVEL   4
FIRST DECISION RESTRICTED TO   2   SM.EXTN

YEAR	DECISION BLD SCRP		CAPY	WORTH	DEMAND 1 LEVEL	STATE	
					INITIAL	10000	0
1	2	0	11000	3509	10100	11000	0
2	0	0	11000	7314	10300	11000	0
3	0	0	11000	10863	10500	11000	0
4	0	0	11000	14172	10700	11000	0
5	0	0	11000	17254	10900	11000	0
6	4	0	13500	16932	11100	11010	0
7	0	0	13500	19372	11300	11010	0
8	0	0	13500	21646	11500	11010	0
9	0	0	13500	23764	11700	11010	0
10	0	0	13500	25736	11900	11010	0
11	0	0	13500	27591	12200	11010	0
12	0	0	13500	29316	12400	11010	0
13	0	0	13500	30936	12700	11010	0
14	0	0	13500	32440	12900	11010	0
15	0	0	13500	42522	13200	11010	0

Table 6.10d

Four Optimum Policies that give
the Strategy for Case 1,
with SM.EXTN as first Decision.

2. Employing Level 2, the best plan is shown by an evaluation between probability levels 2 and 3. A further computer run was carried out with the mean sales and this gave alternative LG.EXTN as the best first decision, with a NPW of 7,970 -- $(NPW)_{P,2}$.

3. Level 3 allows for probabilities, but with a single plan. This corresponds to the first column headed "No yearly correlation NPW". Here, alternative SM.EXTN has the highest NPW, with a value of 7110 -- $(NPW)_{P,3}$.

4. Level 4 represents the choice of the first decision with the highest Expected NPW when different plans are allowed for different probability levels. The results can be extracted from the columns headed "Mean -- with total correlation" of Table 6.9. The best first decision is now alternative LGE.EXTN with an NPW of 7690 - $(NPW)_{P,4}$.

To evaluate the simulated worth $(NPW)_{S,1}...(NPW)_{S,4}$, it is necessary to take the optimum strategy for each first decision involved, in this case SM.EXTN and LGE.EXTN and proceed with a Monte-Carlo simulation. The strategies can be extracted from the intermediate DP search printout. Tables 6.10a-d give the 4 tables that make up SM.EXTN strategy. Table 6.11 summarises the LGE.EXTN strategy.

```
OPERATING STRATEGY
          EXISTING    START/END PAIRS, BY PROBABILITY

1 EXISTING   1        0   5     0   5     0 26     0 26
2 SM.EXTN    0        0   0     0   0     0  0     0  0
3 LGE.EXTN   0        1   5     1   5     1 26     1 26
4 2500TNPLT  0        0   0    14 26     8 26     0  0
5 5000TNPLT  0       10 26     0   0     0  0     0  0
6 NEWPROC    0        5 26     5 26     0  0     0  0
```

Table 6.11

Summary of the Strategy for Case 1,
with LGE.EXTN as the list Decision

In this case, since two decisions are the same,

$(NPW)_{S,1} = (NPW)_{S,3}$ and $(NPW)_{S,2} = (NPW)_{S,4}$

Runs with the PLADE program in the Monte-Carlo evaluation phase gave the Expected NPW's as :

$(NPW)_{S,1} = 6960$

and

$(NPW)_{S,2} = 7050$

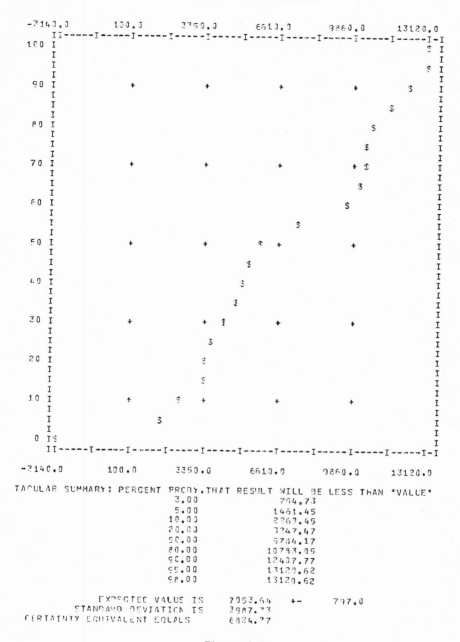

TABULAR SUMMARY; PERCENT PROBY.THAT RESULT WILL BE LESS THAN "VALUE"

3.00	764.73
5.00	1461.45
10.00	2260.45
20.00	3247.47
50.00	5784.17
80.00	10733.05
90.00	12407.77
95.00	13120.62
98.00	13120.62

EXPECTED VALUE IS   7053.64   +-   707.0
STANDARD DEVIATION IS   3987.73
CERTAINTY EQUIVALENT EQUALS   6824.27

Figure 6.6

A Simulation of Case 1
with the Optimum Strategy Associated
with LGE.EXTN.

The program output, showing the distribution, is given as Figure 6.6. We now have all the necessary data to evaluate the effect of the four levels for Case 1. Since $(NPW)_{S,2} = (NPW)_{S,4}$, there is no advantage in using the more sophisticated level rather than level 2 in this case, and the difference between $(NPW)_{S,1}$ or $(NPW)_{S,3}$ and $(NPW)_{S,4}$ is too small to be significant.

We can conclude therefore that there is no saving to be gained by using anything better than the simplest level in this case.

Now let us consider Case 2 in a similar way. Table 6.12 corresponds to Table 6.9 for Case 1, from which the following data can be withdrawn by taking the highest NPW from the appropriate columns.

	Value	Corresponding First Decision
$(NPW)_{P,1}$	131780	LG.PLANT
$(NPW)_{P,2}$	26430	SM.PLANT
$(NPW)_{P,3}$	1600	SM.PLANT
$(NPW)_{P,4}$	20560	PILOTPLT

The evaluation of $(NPW)_{S,1} \ldots (NPW)_{S,4}$ could be made with a probabilistic evaluation of PLADE using the best strategy appropriate to the first decision corresponding to the level in question. However, since the Demand/Year correlation coefficient is 1.0, the evaluation gives the same result as the mean NPW with total correlations given in Table 6.12. The simulation output of PLADE, shown as Figure 6.7, shows this to be the case for the PILOTPLT as the first decision.

Hence, from Table 6.12 :

$(NPW)_{S,1}$ = -32370   (First Decision - LG.PLANT)

$(NPW)_{S,2}$ = 7860   (First Decision - SM.PLANT)

$(NPW)_{S,3}$ = 7860   (First Decision - SM.PLANT)

$(NPW)_{S,4}$ = 20560   (First Decision - PILOTPLT)

Clearly, in this case, there is a loss associated with using a simpler level since $(NPW)_{S,4}$ is greater than $(NPW)_{S,1}$ or $(NPW)_{S,2}$. The first decision is dependent on the level of analysis chosen. Furthermore, the differences in the NPW are significant, -- up to 53,000 on a project with a capital outlay in the region of 200,000. In this case, there is a definite benefit in going to the most detailed level of risk analysis.

Cases 3 and 4 can be treated in a similar way; the corresponding Decision Matrices are shown as Tables 6.13 and 6.14.

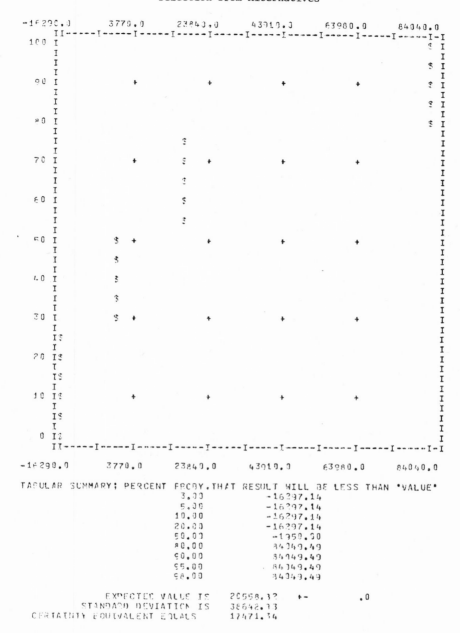

```
TABULAR SUMMARY; PERCENT PROBY.THAT RESULT WILL BE LESS THAN "VALUE"
                          3.00              -16297.14
                          5.00              -16297.14
                         10.00              -16297.14
                         20.00              -16297.14
                         50.00               -1350.00
                         80.00              84049.49
                         90.00              84049.49
                         95.00              84049.49
                         98.00              84049.49

             EXPECTED VALUE IS     20558.32    +-          .0
          STANDARD DEVIATION IS    38642.33
     CERTAINTY EQUIVALENT EQUALS   17471.34
```

Figure 6.7

A Simulation of Case 2
with Optimum Strategy
First Decision: Pilotplant

```
NEW PRODUCT    CASE2              DECISION MATRIX

FIXED DECISION YEAR    1

                 NO YEARLY    I--------WITH TOTAL CORRELATION--------
                 CORRELATION  I    CERT        BY   PROBABILITY
INITIAL          NPW          I MEAN EQUIV    1      2       3        4
DECISION

1 LGE.PLT        -32350    -32370   -86303.  131780    7540   -91100   -177710
2 MED.PLT         -8860     -8890   -32410.   93250   44050   -43130   -129730
3 SM.PLT           1600      7860    -1833.   78330   32440     3650    -82950
4 PILOT PLT       -3180     20560    17480.   84050   16440    -1940    -16280

BASED ON THE ABOVE FIRST DECISIONS,
THE EXPECTED VALUE OF COMPLETE FORECAST INFORMATION IS    20230.
```

Table 6.12

Decision Matrix for Case 2

LIMITED LIFE   CASE 3   DECISION MATRIX

	NO YEARLY CORRELATION	WITH TOTAL CORRELATION		PROBABILITY BY CORRELATION			
				1	2	3	4
INITIAL DECISION	NPW	MEAN	CERT EQUIV				
2 EXTN	1490	5190	4724.	12820	6670	1270	0
3 SM.PLT	720	4100	3771.	10560	5110	750	0
4 2ND SM.PL	0	-500000	-500000.	-500000	0	0	0
5 MED.PLT	1860	4790	4302.	12120	7050	0	0
6 LGE.PLT	290	4380	3841.	13470	4040	0	0

BASED ON THE ABOVE FIRST DECISIONS,
THE EXPECTED VALUE OF COMPLETE FORECAST INFORMATION IS   260.

Table 6.13

Decision Matrix for Case 3

FALLING DEMAND   CASE 4   DECISION MATRIX

FIXED DECISION YEAR   1

INITIAL DECISION	NO YEARLY CORRELATION VPW	I MEAN	I CERT EQUIV	WITH TOTAL CORRELATION BY PROBABILITY 1	2	3	4
3 LARGE	-9610	-11160	-11425.	-5310	-8250	-12170	-18900
4 MED/LGE	-7210	-8750	-8847.	-10810	-5070	-6440	-12680
5 MED/SM	-8680	-9470	-9492.	-8580	-8580	-9240	-11490
6 SMALL	-8840	-10050	-10140.	-14900	-8380	-7370	-9550

BASED ON THE ABOVE FIRST DECISIONS,
THE EXPECTED VALUE OF COMPLETE FORECAST INFORMATION IS   2150.

Table 6.14

Decision Matrix for Case 4

The comparison can be summarised on two tables: Table 6.15 shows the
Cost of Using the Simpler Levels for each case, and Table 6.16 shows the
Error in Predicted Profitability, as evaluated by equations (6.3) and (6.4).

Appreciating results expressed as NPW is rather difficult because one
has to know the magnitude of the project before the NPW is significant.
This disadvantage of NPW was discussed in Chapter I. To overcome this problem, the results in Tables 6.15 and 6.16 are given in the form of a
Benefits/Cost ratio (Section 1.4c) by expressing them as a percentage of the
sum of the mean discounted capital for each project. This may seem a little
obtuse, but the resulting figures are then equivalent to percent changes in
the capital cost. Hence, 1% in the table is no more significant than a 1%
saving in capital cost; 10% is certainly significant and 50% is a serious
matter.

6.3d Discussion of Results

Table 6.15 gives the cost of employing the simpler levels of analysis
for the four test cases. The conclusions drawn are very dependent on the
individual project. In Case 1 there is no cost in using the simpler method
because there is no significant difference between the different first
decisions chosen. There was really no problem of risk. This point became
apparent whilst the synthetic data was being constructed (See Section 6.2b).
Case 4 showed that only Level 1 has a first decision which differed from
the best, and the resulting cost was equivalent to only 6% of the capital
involved. In this case again, during the synthesis of the data it became
apparent that it was not a risky situation. Although this problem effectively was one of choosing a plant capacity in an uncertain market, the
market was falling and much less risk was apparent than in the normal case
where the market   increases   with uncertainty. This is because with a
falling market the high production rates are least uncertain and least
discounted. Hence, the major contribution to the NPW is certain, and the
less significant cash flows are those that are uncertain. It is also
interesting to note that the analysis in Case 4 included the constraint
that it was company policy to install the process in year 1. That all the
figures in the Decision Matrix (Table 6.14) were negative, indicates that
the best decision would really be to make no investment in the new process.

Table 6.15
Cost of Using a Simpler Level

Level of Analysis	Cost of Using Simpler Level (Units:NPW as % of Capital)			
	Case 1	Case 2	Case 3	Case 4
Level 1 (optimistic plan)	0%	30%	14%	6%
Level 2 (mean sales plan)	0%	7%	0%	0%
Level 3 (probabilistic plan)	0%	7%	7%	0%
Level 4 (strategy)	-	-	-	-

Table 6.16
Error in Predicted Profitability

Level of Analysis	Error in Predicted Profitability (Units: NPW in % of Capital) (+ ≡ over estimate)			
	Case 1	Case 2	Case 3	Case 4
Level 1 (optimistic plan)	+40%	+90%	+140%	+8%
Level 2 (means sales plan)	6%	+10%	- 30%	-6%
Level 3 (probabilistic plan)	0%	- 3%	- 50%	0%
Level 4 (strategy)	+4%	0%	0%	0%

Case 2, on the other hand, was an attempt to determine the plant size
for a profitable new product with a very uncertain market. Obviously a risk
risky situation. Figure 6.12 shows, in this case, that the different levels
of analysis do produce different decisions, and Table 6.15 shows that the
cost of these decisions can be considerable. The optimistic approach incurs
an extra cost equivalent to 30% of the capital cost involved. Taking mean
sales or the simple probabilistic approach still involves significant costs,
(7% of the capital), compared with a strategy approach to planning. The
search has shown in this case it is economically the best policy to build a
small plant (alternative PILOTPLT) to supply the early market and then see
how the market develops. Since the simpler probabilistic approach of Level
3 cannot consider different decisions for the different market outcomes, it
could not properly assess the value of a pilot plant, and therefore it
discards this alternative as not being the best first decision.

Case 3 represents the determination of plant capacity when the product
is expected to have a limited life and the market is uncertain. This has
all the ingredients of a risky situation, and in fact, Table 6.15 shows the
situation to be intermediate between 2 and 4. Different levels do recommend
different first decisions and the costs of using the wrong level are equivalent to 7 to 14% of the capital involved.

Table 6.16 shows the error in predicted profitability of the projects,
using the different levels of analysis. Again the results are scaled to
the capital costs involved, so the figures should be treated with the same
seriousness as corresponding errors in the original capital costs estimates.
For instance, +50 is equivalent to an underestimate of 50% in the capital
cost.

Level 1, in most cases, produces extremely serious over-estimates
of profitability, equivalent in one case to 140% under-estimate in capital
cost! This is least evident in Case 4 because the inverted sales forecasts
again have a stablizing influence. This over-estimate of profitability
is not surprising and hopefully no firm ever uses it to obtain an honest
profitability estimate. What is worth noting is the magnitude of the
over-estimate.

Level 2 produces errors in predicted profitability which are sometimes positive and sometimes negative. The one error of -30% for Case 3
shows that sometimes the error can be serious. One would normally expect
this level to produce an over-estimate of profitability, because it
balances low sales against high sales in any one year, whereas the physical

capacity of the plant does not allow this complete balance, and the result
is reduction in profit. This is true if there is only one investment being
investigated, but if the plan involves more than one investment changes in
the timing and size of the following decision can lead to Level 2 sometimes
predicting profitabilities too low, and sometimes too high. Level 3 in fact
represents a corrected form of Level 2 where the Expected value has been
correctly calculated. This always results in the same or lower predicted
profitability than that of Level 2, -- never higher, because the mean production must be less than or equal to the mean sales, but never higher.

What is more interesting is that these results are also the same, or
lower, but never higher than Level 4. The difference can be equivalent to
significant over-estimates of capital cost -- 50% in one case which is so
over-conservative that it may result in reasonable opportunities being lost.

The reason for this conservatism is that Level 4, which we are assuming
is our best estimate of reality, has further degrees of freedom to make
different decisions for different probability levels. Level 4 will therefore be different from Level 3 only if it results in a better profitability.
It is therefore clear that at the worst, Level 4 must equal Level 3 and in all
all other cases Level 4 will show improvements in profitability over Level
3.

Level 4 properly predicts profitabilities when the correlation coefficients are 1.0, -- at least within the definition of the study. However,
when the correlation coefficients are not unity, this represents some
additional uncertainty and plans can be thwarted by the sales drifting to
other levels after irreversible decisions have been made. This results in
a lower profitability than that given by the unity correlation calculation.
Hence, when correlation coefficients are not unity, Level 4 predicts
profitabilities that are high. This is shown in Case 1, where a 4% error
results. This subject is discussed in more detail in Chapter VIII.

Conclusions

This whole study on depth of analysis can be summarised as follows:
The simple optimistic sales level of analysis has nothing to recommend it as it can give very misleading results. The deterministic
analysis, using the mean sales forecast, is adequate when the project
is clearly not representing a risky situation. This can often be
recognised from a preliminary study of the project but it is rather
begging the question since one object of a study is to determine
whether there is a risk problem or not. In a project

involving risk the use of mean sales can result in poor decisions, and in one of the test cases a cost equivalent to 7% of the project capital cost was incurred.

A simple probabilistic analysis, but without the facility for making different future decisions depending on the market development, did not always select the best decision and additional costs of about 7% of the capital cost of the project were incurred. This Level generally predicts profitabilities that are lower than will be realised, -- although this could be considered a good fault. An error equivalent to a 50% capital cost that was observed in one case may be so conservative as to involve losses of opportunity.

The most complete risk analysis which included the allowance for different future decisions depending on the market development could be justified, both in that it gave a better decision than any other level in two of the four cases, and as good a decision in the rest and that it gave a better prediction of the Expected profitability.

6.4 COMPARISON OF DECISION CRITERIA

The study of various levels of analysis have not touched on the question of Decision Criteria introduced in Section 3.7a. Only when one is using a sufficiently advanced level of analysis to produce a distribution of NPW (such as that given by the Decision Matrix, of Level 4), can one begin to discuss Decision Criteria in weighing the effects of risk.

Of the possible Decision Criteria, Chapter 3 shows that Expected Value, Certainty Equivalent, Minimax and Minimum Regret were four reasonably practical, different approaches that could be taken. We can look at the Decision Matrix for our four cases Tables 6.9, 6.12, 6.13, and 6.14, and see whether these different criteria lead to different first decisions.

This can be done in a few minutes with paper and pencil, extracting best first decisions presented by each Decision Criterion for the four Decision Matrices, as shown in Chapter III. Before the Minimum Regret Criterion can be used, the Regret Matrix must be written down, which takes two or three minutes.

The reader might like to confirm the interesting result that, apart from one exception, all the Decision Criteria choose the same First Decision. The exception being the choice of alternative MEDSM instead of MEDLGE for Case 4, but since this represents a change equivalent to less than 1% in the capital cost, it can hardly be cited as a reason for carefully selecting a

Criterion. It is not possible to draw definite conclusions from such a
complex system from only four results, but at least they form some evidence
that perhaps the selection of Decision Criteria is not an important aspect
of studies involving manufacturing capacity. These four cases were carefully
created to represent risky situations and they show insensitivity to the
Decision Criterion chosen. It is no doubt possible to find a project which
is as sensitive to the Decision Criterion as that example given in Section
3.7a, but it must take some finding and it would have peculiar characteristics.

The example cited in Chapter III, which shows so nicely that the decision arrived at depended on the Decision Criterion chosen, does in fact
have curious characteristics. Table 3.1 shows Alternative Action A to have
an equal difference in outcome between States of Nature (1) and (2) and (3),
whereas Alternative Action B shows nearly three times the difference between
(1) and (2) and (2) and (3). That is to say, that these two Alternative
Actions are affected in very different ways by the State of Nature. In
manufacturing capacity proposals, there is a tendency for all Alternative
Actions -- which are the conditional plans -- to be similarly affected by
the State of Nature, and this would explain the observation that the Decision Criterion does not affect the decision in our examples.

It would be wrong to conclude that no effort should be made in choosing the appropriate Decision Criterion for manufacturing capacity decisions.
There may occasionally be combinations of circumstances which make the decision sensitive to the Criterion chosen and so if the appropriate Decision
Criterion is always employed we do not have to worry about identifying the
exceptions. However, it cannot be claimed that idenfication of the right
Decision Criterion is a serious problem in manufacturing capacity decisions.

6.5 APPRECIATION OF THE PLANNING RESULTS

The essence of the planning activity is much more than recommending
the proposal with the highest profitability from a set of proposals and data
supplied by an outside department. It is more a matter of fully understanding the situation and proposing a course of action which can be defended as
being sensible, economic and free from high risk.

The central point is the real understanding of the problem. The careful
definition of a set of projects for a comparative study is a very difficult
exercise because there are so many assumptions to be made where an ill-
considered judgement can so easily change their ranking. Table 6.17 gives

a list of some areas where different assumptions can easily affect the
ranking of projects.

Table 6.17

Important Aspects in Comparing

Alternative Projects

Capacity
Project Life
Tolerated Shortfall
Treatment of Existing Plant
Timing of Capital Outlay
Discount Rate
Inflation
Product Quality
Raw Material Efficiencies
Manpower Requirement
Utility Usage
Overhead Charges
Associated Utility Investment
Demand
Price
Probability Level for Demand and Price
Degree of Correlation Between Demand and Prices/Year
Ability for Later Expansion

Once a study has provided a ranking for a set of projects the planner
should discover the major reasons for this ranking. He should then return
to the source of the data and check that the data defining these major
reasons are, above all correct and secondly, certain enough to support the
final decision of the whole study.

In a planning study associated with process development the planner
can play an additional creative role by recommending exactly the areas on
which to concentrate development.

The need for the planner to return to his data source and ask "is your
data really correct?" Is brought about by the sheer size of projects.
The planner may be comparing the work of different teams and each team will
have its own way of deciding on the data. Although large firms try to stan-
dardise on methods for economic analyses, these are always inadequate and in
fact can themselves lead to poor comparisons. When a company procedure has
been standardised for the sake of consistency, but the basis for the standard does not apply to all projects being compared, some projects will be
unfairly penalised.

To illustrate these rather vague generalities let us take some examples
from Table 6.17.

Example 1

Consider two projects, one capital intensive, the second requiring
comparatively little capital; a selection based on DCF shows the
second to be preferable. In such cases where the major significant
difference in the project is the capital, one must be sure that the
right economic criterion is chosen. If the firm had ample funds and
was short of projects a NPW with a 10-12% discount rate would have been
a more suitable criterion -- even when "company policy" is still out-
wardly to maximise DCF. If discussions could ascertain that NPW was
the more appropriate criterion the ranking of the two projects may well
be reversed.

Example 2

Consider two projects; one to install a duplicate plant immediately
for an existing product, the second to accept a short-fall for two
years (attempt to buy-in product) and then install a new process;
assume the second proposal shows better economics. This comparison
is heavily influenced by the delay in spending capital associated with
the new project. The study is more a comparison of delaying capital
spending than comparing alternative processes. The study should be
repeated assuming the same "buying-in" facilities exist in the existing process to get a true valuation of the new process.

Example 3

Consider two plants being compared for a new product. One a process
purchased from a contractor, and the other a process under development
within one's own research department. Naturally, you research department process is ranked first by the study, the dominating advantage
being the lower raw material costs.

The basis for the reaction efficiencies for the two processes must be
compared. If the contractors efficiencies are based on actual plant
operation and the research department figures are a better-than-average
laboratory result, it would be unwise to accept the result of the study.

Example 4

Consider two proposals for a new product which have been put forward;
one with a lower raw material cost and the second with a lower utility
cost, and assume the second process was ranked first by the economic

study, -- basically because it has lower capital costs: A study of
the capital cost might reveal the high capital associated with the
higher utility project was due to capital expenditure in the boiler
house, because it is company accounting policy to debit the project
which causes a boiler expansion to be initialized with the total
expansion costs. Clearly, it is wrong for a process selection to
be influenced by the coiincidence of service facilities being overloaded and discussion are then necessary to determine whether the
company's "Allocation of Utility Capital rules" can be modified for
this selection study.

Example 5

Consider two proposals for increasing manufacturing capacity; one to
add a small plant to an existing manufacturing facility, and the second
to scrap an existing plant and build a large new plant to give a higher
quality product. Assuming all credit for the single new plant, --
lower manpower, higher efficiency and lower maintenance -- could not
make it more attractive than keeping the old plant.

In this case the product quality may not have been properly considered.
With certain products, quality is of little importance. In others,
product quality specifications are continually becoming tighter, resulting in old plants being unable to achieve a high enough quality to
maintain sales. The forecasting department should be asked to estimate
two future sets of demands and prices, specifically taking into account
the two qualities involved. It may be that to keep the old plant is
tantamount to withdrawing from this product area in about 10 years time.

For further examples let us consider the four case studies made earlier in this chapter. These cases were very carefully prepared and it is
interesting to see how well they answered the planning question that were
posed.

Case 1 (See Section 6.2a, Fig. 6.1 and 6.10a,b,c,d, Table 6.5 and 6.9)

This study was to decide whether to install a new process for an
existing product or simple to expand the existing plant, or to add a second plant with the old process, as a means of increasing manufacturing capacity on a basic chemical with a stable steady growth. Table 6.9 shows
that a plant expansion is the most economic first decision, and inspection
of the individual plans shows that the new process would be attractive in

two years time if the demand turned out to be optimistic. The conclusion
from the study therefore is to expand the existing plant, and prepare the
new process for installation in two years time, since it has a 25% probability
of being required. However, the chance of changing demand levels (correlation coefficient 0.75) really means that there is no point in planning this
product as though it will stay on a fixed demand curve. This is considered
in more detail in Chapter VIII. This modifies our conclusions -- the new
process will not be needed for five years, and so, for the present, development work should stop. Before such a recommendation is made it behoves
the planner to check more carefully the effect of the correlation coefficient, and also to ascertain more exactly from the commercial people what the
most likely values of the coefficient might be.

The cost of going directly to the new process, rather than firstly
expanding the existing plant is about £200,000 in NPW. (See Table 6.9)
It might be possible to justify this line of action if savings associated
with rationalisation were not properly considered when the data was first
collected. One point of study shows very clearly; although the new process
is much better than the old process when compared directly, given the company's present situation it is not very attractive. If the research department has already worked long on it, it should be persuaded to shelve it.
If it is still a new idea, it is showing very considerable promise and
should be developed further.

Case 2 (See Section 6.2a, Fig. 6.2, Tables 6.6 and 6.12)

Case 2 is a study to determine the best initial size of plant to
install for a new product with a very uncertain market. Table 6.12 shows
that the most attractive action is to install a pilot plant and see how the
market develops.

Table 6.6 gives company policy "to satisfy demand only when economic
to do so". Though this might sound a realistic policy for an isolated new
product, the implications of this policy are far from realistic. Table
6.18 gives one of the "Optimum Policies", with PILOTPLT as the first decision. This Table shows that for ten of the fifteen year there is a short-
fall -- sometimes up to 40% of the demand cannot be met! Clearly, such a
marketing policy for a new product is impossible, particularly because the
short-fall would alter the demand forecasts. The fact that plants normally
work uneconomically at first as the market builds up is well known, but the
effect is magnified considerably in this case where a small plants shuts
down when the new one is built.

```
                    PROBABILITY LEVEL    2
          FIRST DECISION RESTRICTED TO    4    PILOT PLT


               OPTIMAL POLICY

  YEAR    DECISION   CAPY     WORTH      DEMAND          STATE
          BLD SCRF                       1 LEVEL


                                              INITIAL   00000 0
    1      4    0    1200    -97175      1200            00010 0
    2      0    0    1200    -83677      1400            00010 0
    3      0    0    1200    -71940      1600            00010 0
    4      0    0    1200    -61734      1810            00010 0
    5      0    0    1200    -52859      2080            00010 0
    6      0    0    1200    -45141      2340            00010 0
    7      2    4    2750   -100836      2580            01000 0
    8      0    0    2750    -84194      2760            01000 0
    9      0    0    2750    -69723      2780            01000 0
   10      0    0    2750    -57139      2984            01000 0
   11      0    0    2750    -46197      2970            01000 0
   12      0    0    2750    -36682      2980            01000 0
   13      0    0    2750    -28408      2990            01000 0
   14      0    0    2750    -21213      3000            01000 0
   15      0    0    2750     16441      3000            01000 0
```

The shortfall is the difference between columns DEMAND and CAPY.

Table 6.18

Optimal Policy for Case 2

Clearly, further studies should be done assuming no short-fall is allowed, or assuming say only 5% short-fall could be tolerated. This bound can easily be inserted in the search for optimal plans, and when this was done it resulted in a general lower level of profitability -- an indication of the incentive for proposing some form of warehousing of products to enable investment to be delayed by supplying product from long term stock. It also showed that installing a 2000 ton/year plant was equally attractive to the 1200 ton/year PILOTPLT as a first decision.

Two of the four conditional plans starting with PILOTPLT recommend shutting down the small plant and replacing it with one two or three times as big. This is useful guidance, but probably not a sensible decision. With a 50% chance that the plant will be shut down within a few years it is worth going back to the engineers and asking them for the cost of an

expandable pilot plant, together with a range of expansions for it. It is
often easier to forget a small pilot plant and replace it by a larger, improved unified plant, but if the size difference is only two to three, as in
this case, it may be worth treating it as the first stream of a two or
three-stream plant.

Supposing that the engineers say that making the first 1200 ton/year
plant an extendable plant will increase its costs by 20% but up to 4 further
1200 ton/year plant could be added at a cost of only 90% of the original
cost. Re-running the PLADE program gave a Decision Matrix which showed that
profitabilities were, in fact, overall worse, and risks increased. The
lower forecast levels were worse because the single plant cost more, and
the higher levels were worse because 3 or 4 small plants were, in this case,
very much more expensive to run than one large plant.

In 2 iterations we have learnt much more about the project. Due to
gross uncertainty it is worthwhile delaying the real investment by installing a 1200 ton/year plant which can later be scrapped if the market turns
out to be good. The real profitability of the project comes from the
ability to instal the larger plants -- operation of small plants is not
economic. The maximum risk in the project is of the order of 15% of the
capital investment. There is a considerable incentive for having a more
accurate forecast, and a long-term warehousing policy could produce
considerable savings. Finally, the expected DCF for the project will be a
little below 15%

Case 3 (See Section 6.2a, Fig. 6.3. Tables 6.7 and 6.13)

This study considers the expansion of manufacturing capacity for a
product with a growing market, but which is expected to be replaced eventually by other products.

Once more, it is a problem of choosing the plant size and because the
market is expected to decline, the company feels no responsibility to meet
the demand if it is uneconomic to do so.

The Decision Matrix, Table 6.13, indicates that a small extension of
the existing plant as a first decision is the most economic decision to
make. Table 6.19 shows an individual Optimum Policy for the most optimistic
demand level, with EXTN as the first decision. Once more we have an unrealistic picture of a short-fall up to 20% with the market being fully
available immediately the plant is built. A realistic policy would be
either to ignore all demand over installed capacity if the overall investment
was uneconomic, or to have no short-fall if a further plant was going to

PROBABILITY LEVEL 1
FIRST DECISION RESTRICTED TO 2 EXTN

OPTIMAL POLICY

YEAR	DECISION BLD	SCRF	CAPY	WORTH	DEMAND 1 LEVEL	STATE	
					INITIAL	10000	0
1	0	0	1000	5146	860	10000	0
2	0	0	1000	10272	960	10000	0
3	2	0	1100	13624	1060	11000	0
4	0	0	1100	18083	1220	11000	0
5	0	0	1100	21960	1440	11000	0
6	6	0	2400	16089	1590	11000	1
7	0	0	2400	21161	1960	11000	1
8	0	0	2400	26192	2180	11000	1
9	0	0	2400	30910	2320	11000	1
10	0	0	2400	35183	2450	11000	1
11	0	0	2400	38898	2400	11000	1
12	0	0	2400	42064	2360	11000	1
13	0	0	2400	44705	2280	11000	1
14	0	0	2400	46880	2180	11000	1
15	0	0	2400	57561	2070	11000	1

Table 6.19

Optimal Policy for Case 3

be built. It is therefore up to the planner to predict the outcome and decide whether the policy should be to meet the market or to ignore it. This first run recommends that not even EXTN should be installed for the pessimistic level; the EXTN only should be installed (in year 4) for demand level 3; but for demand levels 1 and 2, further runs must be carried out to determine which plants are economic, when they are installed early enough to prevent a short-fall. Until these runs are made, one of the major objectives of the planning study -- to indicate to what level it is economic to manufacture when the market will later fall -- has not been achieved.

Case 4 (See Section 6.2a, Fig. 6.4. Tables 6.8 and 6.14)

Case 4 represents a study of a product in its decline but still with a long expected life for which an improved process has become available.

Should the two existing plants be scrapped and replaced by the new process? The Decision Matrix, Table 6.14, shows that all types of new plant are less profitable than leaving the old plants operating. This is indicated by all NPW values in the table being negative. However, should an installation be unavoidable, then a medium/large plant has the least Expected loss and is one of the least risky.

The relation between NPW and the proposed plant size is interesting in that it shows the intermediate sized plants to be better than either the large or the small. Also, the large and small plants have higher maximum losses than the intermediate sized. Obviously, once the cost of installing a new plant is incurred one should put in adequate capacity to benefit from lower running costs.

However, the study so far has only considered four sizes and the best size will probably be somewhere between 12,000 and 18,000 tons per year -- requiring further analyses.

Besides carrying out these extra studies the planner should go back to the product department and discuss why, or if, they are interested in this loss-making project. It may be essential to consume some by-products, in which case, have they taken a reasonable transfer price for it? It may be only some passing idea from research department and so had best be dropped.

The object of these rather tedious descriptions has been to show that planning is basically an activity which cannot be programmed -- every case is different and requires deep study to appreciate its characteristics.

The role of computer programs such as PLADE is to provide quick answers to questions posed by the planner. The programs themselves do not yield the answers to the planning study in the single run, and the completeness of any study is governed entirely by the competence and creativity of the planner himself.

It is the selection from alternatives that poses the greatest difficulty in the planning activity. Choice of plant capacity and the economic evaluation, -- discussed in Chapters VII and VIII -- pose much less difficulty because these stages are less individualistic and analyses can more easily follow a common procedure.

6.6 THE INTRODUCTION OF QUALITATIVE FACTORS

Up to this point consideration has only been given to obtaining a fair comparison of the economics of the various alternatives. Every attempt should have been made to express the characteristics of each alternative in terms of money and include this in the economics. It is surprising how much can be expressed quantitatively given a little thought and imagination. It is often possible to determine the cost of any disadvantage by enacting the procedure required to overcome the disadvantage. Chapter I describes the technique of Cost/benefit Analysis which is a fairly well advanced technique for determining the money equivalent of qualitative factors, and Chapter II describes methods of quantifying uncertainty.

However there may always remain some difference between the alternatives that cannot be quantified in money terms. One project may fit better into a desired future policy than another, or one may cement or destroy relationships with an outside firm with which it is intended to develop collaborative projects, for example.

After the economic analysis has been completed the profitability and all non-quantitative factors should be considered together before the final decision is taken.

Decision-taking in Multi-criterion situations was discussed in Chapter III Section 3.3 which describes unweighted selection methods and weighted selection methods. The weighted methods are generally preferred because they provide extra flexibility and the economics often play a more important part in the decision than any other single factor.

6.7 THE VALUE OF FURTHER INFORMATION

6.7a The Value of Perfect Information

The presentation of the Decision Matrix affords a method of determining the value of better forecast information. Forecasting can be expensive and so forecasting studies should only be carried out when an economic justification has been presented. It often happens that widely differing sets of data result in exactly the same decision being made. There is no point in paying for better information to be gathered if the only result is to predict more accurately the outcome of the decision; it is only worth paying for better information when this could change the decision about to be made.

The value of Perfect Information can easily be extracted from the Decision Matrix, in the way described in Section 3.5, as follows:

Taking Case 2, we can see from the Decision Matrix (Table 6.12) that:

If we knew the outcome would be Probability 1,
we would choose Alternative 1 with a NPW of   131,780

If we knew the outcome would be Probability 2,
we would choose Alternative 2 with a NPW of   44,050

If we knew the outcome would be Probability 3,
we would choose Alternative 3 with a NPW of   3,650

If we knew the outcome would be Probability 4,
we would choose Alternative 4 with a NPW of   -16,280

Since the probability of each level is 0.25, we can say the Expected Net Present Worth from the project with Perfect Information would be:

$$(131,780 + 44050 + 3650 - 16280)/4 = 40800$$

and we would install either Alternative 1 or 2 or 3 or 4, depending on our Perfect Information. However, without information, as we have already seen for Level 4, we would choose Alternative 4 which has an Expected NPW of 20,560.

The difference between these two figures:

$$40,800 - 20,560 = \underline{20,240}$$

represents the Expected Value of having Complete Information. Expressed as a fraction of the discounting capital involved it is 11%, which is considerable.

Table 6.20 shows the Value of Complete Information for the four case study examples. It shows, in all cases, that there is a gain to be made in having more information, and that this gain, equivalent to between 2 and 11% of the capital involved, is certainly not negligible.

It is not surprising that there is some definite value in having complete information in the four examples here, since they were originally chosen to be projects involving risk. That they were synthesized to appear as Figure 6.5 guaranteed that there will be a value for Perfect Information. However, this is not always the case, and when one alternative is the best first decision for all probability levels, the Value of Perfect Information is zero. In practice, this is likely to occur quite frequently and in Operations Research literature this is referred to as a "Dominant" first decision.

Table 6.20

Value of Perfect Information for the 4 Case Studies

(expressed as a percentage of the discounted capital involved)

	Value of Information	
	Numerical Value	Expressed as % of Capital
Case 1	370	2%
Case 2	20,200	11%
Case 3	260	4%
Case 4	2,150	3%

6.7b The Value of Imperfect Information

Although it is useful to have the Value of Perfect Information as a yardstick, it does represent an impossible situation, and so it is interesting to be able to calculate the Value of Imperfect Information which has only a probability of being correct. This can be obtained from the Decision Matrix using the same treatment as described for Decision Trees in Section 3.6.

Consider that, from past experience of the forecasting method, probabilities can be assigned to the accuracy of the forecast, as shown by Table 6.21:

Table 6.21

Probabilistic Outcome of a Forecast

	Level 1	Level 2	Level 3	Level 4
If Level 1 is forecast the probability distribution is:	0.8	0.2	0.0	0.0
If Level 2 is forecast the probability distribution is:	0.2	0.6	0.2	0.0
If Level 3 is forecast the probability distribution is:	0.0	0.2	0.6	0.2
If Level 4 is forecast the probability distribution is:	0.0	0.0	0.2	0.8

Now, taking Case 2 once more as an example, and taking the appropriate data from the Decision Matrix (Table 6.12), we can say:

If Alternative 1 (LGE.PLT) were chosen, and

if Level 1 were forecast, our Expected NPW would be
$$0.8 \times 131780 + 0.2 \times 7540 \qquad = \quad 106932$$
if Level 2 were forecast, our Expected NPW would be
$$0.2 \times 131780 + 0.6 \times 7540 + 0.2 \times (-91100) \qquad = \quad 12660$$
if Level 3 were forecast, our Expected NPW would be
$$0.2 \times 7540 + 0.6 \times (-91100) + 0.2 \times (-177710) = -88694$$
if Level 4 were forecast, our Expected NWP would be
$$0.2 \times (-91100) + 0.8 \times (-177710) \qquad = -160388$$

This can be repeated for all the other possible first decisions and a further Decision Matrix constructed where the State of Nature is now the outcome of the forecast, and not the probability level. This is displayed in Table 6.22.

Table 6.22

Decision Matrix for Imperfect Forecast - Case 2

Expected NPW

Initial Decision	Probability Level forecast by proposed Survey			
	1	2	3	4
1. LGE.PLT	106,932	12,660	- 88,694	- 160,388
2. MED.PLT	83,410	36,454	- 43,014	- 112,410
3. SM.PLT	69,152	35,860	- 7,912	- 65,630
4. PILOT PLT	70,528	26,286	- 1,132	- 13,412

Now, after the forecast, the decision will be either to install Alternative 1, or 2, or 3 or 4. The selection will be made after the result of the survey becomes available by choosing the decision with the highest NPW (underlined in Table 6.19).

Since this analysis is to be carried out before the survey, our only "a priori" information is that all levels have a 0.25 probability. We can therefore only discuss the Expected NPW after the survey, and this will be:

$$0.25 \times 106,932 + 0.25 \times 36,454 + 0.25 \times (-1132) + 0.25 \times (-13412) = \underline{32,210}$$

This value then is the Expected Value if an Imperfect Survey is carried out. If no survey is carried out, our decision would be to install 4, with an Expected Value of 20,560. The difference -- 11,650 -- is therefore the Expected Value of our Imperfect Information.

If this approach were to be used a sufficient number of times within an organisation, this use of Expected Value would maximise the benefit to the firm.

This approach gives a quantitative basis for evaluating the worth of a realistic forecasting study so that marketing departments can judge the value of embarking on expensive surveys.

Admittedly, it is far from common practice and it requires a high degree of experience of previous surveys because a probability distribution for the accuracy of the survey must be estimated. Furthermore, it relies heavily on Expected Values, and so it must be used frequently for any benefit to be guaranteed; but at least the method gives an answer to a difficult problem, which is certainly valuable as guidance, and it is very simple to apply once one has the Decision Matrix to hand.

6.8 CONCLUSIONS

This Chapter has been concerned with choosing one from a number of basically different manufacturing procedures.

The Chapter began by generating four synthetic projects which typified various marketing areas within the chemical industry and then used these projects to determine to what level the analysis should be made in order to obtain valid conclusions.

The cases showed that it was often important to consider an uncertain market probabilistically and not merely represent it by a mean deterministic study. Furthermore, it was sometimes important to incorporate the possibility of different future decisions depending on the market outcome in order to make a decision which involved lowest total cost. In short, justification is presented for analysing all projects in great detail where a difficult decision is anticipated. The deeper levels of analysis also provide a range of outcomes which is a form of risk analysis and is therefore of importance in arriving at a good decision.

It appears, from the four case studies, that whatever decision criterion is employed for selection when outcomes are uncertain -- MINIMAX, Minimum Regret, Certainty Equivalent or simply Expected Value -- the same alternatives are selected. This rather surprising conclusion appears to be a characteristic of manufacturing capacity investment problems because the state of nature affect all alternatives in the same way. However, this generalisation cannot always be true and so some effort should be made in

choosing the appropriate decision criterion. It is reasonable to conclude
that this cannot be considered to be a major problem area.

The four case studies well illustrate the major feature of planning
work -- that extreme care must be taken when creating the alternative
projects. A planning study must be an iterative process to enable a planner
to understand fully the difference between his alternatives and to be sure
that the difference in the economics of the alternatives has been brought
about by real differences resulting from the major characteristics of the
alternatives. This understanding may require the planner to go back to
other departments to check on the data or to obtain more data.

The deeper levels of analysis which also produce risk analyses can
be used to estimate the worth of more certain data. This can be of use
in discussions between planner and forecaster in deciding the value of
embarking on expensive forecasting exercises.

CHAPTER VII

THE SIZE OF A CAPACITY INCREASE

> Eis, zwei, drei, vier, feuf, sächs,
> sibe,
> myni Mueter chochet Rüebe,
> myni Mueter chochet Speck,
> und due muesch weg.

7.1 INTRODUCTION

The most common planning problem is that of choosing the size of a capacity increase (2,8,13,14). The whole of this chapter is therefore devoted to this problem. Initially, a number of idealized studies will be discussed and these will then be followed by a number of more realistic examples.

An introduction to the general problem is to make a primary study of the choice of capacity when the forecast demand shows a steady increase into the foreseeable future. This problem was analysed in detail in 1952 by Chenery (3), who was looking for reasons to explain why industries always appeared to have excess capacity. He produced an analytic expression which gave the optimum capacity to install for a simplified situation with an arithmetically increasing, known demand. Later workers have expanded his treatment to give analytic expressions for different demand patterns (1,16,17), for the inclusion of a loss when the market is not satisfied (7,10), and for when the demand is uncertain (10).

The situation of an unlimited arithmetic increase is the idealized case and therefore, in many ways, the most informative. This case, as reported by Manne (10), will be developed here, after which the more complex cases will be discussed.

The problem of locating the optimum plant size is introduced entirely by the "Economy of Scale" factor. In general, one can say the cost of a plant can be approximately related to its capacity by equation 7.1:

$$C = k_c(V)^a \qquad (7.1)$$

where C = capital cost

k_C = constant

V = plant capacity

a = a constant, the Economy of Scale factor

The value of the Economy of Scale factor is generally found to be
between 0.5 and 1.0. Values near 1.0 occur when there is no "economy of
scale." This is so with electrolytic cells for chlorine production, or
other plants where the only way of increasing capacity is to duplicate the
existing equipment. Values near 1.0 can also occur, even when a larger
plant is built, if the size of the larger plant is such that extra difficulties are encountered as a result of the increased size. Some of the larger
ethylene plants had disappointing economics because of on-site stress-
relieving and compressor problems introduced by taking equipment sizes
beyond normal experience.

In practice, the cost is not a smooth function of capacity, as given
by equation 7.1, but some discontinuous function which depends on the
construction techniques available; but for a generalized treatment equation
7.1 suffices.

A popular value for the Economy of Scale factor in the chemical
industry is 0.6--the "two-thirds power rule." This was first suggested by
Chilton (4,5) as being the mean relationship found in a survey of 36
chemical plants. However, individual factors varied between 0.33 and 1.39.
In particular, multistream units tended to have factors nearer 1.0 than
0.6. Clearly, for accurate work, the value of the factor should be determined from cost studies for the particular process in question.

If this value of the Economy of Scale factor was 1.0, that is, there
was no "economy of scale" and there was no discounting, it would be immater-
ial when the plant expansions were made and how big they were, as long as
demand was always met. However, since cash flows are discounted, this
leads to an advantage in spending money as late as possible. When the
Economy of Scale factor equals 1.0, the optimum policy is to make the
smallest increments possible when demand meets capacity. In the extreme,
this represents a continuously expanding plant. However, because the
Economy of Scale factor is generally less than 1.0, there is a capital cost
disadvantage in installing small increments--hence we have the optimisation
problem.

7.2 PLANT SIZE FOR ARITHMETICALLY INCREASING DEMAND WHICH MUST BE MET

Consider the case of a steadily increasing demand, with plant capacity increased in steps whenever demand meets capacity, as shown by Figure 7.1. Also assume that this situation exists without limit, with the demand increasing at the same rate for an infinite time.

It can be proved that in this case the optimum policy requires a constant cycle time -- i.e. the plant capacities for all the expansions are the same -- as follows:

> Consider the problem as seen from the first plant on Figure 7.1 (point A): The optimum capacity corresponding to the minimum discounted total cost, will be dependent on the plant cost, the growth rate and the future beyond point A. The second plant is installed at point B. At this point the growth rate is the same as point A, the plant costs are the same, and the future is identical because we are considering an infinite future. Point A appears identical to point B; hence the optimal size for point A must also be the optimum size for point B. This means that at optimality, all plants will be the same size.

Figure 7.1 can be re-drawn to show the time path of excess capacity, presented as Figure 7.2 The point at which excess capacity is zero is called the point of regeneration and because of the constraint that demand must always be met, a new plant is required at each point of regeneration. Because all plants and cycle times must be identical at the optimum point, Figure 7.2 will have a constant, infinite, saw-tooth pattern.

Since demand will always be met, the income from sales will be constant, whatever plant size is chosen. Similarly, total variable costs will be independent of plant size. Let us also assume for the present that fixed costs are small enough to be neglected. With these assumptions, our optimum plant size is simply the size that results in the lowest discounted total capital cost, since all other cash flows are independent of the plant size.

The total discounted capital costs from the point of regeneration A (T_A) are the capital cost of the plant installed at point A (C), plus the total costs at the next point of regeneration, point B (T_B), discounted to point A. Using the continuous discounting formula, (section 1.4b), which is analytically more convenient than the dicrete formula used throughout the majority of this text, we can write down the cost at each point A as:

$$T_A = C + T_B \, e^{-rx} \qquad\qquad (7.2)$$

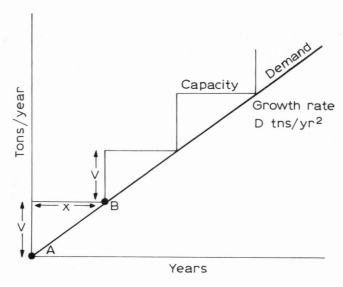

Figure 7.1

Arithmetically Increasing Demand which must be met

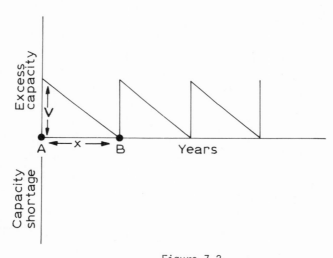

Figure 7.2

Time Path of Excess Capacity for Figure 7.1

where

r is the fractional discount rate and

x is the time between points of regeneration

Since we are dealing with an infinite series:

$$T_A = T_B \tag{7.3}$$

and so we can obtain the recursive equation 7.4 by substituting 7.3 and 7.1 in 7.2:

$$T_A = k_C V^a + T_A e^{-rx} \tag{7.4}$$

Re-arranging, to isolate the total discounted cost and re-expressing plant capacity, V, in terms of cycle time (x) and growth rate (D) , we get

$$T_A = \frac{k_C(xD)^a}{1 - e^{-rx}} \tag{7.5}$$

The optimum cycle time can be conveniently obtained by taking logarithms, differentiating $\ln(T_A)$ with respect to x, and equating to zero.

$$\ln\left\{\frac{T_A}{k_C D^a}\right\} = a\ln(x) - \ln(1 - e^{-rx}) \tag{7.6}$$

differentiating

$$\frac{d(\ln(T_A))}{dx} = \frac{a}{x} - \frac{re^{-rx}}{1-e^{-rx}} = 0 \tag{7.7}$$

and re-arranging,

$$a = \frac{rx}{e^{rx}-1} \tag{7.8}$$

Equation 7.8 is a remarkably simple relationship which defines the optimum cycle time. For our idealised situation, the cycle time is independent of the level of capital investment, sales income and variable cost. It depends only on the Economy of Scale factor and r, the discount rate.

It must be remembered that this simple relationship has been obtained by assuming the criterion was the minimisation of total discounted costs. This corresponds to taking NPW as the economic criterion and not DCF. If DCF is the criterion, then no such simple derivation is possible, and all investigations have to involve trial error with an evaluation computer program.

Figure 7.3 plots values of x against discount rate with the Economy of Scale factor as the parameter. These curves are excellent initial guidance to capacity expansions since they summarise the parameters of major importance for <u>arithmetic growth when the demand must be met</u>.

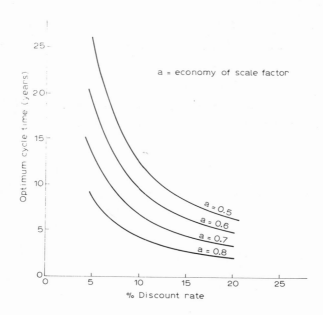

<u>Figure 7.3</u>

<u>Optimum Cycle-Time for a certain, arithmetically increasing</u>
<u>demand which must be met</u>

We can now look at the effect of tax, inflation and fixed costs on the optimum time. Tax rate has an effect on income (which does not affect optimum plant size), and the effect of lowering capital investment by a constant proportion, (see Capital Tax Factor Calculation in Section (1.6b)). Hence Equation 7.1 becomes:

$$C = fk_C(V)^a \qquad\qquad (7.9)$$

where f is the tax factor which allows the total tax rebate on depreciation. Hence k_C in the previous analysis is replaced by fk_C. However, since k_C does not appear in equation 7.8, which defines the cycle time, we can conclude that tax rate changes do not affect the optimum plant size, though of course they do affect the economic picture of that plant and the level of the total discounted costs (T_A) at the optimum.

Inflation, if treated as described in Chapter I, -- i.e. taking a constant value money basis and assuming it applies equally to all costs, -- affects only the credit returned as depreciation tax allowance. This, like tax, has an effect on the capital cost. Therefore, like tax, we can conclude that changes in inflation rate do not affect the optimum plant size. This presupposes that if money is not invested in a plant it is stored in a way which is protected against inflation.

Annual Fixed Costs are generally proportional to plant capacity in the same way that capital costs are:-

$$F = k_f(V)^b \qquad\qquad (7.10)$$

Hence, the total fixed costs for one cycle of x years is:

$$\int_{t=0}^{x} k_f(V)^b e^{-rt} dt = k_f(V)^b \frac{1}{r} (1 - e^{-rx}) \qquad\qquad (7.11)$$

The recursive equation 7.4 then becomes:

$$T_A = k_c(V)^a + k_f V^b \frac{(1 - e^{-rx})}{r} + T_A e^{-rx} \qquad\qquad (7.12)$$

Substituting Dx for V, and re-arranging, we get:

$$T_A = \left(k_c D^a x^a + k_f \frac{D^b x^b}{r} (1 - e^{-rx})\right) \frac{1}{1 - e} -rx \qquad\qquad (7.13)$$

in place of equation 7.5.

From equation 7.13 it can be seen that neither the plant costs k_c nor the growth rate, D, will disappear from the expression for the minimum point, and so now our optimum is influenced by these two factors. The optimum will also be affected by the fixed costs factor, k_f, the fixed costs Economy of Scale factor, b, as well as the capacity Economy of Scale factor, (a), and discount rate, (r) of equation 7.5.

The optimum point for equation 7.13 can be obtained numerically for any particular set of data using a single variable optimisation routine (13). Alternatively, the optimum can be obtained for any specific set of data using a computer program such as PLADE, in the way described in Section 7.9. As an example, PLADE was used to determine the optimum cycle time of a project with high fixed costs (around 40% annually of the capital cost), and these fixed costs had no Economy of Scale (i.e. b = 1.0). The optimum cycle time was found to be 15 years. This is considerably different from the optimum cycle time of 5 years when fixed costs are excluded. If a = b or k_c is large compared with k_f/r it is safe to apply the results of the simple treatment summarised by Figure 7.3. But clearly in situations when the

fixed cost is independent of the plant size, -- in chemical production
supervision for instance, -- the optimum plant size can differ significantly
from the results of the simpler treatment.

Concluding this section on arithmetically increasing demand when demand
must be met, we can say that the optimum cycle time between investments is
dependent only on the Economy of Scale factor, the discount rate and fixed
costs unless the fixed costs do have an identical Economy of Scale factor
as the capital cost. The sales price, variable cost, tax rebate, inflation
rate and capital costs, and demand growth rate do not influence the optimum
cycle time. The influence of fixed costs can be very significant, particularly if the total discounted fixed costs that are incurred are similar in
magnitude to the capital costs.

7.3 PLANT SIZE FOR ARITHMETICALLY INCREASING DEMAND WHICH NEED NOT BE MET

Manne extended Chenery's original work to allow for demand to be
greater than capacity. Allowing demand to outstrip capacity is a realistic
situation, particularly when one is considering national production of
goods when importation is a perfectly valid activity. For the country
concerned there is an additional cost because the cost of the imported goods
will be higher than the marginal cost of production at its own plant. The
cost of not supplying demand will simply be (import price -- marginal plant
costs) per ton not supplied. Since costs are incurred from supplying
over-capacity in early years to meet the demand at the end of the investment
period, we have an optimisation problem to determine the economic amount of
shortfall to allow before the next plant is built.

In company, rather than national activities it is more difficult to
discuss the costs of shortfall. For a company this cost has two parts;
firstly, there is simply the loss in profit, as with the national case
described above; secondly, because of competition, there is a further
"penalty cost" to consider. This is a cost equivalent to the bad will and
permanent transfer of customers to competitors. This second cost is very
difficult to estimate, and since this penalty cost can be one of the most
significant parameters in the optimisation study, a rough guess is very
dangerous.

In the following analysis the cost incurred through not meeting demand
is taken as the loss in profit. The second penalty cost should only be
included if it can be reliably estimated.

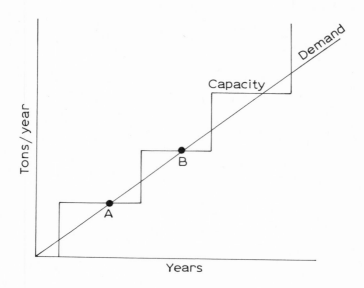

Figure 7.4
Arithmetically Increasing Demand Which Does Not Have To Be Met

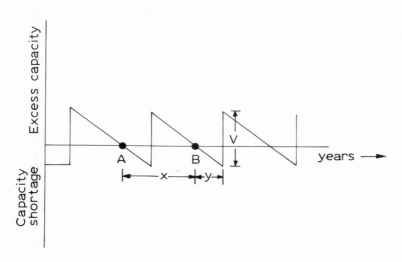

Figure 7.5
Time Path of Excess Capacity for Figure 7.4

Consider Figure 7.1 but now assume that shortfalls are admissable; this results in Figure 7.4, and the corresponding time path of excess capacity is given by Figure 7.5 Again this is a repeated infinite series, and so we can make an analysis similar to the first case to determine total costs by developing a recursive equation. If we again consider one cycle from recursive point A to recursive point B, the discounted costs at point A, T_A, are given as the discounted cost of installing the plant at time Y on Figure 7.5 (defining point A as zero time), plus the summed discounted cost of not meeting the demand, as shown by the area under the line in Figure 7.5, plus the discounted cost at point B to infinity, (T_B) which, as it is an infinite series is equal to T_A.

Hence,

$$T_A = k_C(Dx)^a \, e^{-ry} + \int_{t=0}^{t=y} pDte^{-rt}dt + T_A \, e^{-rx} \tag{7.14}$$

where: t refers to time, and p to cost/ton penalty of not meeting demand. Other symbols correspond to the treatment given in Section 7.1. By re-arranging and integrating we get:

$$T_A = \frac{1}{1 - e^{-rx}} \left(\frac{pD}{r^2} (1 - e^{-ry} (1 + ry)) + e^{-ry}k_C(Dx)^a \right) \tag{7.15}$$

by dividing through by $k_C D^a$

$$\frac{T_A}{k_C D^a} = \frac{1}{(1 - e^{-rx})} \left(\frac{pD}{k_C D^a r^2} \quad (1 - e^{-ry} (1 + ry)) + e^{-ry}x^a \right) \tag{7.16}$$

This equation is more complicated than the "no-shortfall" case, equation 7.5, and it does not really lend itself to easy differentiation to find the optimum values of x and y, -- the cycle time and time for which shortfall is allowed.

Manne and Erlenkotter have both evaluated minimum points for this equation and published their results as tables. The following parameters are important; the Economy of Scale factor, (a) and Discount Rate (r) as in the no-shortfall case, the factor $(pD)/k_C D^a$ which relates the total shortfall cost to the plant capital costs.

Some of their results are displayed in Table 7.1. These show the effect of the changes in the ratio $pD/k_C D^a$, for different interest rates on capital. An Economy of Scale factor of 0.6 has been assumed and the demand normalised to unity.

Table 7.1

Effect of Cost of Short-fall on Optimum Cycle Time

(x*) and Years of Short-fall (y*)

Minimum point of Equation 7.16, (a = 0.6)

Discount Rate r		Cost Function $pD/k_C D^a$				
		0.05	0.10	0.25	0.50	∞ demand satisfied
0.05	x* =	25.0	21.7	20.0	19.5	19.0
	y* =	6.9	3.1	1.2	0.6	0
0.10	x* =	20.3	13.7	10.9	10.2	9.5
	y* =	12.2	4.3	1.7	0.8	0
0.15	x* =	25.5	12.5	8.2	7.2	6.3
	y* =	21.0	6.9	2.1	1.0	0
0.20	x* =	39.0	13.5	7.0	5.7	4.7
	y* =	36.0	9.7	2.6	1.1	0

Before Table 7.1 is useful, we must understand the significance of the cost function, $pD/k_C D^a$.

Consider a product with a market increasing at the rate of 100 tons a year. The capital cost of a 500 ton-plant is £1,000,000, hence from equation 7.1.

$$k_C = 1 \times 10^6 / 500^{0.6} = 24022$$

and so

$$k_C D^a = 24022 \times (100)^{0.6} = 0.380 \times 10^6, \text{ (assuming } a = 0.6)$$

Let us further assume that the process has a marginal income (selling price - variable cost) of £500 per ton.

Hence pD = 500 × 100 = 50,000

This gives the value of the cost function as:

$$pD/(k_C D^a) = 50,000/(0.380 \times 10^6) = 0.131 \qquad (7.17)$$

Assuming also that the discount rate is 10%, then Table 7.1 shows the optimum cycle time to be about 13 years, with the plant being unable to supply demand for four of those years.

This may be the mathematical optimum but a firm that is content to turn customers away for one third of the time for a product giving a 25% return of invested capital is not seriously in the business, and customers may not return when the expanded plant becomes available. In other words, this policy may change the original demand curve -- Figure 7.4.

If it is considered that to turn one customer away unsatisfied has a repercussion equal to perhaps four times that sales loss, then C becomes four times bigger and our cost function is 0.53. Table 7.1 now shows the optimum cycle to be ten years, with 0.8 years in which demand will not be met. Table 7.1 is interesting in that it summarises the economic incentive, or otherwise, of meeting demand. However, it must be remembered that it is based on idealised conditions -- an arithmetically increasing market to infinity, -- and there is considerable difficulty presented in determining the proper cost of not meeting demand.

7.4 PLANT SIZE FOR GEOMETRICALLY INCREASING DEMAND WHICH MUST BE MET

Srinivasan (16) considered the case where demand growth was geometric, not arithmetic as described in Section 7.2. He showed that, as with the arithmetic case, it was optimum to construct plants at equal time periods, which means that each plant is bigger than the preceding plant. With the time period x, the geometric yearly increase g, and the first plant size V, then the following plant sizes must be:

$$Ve^{gx}, \quad Ve^{2gx}, \quad Ve^{3gx}$$

installed at times:

$$0, \ x, \ 2x, \ 3x.$$

The total discounted capital cost, T, is therefore given as:

$$T = \sum_{n=0}^{\infty} e^{-nrx} \, k_c (Ve^{ngx})^a$$

$$= k_c V^a \sum_{n=0}^{\infty} e^{-n\,(r-ag)x}$$

$$= \frac{k_c V^a}{1 - e^{-(r-ag)x}} \tag{7.18}$$

The value of x which results in minimum costs, T, as given by equation (7.18), is the optimum investment cycle time when growth is geometric. Srinivason tabulates some numerically determined optimum values (16) for

10% growth rate (g=0.10) and compares them with the optimum investment cycle time for arithmetic growth.

For example; when a = 0.6, and r = 0.10 the optimum geometric growth cycle time is 7.6 years, whereas the corresponding arithmetic growth cycle time is 9.5 years. What is more interesting is that the error involved in employing 9.5 instead of 7.6 is only equivalent to 1.1% of the capital cost. In short, it is often adequate to employ the optimum time period for the arithmetic growth, even if it is suspected that the growth may be geometric, at least with growth rates of less than 10% per annum.

Berretta and Mobasheri (1) combined Srinivasan's geometric model with Manne's model allowing short-fall, and applied it to a study of Aluminium Production in Argentina. The paper is an interesting study of the application of these methods but it is clear that they are more appropriate to determining national economic policy for basic industries in developing countries than to providing a technique to be applied by individual companies in competitive situations.

7.5 PLANT SIZE FOR A CHARACTERISTIC GROWTH CURVE

Most products do not exhibit steady growth to an infinite horizon, but follow a characteristic S curve which flattens out as the market saturates (Section 2.2b). In most cases therefore, the problem is that of determining whether there should be one, two, three or more production expansions to reach the saturation level, rather than of determining the cycle time for an infinite series.

Consider a product with sales arithmetically increasing to a plateau of P tons per year, reached over N years. Assume also that it has been decided as a matter of policy to meet the capacity P by a number of identically-sized plants which will be so installed that the sales demand will always be met. The problem reduces itself to finding out how many steps should be taken over the life of the project. This situation is summarised in Figure 7.6.

Since sales demand must be met, we can ignore income and the optimum is given by the proposal with the least cost. Assuming that fixed costs are directly proportional to capital costs, and equation (7.1) holds for capital cost, then we can write down the total discounted capital costs, T_i, for the three cases using continuous discounting formula, as follows:

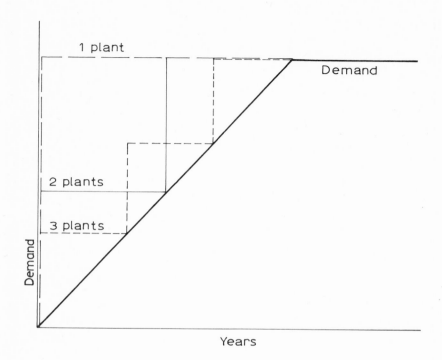

Figure 7.6

Sales Linearly Increasing to a Plateau -

Satisfied by 1, 2 or 3 equal-sized steps

in Plant Capacity

Case 1 :   1 plant

$$T_1 = k_c P^a \tag{7.19}$$

Case 2 :   2 plants

$$T_2 = k_c \left(\frac{P}{2}\right)^a + e^{-rN/2} k_c \left(\frac{P}{2}\right)^a = \frac{k_c P^a}{2^a} (1 + e^{-rN/2}) \tag{7.20}$$

Case 3 :   3 plants

$$T_3 = \frac{k_c P^a}{3^a} (1 + e^{-rN/3} + e^{-2rN/3}) \tag{7.21}$$

Since $k_c p^a$ is common to all the equations, the total costs for the three cases are proportional to

$$1, (1 + e^{-rN/2})/2^a, \text{ and } (1 + e^{-rN/3} + e^{-2rN/3})/3^a$$

These expressions have therefore to be evaluated and that giving the lowest cost identifies the number of expansion steps required. Fairly understandably, the plateau level does not influence the result; the number of steps depends primarily on N, the number of years to reach the plateau, and a and r, the Economy Scale factor and discount rate.

Taking as an example a 15% discount rate and a 0.6 Economy of Scale factor, it is best to install only one plant if the growth period is less than 9 years and two plants if it is not more than 15 years. For slower growth rates, more than two plants are sensible, -- the exact number can be obtained by a simple extension to the treatment here.

This analysis is recommending that plants should not be installed to cover less than five years growth or more than nine; this is in accord with the analysis assuming unlimited growth, which gave 6.3 years as the optimum interval between plants under these conditions, (see Figure 7.3).

Table 7.2 summarises the number of plants required for different values for the Economy of Scale factor, discount rate and growth period.

Table 7.2

Number of Plants for an Arithmetic Growth followed

by a Plateau

Economy of Scale Factor 'a'	(Years of Arithmetic Growth)x(Discount Rate) Nr				
	0.5	1.0	1.5	2.0	3.0
0.5	1 plant	1 plant	1 plant	2 plants	2 plants
0.6	1 plant	1 plant	2 plants	2 plants	
0.7	1 plant	2 plants	2 plants		

Coleman and York (6) described a more sophisticated treatment of a growing market in which they assumed a geometric growth rather than linear growth up to the plateau. This is justified in that it more nearly represents the Gomperz S-shaped growth curve which characterises demand

growth (See Chapter II, Section 2.2a). This is certainly reasonable for
the beginning part of the curve, but does introduce errors in that, where
the Gomperz curve is tailing off to the plateau, the geometric growth is at
its maximum, as shown by Figure 7.7.

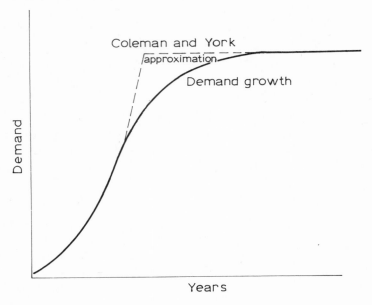

Figure 7.7
Coleman and York's Approximation to the Growth of a Product

They developed an analysis for the situation similar to the one already
described for the linear growth case, but, since equal-sized plants are
wrong in a geometric growth situation, their treatment also optimised the
size of each step, maintaining the conditions that the market must be met
and that the final capacity must equal the plateau capacity. The method
consisted of writing down the equations corresponding to (7.19), (7.20),
and (7.21), which now includes different capacity plants in the same
expression. By equating partial derivatives to zero, and solving the
resulting simultaneous equations, they obtained expressions for optimum
plant sizes which could then be fed back into the total cost expressions.
The final step was to choose the number of expansion steps that gave the
lowest cost -- exactly as the previous example.

This will not be described in more detail here because the method is
tedious to apply and the major achievement of the work -- to optimise each
individual plant size -- is unnecessarily pedantic.

By assuming equal cycle times, as recommended in Section 7.4, 2-,3-,
and 4-stage expansions can be proposed by inspection and evaluated by any
evaluation program. Using this technique for the example quoted in Coleman
and York's paper, with the PLADE program, results displayed in Table 7.3
were obtained.

Figure 7.3
Comparison of Coleman and York's Method

Proposal	Sizes	Total Discounted Costs
Coleman/York "optimised" proposal	1900, 1000, 1500, 1600	139,430
4 equal cycle times	1400, 840, 1490, 2270	138,169
3 equal cycle times	1640, 1530, 2830	139,865
2 equal cycle times	2280, 3720	146,140

There is agreement in that both techniques propose that a 4-stage
expansion policy is better than a 3-stage one, but the simple constant
cycle time assumption produces a proposal with 0.9% lower costs. Clearly,
differences are so small that it is not worth introduing complex techniques
to locate the exact optimum. It does suggest, however, that there may be
a mistake in the Coleman and York paper.

7.6 PLANT SIZE FOR AN UNCERTAIN GROWING DEMAND

Manne extended his idealised arithmetic infinite growth case to
include uncertainty. To do this he used a Markov Process to give a probabilistic growth pattern which gave a mean linear growth and an attended
distribution to describe future demand (10).

His overall conclusion was that uncertainty increases the size of the
optimum capacity increment, and that when the market must be satisfied, the
discounted costs at the optimum are higher than the deterministic case.
Manne further developed the analysis to allow demand not to be met by
introducing a loss or demand not satisfied, which resulted in a complex
expression that could only be solved numerically. The conclusion in this

case was that, as before, uncertainty increased the size of the optimum capacity increment, but now the system had become too complicated to enable a simple statement to be made concerning the total cost at the optimum.

Manne can be credited as the first investigator of this problem, but his approach gives rise to difficulties in that it requires parameters to define the penalty cost and the Markov chain, which are difficult to obtain, and also the treatment is far from simple. Section 7.9 will show that the effect of uncertainty can be adequately evaluated more conveniently with the PLADE program without confusion over the meaning of the parameters.

7.7 THE SENSITIVITY OF THE OPTIMUM

In practice, it is as important to know the sensitivity of the optimum as to know the optimum itself. All decisions are a compromise of quantitative and non-quantitative factors, and to know whether the optimum is flat or sharp affects the weighting of the non-quantitative factors in the balance leading to the decision. For example, the advantage of common design, common spares and common operating experience associated with two identical plants may well be worth 5% of the plant capital cost. If a study proposes two different plant sizes as the optimum solution to a two-stage expansion, it may still be worth installing identical sizes, if the optimum is sufficiently flat for the cost incurred above the minimum to be small.

Taking the "linear growth to infinity with the market being satisfied" case as an example, the total discounted cost for any plant size is given by equation (7.5). This equation can therefore be used to evaluate the cost of any proposed plant size. The results from this case at least give some guidance for cases not involving infinite growth, or not linear growth.

Equation (7.5) has been evaluated over a range of plant sizes, and, for demonstration purposes, the results have been scaled to show increased costs as a percent of capital, plotted against the ratio of plant size to optimum plant size. This is shown in Figure 7.8.

This optimum is very "flat"; for example, a 10% deviation from the optimum capacity produces an extra cost of only 0.16%! Clearly, the decision-maker has very considerable scope for adjusting the results of any optimisation in a growing market to fit qualitative considerations such as rationalisation of plant size, availability of existing equipment and so on.

Figure 7.8 shows that, in a steady growing market, doubling or halving the plant size recommended by the optimum incurs total cost increases of less than 10% of the capital cost.

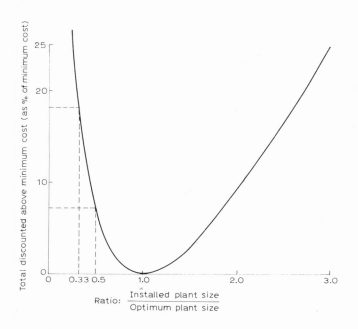

Figure 7.8

Sensitivity of Discounted Total Capital Costs to Plant Size
in an infinitely growing market

The sensitivity of the total cost to errors in the Economy of Scale
factor (a) and assumed applicable discount rate (r) are also of interest.
Assume an optimum plant size was determined, but the Economy of Scale factor
was 0.75 instead of 0.6; what would the cost of this error be?

Similarly, an error in the assumed applicable discount rate would
introduce an additional cost. Table 7.4 shows the % increase in cost
through an error in the estimation of (a) or (r), determined again from
equation (7.5).

Again, the costs are not very sensitive to errors; although big
differences are given for the optimum plant size, these correspond to very
small increases in cost because of the shape of Figure 7.8.

Table 7.4

Sensitivity of Cost of Decision to Errors in
Discount Rate and Economy of Scale Factor in
a Growing Market

Mean Point: r = 0.15, a = 0.6.

	Actual Optimum Plant Size	Size obtained by optimisation using data with error	Increase in capital cost through using non-optimum plant size (as % of optimum capital cost Figure 7.8)
0.05 error in discount rate (0.15→0.10)	6.32	9.47	3.0%
0.15 error in scale-up factor (0.6→0.75)	6.32	3.7	4.0%

7.8 REDUCTION OF PLANT SIZE AS A FORM OF INSURANCE

This insensitivity of the optimum provides a technique for minimising
risk by building small plants and later expanding if demand is maintained.

For example, if a small company were considering a plant for a rapidly
growing product which had a 90% chance of a good future, but there remained
a 10% chance that the produce would be replaced by something completely
different, then it might not be in a position to accept the 10% risk of a
complete failure. With an Economy of Scale Factor of 0.6 and an interest
rate of 15%, the optimum plant size would be to satisfy 6 years' demand.
This 10% chance of a disastrous loss could be effectively removed by
building for only 3 years' demand, then adding the second 3 year plant if
the market requires it in 3 years' time. Figure 7.8 shows that the cost
of being off the optimum would be equivalent to an increased capital cost
of 7%.

This can be considered to be the cost of a single payment insurance
policy against the 10% chance of the product becoming superseded. Looking
at it in this way, the manager can decide subjectively "would I pay 7% of
the capital cost as an insurance to cover that risk?" Alternatively the
decision could be taken much more analytically using Utility Functions.

The answer to this question is dependent on the particular situation.
If it is a new product with a high profit margin, the 7% premuim is a good

buy. If it is a more basic product this 7% may be an appreciable fraction
of the expected profit. It must always be remembered that the end result
is a 90% chance that the market will be maintained, and the conservative
firm will be left with a series of small plants which are less efficient
than the competitors, -- which is in itself a risky situation.

7.9 THE PLADE PROGRAM FOR PLANT CAPACITY STUDIES

One obvious way to determine the optimum capacity for a steadily
increasing demand is to take a number of possible plant sizes, determine
their capital and fixed costs, propose them as separate projects, and
evaluate these projects. Graph paper and pencil will then enable the optimum
to be determined by interpolation.

This offers a series of advantages:

a) Any forecast Demand Curve can be considered:
    it need not be arithmetic, geometric or a Gompertz
    curve, and there is not the problem of choosing
    an approximate form and afterwards worrying about
    the errors introduced by the approximation.

b) Any price and costs profiles can be included.

c) The case can be evaluated for the more realistic
    non-infinite lifetime situation.

d) The graph is a sensitivity analysis about the
    optimum point.

e) The profitability of the project has been
    evaluated at the same time.

The evaluation of each project can be made by any evaluation program once
the year of installation of each extension is given.

If demand does not have to meet capacity, or if probabilistic demand
data is involved, a problem arises because it is no longer possible to
determine the year of each expansion by inspection of the Demand curve.
However, the PLADE planning phase is capable of generating the years for
each capacity extension -- the optimum plans -- and so, by employing PLADE,
we can use the same technique for determining the optimum plant size. In
fact, by using the Keyword for repeated examples, the optimum plant size
can be determined by one 20 second computer run.

Whereas the methods described earlier in this chapter are good for
giving an appreciation of the factors involved, and for generating rules
of thumb, when the optimum size for a specific project is required, this

is best done as a specific study. Besides the five advantages enumerated at
the beginning of this section for deterministic forecasts where the demand
will be met, the following further advantages can be listed if this specific
study is extended to uncertain sales cases or to cases where there is a
finite cost associated with a shortfall:

    f)   Probabilistic data can be defined in a meaningful way,

    g)   Cost of not meeting demand is clearly calculated by
        the procedure and no difficult-to-define ratios are
        involved.

    h)   The same procedure can handle every case described
        in Sections 7.2 to 7.7, which leads to greater
        overall consistency.

Example:

Consider looking for an expansion policy for a product with a forecast
uncertain demand, shown by Figure 7.9. This has a mean sales forecast of a
4%/year geometric growth, but with the optimistic forecast showing a distinct
tendency to level off. A price forecast is available, -- there is certain
to be a fall over the next 20 years in constant value money terms as shown
by Figure 7.10. The cost of capital is 15%, and the Economy of Scale
factor is 0.7.

We must now produce some projects, each with a distinct expansion
policy. With a = 0.70 and r = 0.15, Figure 7.3 suggests a cycle time of
4.5 years might be the right order of magnitude. Section 7.4 concludes
that we can take this cycle time for the geometric case, and this gives
a series of increases in capacities. This example differs from the
idealised case in that probabilistic data are presented, fixed costs are
included, and the demand does not match the idealised assumptions.

The importance of these individual effects is difficult to assess,
and rather than modify the figure of 4.5 for each effect, it is better
to simulate the alternative proposals with PLADE. It would seem reasonable
to propose cycle times 3.0, 4.0, 5.0,6.7 and 10 years with respect to the
mean sales forecast for these studies. Notice that the larger plants have
been chosen to given an integer number of extensions in the 20 year life.
Though this is not essential, it produces a smoother curve to analyse.
Without this constraint the results show a small irregular additional cost
whenever excess capacity is present at the end of the project.

Table 7.5 summarises the proposals and the associated Capital and
Fixed Costs.

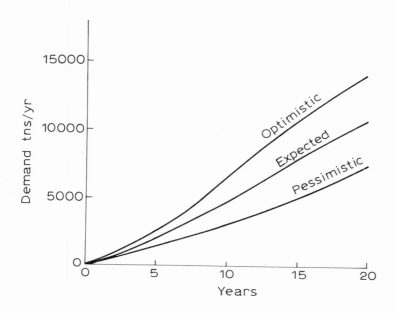

Figure 7.9
Demand Forecasts for Example of PLADE for determining Plant Size

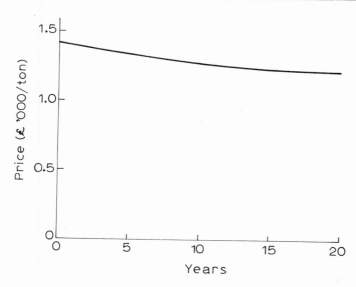

Figure 7.10
Price Forecast for Example of PLADE for determining Plant Size

Plant and Extension Name	Capacity	Capital Cost	Fixed Costs
Proposal 1; 3yr cycle			
3yr,1	1050	2850.8	335.4
3yr,2	1450	3573.5	420.4
3yr,3	1600	3828.5	450.4
3yr,4	1950	4397.1	517.3
3yr,5	1800	4157.5	489.1
3yr,6	1800	4157.5	489.1
3yr,7	1050	2850.8	335.4
Proposal 2; 4yr cycle			
4yr,1	1500	3659	430
4yr,2	2050	4553.7	535.7
4yr,3	3100	6082.6	715.6
4yr,4	1800	4157.5	489.0
4yr,5	2250	4860.3	571.8
Proposal 3; 5yr cycle			
5yr,1	2000	4475.7	526.5
5yr,2	2750	5593.4	658.0
5yr,3	3100	6082.6	715.6
5yr,4	2850	5734.9	674.7
Proposal 4; 6.7yr cycle			
6.7yr,1	2830	5707	671.4
6.7yr,2	4020	7296.2	858.4
6.7yr,3	3850	7078.9	832.8
Proposal 5; 10 yr cycle			
10yr,1	4750	8200	965
10yr,2	5950	9601	1129

Table 7.5

Example of Use of PLADE to determine Plant Size
- Proposed Plant Sizes and Costs

We now have all the necessary data to run PLADE and obtain the results. We know from Section 7.4 that for each project the plants will be installed in the inverse order of their capacities. The PLADE Planning Search does in fact verify this if it is allowed to try all possible combinations, but, to reduce computer time, we will define the order in which they will be installed by defining an extension structure. This drastically reduces the

```
HEAD
EXPANSION POLICY EXAMPLE
DISCOUNT
15
TAX
25 10
LIFE
20 10
PROBABILITY
0.33 0.34 0.33
DEMAND
   200   800  1400  2000  2600  3300  4000  4800  5500  6400
  7400  8300  9100  9900 10700 11500 12300 12900 13500 14300
   200   600  1050  1500  2000  2530  3000  3550  4100  4750
  5400  6050  6650  7250  7810  8450  9100  9550 1.155 10700
   200   400   700  1000  1400  1700  2000  2330  2700  3100
  3400  3800  4200  4600  5050  5400  5900  6400  6800  7400
PRICE
1.40 1.38 1.37 1.35 1.34 1.32 1.31 1.30 1.29 1.28
1.28 1.27 1.26 1.25 1.24 1.24 1.23 1.23 1.23 1.23
```

```
CORRELATION
0.5  0.5
BOUNDARIES
1.0  2.0
ALTERNATIVES
3YR,1     1050 2850.8 335.4 0.25
3YR,2     1450 3573.5 420.4 0.25
3YR,3     1600 3823.5 450.4 0.25
3YR,4     1950 4397.1 517.3 0.25
3YR,5     1800 4157.5 489.1 0.25
3YR,6     1800 4157.5 489.1 0.25
3YR,7     1050 2850.8 335.4 0.25
END
EXTNS
3YR,2     3YR,1
3YR,3     3YR,2
3YR,4     3YR,3
3YR,5     3YR,4
3YR,6     3YR,5
3YR,7     3YR,6
END
PLAN
1
END
ALTERNATIVE
4YR,1     1500 3653 430 0.25
4YR,2     2050 4553.7 535.7 0.25
4YR,3     3100 6082.6 715.6 0.25
4YR,4     1800 4157.5 489 0.25
4YR,5     2250 4860.3 571.8 0.25
END
EXTNS
4YR,2     4YR,1
4YR,3     4YR,2
4YR,4     4YR,3
4YR,5     4YR,4
END
PLAN
1
END
ALTERNATIVES
5YR,1     2000 4475.7 526.5 0.25
5YR,2     2750 5593.4 658.0 0.25
5YR,3     3100 6082.6 715.6 0.25
5YR,4     2850 5734.9 674.7 0.25
END
EXTNS
5YR,2     5YR,1
5YR,3     5YR,2
5YR,4     5YR,3
END
PLAN
1
END
ALTERNATIVES
6.7YR,1   2820 5707 671.4   0.25
6.7YR,2   4020 7296.2 858.4 0.25
6.7YR,3   3850 7078.9 832.8 0.25
END
EXTNS
6.7YR,2   6.7YR,1
6.7YR,3   6.7YR,2
END
PLAN
1
END
```

```
ALTERNATIVES
10YR,1    4750 8200 965 0.25
10YR,2    5950 9601 1129 0.25
END
EXTNS
10YR,2    10YR,1
END
PLAN
1
END
STOP
```

Figure 7.11

Example of PLADE
to Determine Plant Size

number of possible combinations to be evaluated and so cuts down computation
time by a factor of 10. This is done by defining, for example, Plant 4yr5,
as an expansion of 4yr4 which is an extension of 4yr3, which is an extension
of 4yr2, which is an extension of parent plant 4yr1 for the 4-year cycle
case. This technique is also worthwhile when the extensions are of identical
size.

Figure 7.11 shows the Data Input for the computer run. Notice that by
the arrangement of Alternative and Plan Keywords, all four projects can be
evaluated in one run. The five Expected NPW's from the five projects are
plotted on Figure 7.12. The curve drawn through the points locates the
optimum cycle time as 4.8 years, -- the major objective is accomplished.

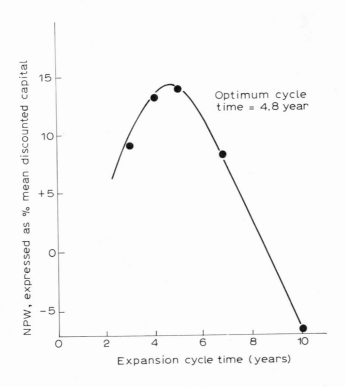

Figure 7.12
Use of PLADE to determine Plant Size -
Location of Optimum Cycle Time

Figure 7.12 is again presented as NPW/Discounted Capital because it then shows immediately the significance, as a percentage of the capital cost, of being away from the optimum. This figure clearly locates the optimum at 4.8 years and shows the shape of the curve. Notice that a 20% error in cycle time produces an additional cost of only 1.5% of the capital cost. This sensitivity has already been discussed in Section 7.6 and shown on Figure 7.8.

Any comparison of the result of 4.8 with the original estimate of 4.5 is difficult because this example, to show how a practical case can be evaluated, has departed from the ideal situation in so many respects. The overall effect, by chance, is very close to the original estimate. The same technique can be used to identify the contribution of the individual points -- fixed costs, uncertainty, non-ideal growth -- but space does not allow a detailed discussion here.

7.10 OPTIMUM SIZE OF PLANT WHEN LONG TERM STORAGE IS ALLOWED

Consider manufacturing a product when it is company policy always to meet demand and rented warehousing space is available. Clearly, if product made when capacity is greater than demand is stored, then this store can be fed into the market when demand exceeds capacity. This will delay the date of the next investment in capacity and therefore will result in appreciable savings association with the interest payments on the capital. This saving will be £Cr per year where C is the capital being delayed and r is the interest rate. Figure 7.13 indicates this diagramatically, assuming arithmetic growth.

The demand increases at the rate of D tons per/yr^2 and a delay of y years in capital investment has been secured by warehousing. Assuming that the warehousing space can be rented in stages, at a cost of £s per year for space per 1 ton of product, and letting the working capital per ton of product be £w, the total cost of warehousing 1 ton for 1 year is therefore £(s+wr).

Consider an element of time dy at time y (time zero being defied as the time when capacity meets demand). The product quantity met from storage is Dy dy and this has had to be stored for 2y years, so occurring a storage cost dT_s given by:-

$$dT_s = 2y(s+wr)Dy \, dy \qquad (7.22)$$

The optimum length of time to warehouse is given by the point where the rate of costs/per year of warehousing becomes equal to the rate of

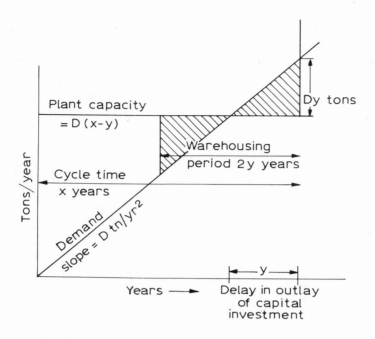

Figure 7.13

Demand and Capacity with long-term Warehousing

interest charges. At this point:-

$$2y^2D \ (s + wr) = Cr \tag{7.23}$$

therefore $\qquad Dy = \sqrt{\dfrac{CrD}{2(s + wr)}} \tag{7.24}$

Equation 7.24 gives the optimum quantity of product (Dy) to warehouse in order to delay investment costs. The optimum delay is y years. This equation applies until half the total cycle time. y greater than half the cycle time is physically impossible, but this constraint is not recognised by equation 7.24.

A more difficult problem is to determine the optimum plant size when one also considers delaying the investment by renting warehouse space. If we re-draw Figure 7.13 in terms of excess capacity against time, we obtain Figure 7.14

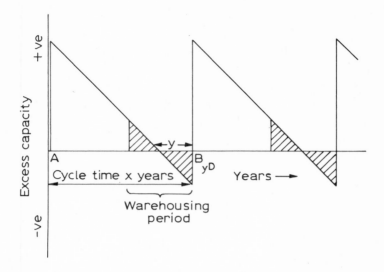

Figure 7.14

Excess Plant Capacity vs Time with long-term Warehousing

    This figure resembles Figure 7.5, which represents the plant operation
when a penalty cost p is incurred for every ton demanded but not supplied.
The tons not supplied per cycle were the hatched area under the axis
-- $\frac{1}{2} y^2 D$ tons, and the incurred cost $£\frac{1}{2} y^2 Dp$.

    The case we are now considering assumes demand must be met, but now the
area under the axis has incurred a warehousing cost. This warehousing cost
incurred at time x (point B) due to this warehousing policy is:-

$$\int_0^y 2 D (s + wr)y^2 dy = \frac{2}{3} D (s + wr)y^3 \qquad (7.25)$$

    As done in section 7.2 and 7.3 we can now write down the recursive
equation for an infinite series as:-

$$T_A = k_c (D(x-y))^a + \frac{2}{3} D (s+wr)y^3 e^{-rx} + T_B e^{-rx} \qquad (7.26)$$

and since

$$T_B = T_A,$$

$$T_A = [k_c (D(x-y))^a + \frac{2}{3} D(s+wr)y^3 e^{-rx}]/(1-e^{-rx}) \qquad (7.27)$$

    The minimum point for T_A with respect to both x and y gives the optimum
plant size and warehousing policy. Any analytic solution to the equation 7.27

Table 7.6

The effect of including Product Warehousing on the optimum Plant Size

Data	Product A (expensive product)	Product B (cheap product)
Growth rate (D) tn/yr.	20 tn/yr	10,000 tn/yr.
Economy Scale Factor (a)	0.6	0.6
Interest rate (r)	0.15	0.15
Working capital in product (w£/tn).	£500	£5
Warehouse rent costs (s£/tn year).	£10/tn year	£10/tn year
Capital cost of new plant	£200,000 for 100 tn/plant	£1,000,000 for 50,000 tn/plant
therefore:	k_c = 12619	k_c = 1515
Results: Case 1 (data as above)		
x	4.7	2.0
y	2.35*	1.0*
Case 2 (s = £20/tn year)		
x	4.5	5.25
y	2.25*	0.80
Case 3 (s = £1000/tn year)		
x	2.0	6.2
y	1.0*	0.12
Case 4 (as case 1 but using equation 7.24)		
when x	5	5
y	2.5*	0.8

* these cases, when $y=\frac{1}{2}x$, imply <u>always</u> working the plant at full output and storing the unsold product.

x Optimum Total Cycle time in years.

y Optimum Number of years when capacity is met partially from stored product.

```
      PROGRAM LAGER (INPUT,OUTPUT,TAPE1=INPUT,TAPE2=OUTPUT)
      DIMENSION X(3), H(3), W(3,7), CL(3), CH(3)
      COMMON/LAG/B(10)
      EXTERNAL FUNCT
C     B(1)=D, B(2)=A, B(3)=R, B(4)=W, B(5)=S, B(6)=KC
1     READ(1,50) B(1),B(2),B(3),B(4),B(5),B(6)
50    FORMAT(8F10.0)
      IF(B(1).LT.0.00000001) STOP
      WRITE(2,40) (B(I),I=1,6)
40    FORMAT( 29H INPUT DATA    D,A,R,W,S,KC         ,/6F15.4)
      X(1)=10.0
      X(2)=2.
      X(3)=X(1)-    X(2)
      M=250
      N=2
      NN=7
      IW=2
      IPR=0
      DO 70 I=1,3
      H(I)=0.2
      CL(I)=0.00001
70    CH(I)=25.
      NC=3
      CALL RSNBK(N,NC,X,H,CL,CH,M,IPR,F,FUNCT,W,NN,IW)
      Z=X(1)+ X(2)
      WRITE(2,100) F,Z,X(2)
100   FORMAT(//13H TOTAL COST=    ,F15.4,  13H CYCLE TIME=    ,F6.2,/
     1 24H EXTENSION BY STORAGE=    ,F6.2///)
      GO TO 1
      END
      SUBROUTINE FUNCT(N,X,F )
      DIMENSION X(N)
      COMMON/LAG/B(10)
      Z=X(1)+X(2)
      X(3)=X(1)-X(2)
      IF(X(1).LE.0.0.OR.X(2).LE.0.0.OR.X(3).LE.0.0)  GO TO 250
      A=EXP(-B(3)*Z)
      F=B(6)*(B(1)*X(1))**B(2)
      F=F + (2.0*B(1)*( B(5) + B(3)*B(4)) *    X(2)    **3.0*A) /3.0
      F=F/(1.0-A)
      RETURN
250   F=10000000000.0
      RETURN
      END
20.        0.6      0.15    500.    10.      12619.
```

Figure 7.15

Listing of Program LAGER for determining optimum

Investment delay by Long-term Warehousing

is difficult to envisage, but the minimum point can be located by a numerical method using a computer subroutine for constrained non-linear optimisation. Figure 7.15 gives the program for use with a Rosenbrock optimisation routine (13).

Table 7.6 gives examples of the optimum warehousing policy for two products -- product A being a small tonnage, expensive product, and B being a large tonnage, cheap product -- for a range of warehousing costs.

The most interesting feature of this table is the optimum storage policy. Operation of plant in the early years is so uneconomic that considerable warehousing is justified. For the small tonnage case it was always justifiable to operate the plant at full output and store all its product ($y = 0.5 \ x$). This policy was also optimum for the large tonnage when storage costs were low, resulting in suprisingly small expansion increments.

Clearly this study should include uncertainty since a policy decision to store product many years before it will be sold cannot be separated from the probabilistic aspects of the demand forecasts. This could be done, for instance, by determining the optimum policy for the mean sales forecast from equation 7.27, and then checking this policy for the case of the pessimistic sales outcome. Long term storage is advantageous when the demand is optimistic, but it adds further costs should the demand turn out to be less than the mean forecast. This policy, therefore, is increasing the risk associated with the project.

A second interesting point shown by Table 7.6 is that the effect of warehousing is to reduce the optimum "no-warehousing" cycle of 6.3 years down to, in some cases, as low as 2 years.

Equation 7.27 has two local optima, -- as can be seen by comparing Case 1 and Case 2 of product B on Table 7.6. Even when the optimum indicates a large plant, the second optimum of full warehousing and a small plant may only be slightly more costly , and it will probably offer reduced risk. In this way, long-term warehousing may be a way of reducing risk.

7.11 PLANT SIZE FOR ZERO DEMAND-GROWTH MARKETS

So far this chapter has considered only the situation of a growing market and has discussed the optimum step size to cover the demand. A not entirely uninteresting problem is the choice of plant size when the demand is expected to be constant.

If the demand, V tons/year, could be forecast with certainty, the plant could be so accurately designed that its capacity could be assured, the

plant could be certain not to break down and the product could be dispatched
from the plant at a constant rate, then the plant capacity should be V tons
in 8760 hours, and that is the end of the matter. However, these conditions
generally do not hold, and this gives rise to problems in determining the
plant capacity.

Since plants are not 100% reliable, historical data has to be used to
determine the availability to be expected from the plant. The availability
is generally between 90 and 99%, depending on the type of plant and the
philosophy used for the design of the plant. The availability is used to
determine the number of operating hours in the year. For instance, a plant
with an availability of 95% would be assumed to operate for

$$0.95 \times 8760 = 8322 \text{ hours}$$

Hence our plant would be required to produce V tons in 8322 hours.

7.11a Uncertainties in Demand and Capacity

Now consider that the actual demand level is not known with certainty.
The Expected Value for the demand is V, but it has a distribution with a
standard deviation σ_n. A plant of capacity $V + 2\sigma_n$ would have a 98% chance
of covering the demand -- that is, of being too big. But for a marginally
profitable project, where the economics are so poor that no excess capacity
can be tolerated, then a size $V - 2\sigma_n$ would be more sensible, since this
would give a 98% probability that excess capacity was not installed.

Clearly the choice of plant size when the demand is probabilistic
is related to the economics of the process.

Now consider that the actual plant capacity is not known with certainty
but this too has a distribution with a standard deviation σ_p. This case is
synonymous with the distribution in the demand case. A very profitable project
would justify a safety factor of $2\sigma_p$, whereas a break-even one, which was only
profitable when capacity was fully made use of, might only justify one of $-2\sigma_p$.

Section (7.12) discussed these effects and shows how plant safety
factors, to cover combined uncertainties in designed and market forecast,
can be determined. Although this treatment includes a growing market, it is
generally only in the case of the steady market that the matter is important.

7.11b Fluctuations in Demand

Dispatch of product from the plant is never steady. Peak daily shipments of product can easily be 3 times the mean yearly production rate. Such
a variation could be handled by building a plant 3 times the mean rate, but

it is more economically handled by building in <u>buffer storage</u> between the
plant and the customers. The bigger the storage, the greater the confidence
that these random fluctuations of demand can all be met from the stored
material.

Instead of building bigger and bigger storage capacities which tie
up more and more working capital to increase the confidence of meeting
demand, clever commercial people have been known to ask -- why not make the
plant a little bigger, so that it can more easily handle peak demands?

There is obviously a relationship between plant size and storage
capacity and there there must be an optimum arrangement for any particular
service confidence level.

Assume that the mean daily demand is d tons and this has a standard
deviation of σ_d tons -- see figure 7.16.

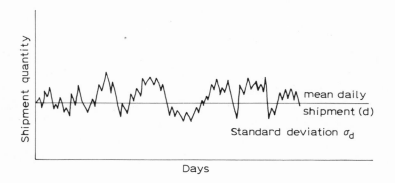

Figure 7.16
Pattern of Daily Shipments

The presence of buffer storage reduces the variance of the demand
put on the plant, as shown by figure 7.17

Consider that we are designing to a 98% immediate delivery to the
customer basis -- this implies that we must be able to supply $d + 2 \sigma_d$
tons on any one day. In general, the expression is $(d + n_\sigma \sigma_d)$ where n_σ
is the No. of standard deviations to achieve the required confidence
level. Values of n_σ are given in Appendix 2.   By having a storage vessel
of n days capacity the standard deviation of the demand on the plant σ_s

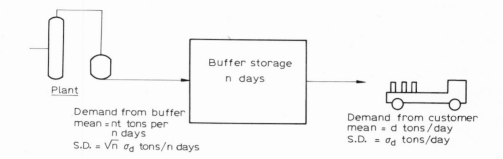

Buffer storage
n days

Demand from buffer
mean = nt tons per
n days
S.D. = $\sqrt{n}\ \sigma_d$ tons/n days

Demand from customer
mean = d tons/day
S.D. = σ_d tons/day

Figure 7.17

Effect of Buffer Storage between Plant and Customer

is reduced since,

$$\sigma_s^2 = n\ \sigma_d^2$$

i.e. $$\sigma_s = \sigma_d\ \sqrt{n} \qquad\qquad (7.28)$$

To maintain this customer service level the plant must supply
$(nt + n_\sigma\ \sqrt{n}\ \sigma_d)$ tons in n days, or $(d + n_\sigma\ \sigma_d/\sqrt{n}\ )$ tons/day.

The cost of this storage is the capital cost of the storage equipment
plus the working capital frozen in the stored product.

Assuming the capital cost of storage equipment is £C_s per ton stored,
and the working capital £w/tons stored, then the total cost for the storage
is

$$£(C_s + w)nd$$

Assuming the Economy of Scale Factor is a, then the capital cost of
the plant is given by

$$C_p \left(\frac{d + n_\sigma\ \sigma_d\ n^{-\frac{1}{2}}}{d} \right)^a$$

where C_p is the capital cost for a plant of capacity d.

The total cost for the plant plus storage system (T) is given by,

$$T = (C_s + w)nd + C_p \left(1 + \frac{n_\sigma\ \sigma_d}{d\ \sqrt{n}} \right)^a \qquad\qquad (7.29)$$

The minimum of cost (T) with respect to n gives the optimum storage size (nd tons) and plant size (d + $n_\sigma \sigma_d / \sqrt{n}$) to maintain our required customer service level.

Differentiation and equating to zero yields an expression too complex for analytical solution, and so the minimum of equation (7.29) is best obtained using a standard single variable search computer subroutine.

Example

Assume there is a market for 10tons/day of product with an expected daily standard deviation of 5 tons. A 10 ton/day plant would cost £1,000,000. The storage costs are £30/ton for capital equipment and £40/ton working capital. The simple computer program shown as Figure 7.18 uses a Coggins search subroutine (13) and yields a minimum value of T at n = 48 days for a 95% service confidence level.

In this example 48 days storage should therefore be provided and the plant capacity should be 12.4tons/day.

Table 7.7 shows the number of days storage for different degrees of customer service level.

Table 7.7
Optimum storage and plant sizes for different degrees of Customers service level
Date from example in text

Customers service % confidence that order can be met directly	No of SD n_σ	No. of days storage n	Required plant size Tons/Day	Total Investment + storage costs £
50%	0	0	10.0	1,000,000
80%	0.87	32	11.54	1,068,000
90%	1.26	41	11.97	1,088,000
95%	1.65	48	12.38	1,103,000
99.0	2.33	60	13.00	1,130,000
99.9	3.10	73	13.63	1,156,000

7.11c Seasonal demands

Some products have seasonal demands -- agricultural products, anti-freeze or salt for defrosting roads etc., -- and a choice must therefore be made between seasonal production from a large plant operated for a

```
       PROGRAM DAILY    (INPUT,OUTPUT,TAPE1=INPUT,TAPE2=OUTPUT)
       COMMON/D/B(7)
       EXTERNAL  FUNCT
C   B(1)=CS,  B(2)=W,  B(3)=D,  B(4)=CP,  B(5)=SIGD,  B(6)=A,  B(7)=NSIG
1      READ(1,100)B(1),B(2),  B(3),  B(4),  B(5),  B(6),  B(7)
       IF(B(1).LT.0.000000001)  STOP
100    FORMAT(F10.4,6F10.0)
       WRITE(2,50)  (B(I),I=1,7)
50     FORMAT(5H1CS= ,F10.4,  3HW=  ,F10.4,  3HD=   F10.4,  4HCP=      ,
      1F15.2,  5HSIGD=  ,F6.3,   3HA= ,F6.3,  5HNSIG= ,F6.3  )
       X=30.
       H=1.
       E=0.01
       M=100
       IPR=1
       CALL  COGGIN(X, H, E, M, IPR, F, FUNCT, 2)
       WRITE(2,60)F, X
60     FORMAT(  13H TOTAL COST=       ,F15.4, 15H DAYS STORAGE=   ,F5.1)
       GO TO 1
       END
       SUBROUTINE FUNCT(X,F)
       COMMON/D/B(7)
       F=(B(1) + B(2))*  B(3)*ABS(X) + F(4)*(1.0 + B(7)*B(5) /(B(3)*
      1   SQRT(ABS(X)))) **B(6)
       RETURN
       END
30.        40.0        10.0      1000000.      5.0        0.6        2.0
```

Figure 7.18
Listing of Program DAILY for determining
Optimum Buffer Storage Capacity

fraction of the year, or continuous production from a small plant together
with a considerable warehouse capacity or, in fact, a compromise between
these extremes. Figure 7.19 shows idealised demand, production and warehousing patterns for various production policies.

The choice of plant size in this situation should be made after
consideration of both plant costs and warehousing costs. The minimum
"total cost" plant size can be determined as follows. Assume the plant
operates for a fraction f_p of the year, and the demand can be taken as
another fraction f_d of a year. Assume that the capital cost of the
continuously running plant is $£C_p$ and that the Economy of Scale Factor is a.

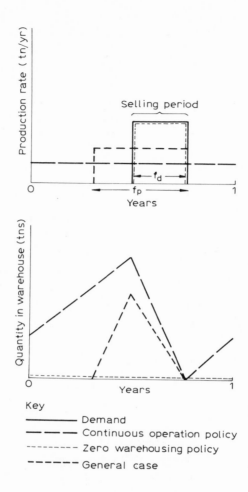

<u>Figure 7.19</u>

<u>Operating Policies for Production of a Seasonal Product</u>

The capital cost of a plant which operates for a fraction of f_p of the year is therefore $C_p(1/f_p)^a$.

The maximum quantity to be warehoused is V tons (the total yearly demand) minus the quantity that the plant will produce during the demand period. The plant capacity is equivalent to an annual rate of V/f_p tons/year and so the maximum warehouse capacity is

$$V - Vf_d/f_p$$

$$= \quad V\left(\frac{1-f_d}{f_p}\right)$$

The mean annual warehouse quantity, taking into consideration that the warehouse is only in use for fraction f_p of the year, is

$$\frac{f_p}{2} \; V\left(1 - \frac{f_d}{2}\right)$$

With an annual warehousing requirement, it is unrealistic to assume that warehouse space can be rented. With such frequent use a specific warehouse would have to be built. Assuming warehouse capital costs to be $£C_s$/tons capacity, and the working capital associated with the product is $£w$/ton, then the total capital involved (T) is the sum of the warehouse and plant investment costs, plus the mean working capital:

$$T = \frac{C_p}{(f_p)^a} + C_sV\left(1 - \frac{f_d}{f_p}\right) + f_p \; \frac{wV}{2} \left(1 - \frac{f_d}{f_p}\right) \qquad (7.30)$$

Differentiation and equating to zero to determine the minimum point, as usual, produces an equation with no easy analytical solution. The minimum cost point of this equation is therefore again best found numerically. The program to do this (SEASON) is shown as Figure 7.20. In fact, equation (7.30) is of such a form and f_p has to be so constrained to be >0 and <1.0 that, rather than locate the optimum by a search method, it is simpler to evaluate over the range of f_p, and then choose the minimum cost point by inspection.

Taking as an example a plant costing £1,000,000 for an annual tonnage of 3,000 tons/year, with an Economy of Scale Factor of 0,6 with a seasonal demand for 3,000 tons over a 3 month period, then the program SEASON can be used to locate the optimum fraction of the year in which to produce product.

Table 7.8 gives the production period for a range of warehousing costs. The last column of Table 7.8 shows a ratio of capital cost to warehousing costs which can be used as a general guide to the policy that can be expected to be near the optimum.

```
      PROGRAM SEASON   (INPUT,OUTPUT,TAPE1=INPUT,TAPE2=OUTPUT)
      DIMENSION  B(7)
C     B(1)=CP,  B(2)=V, B(3)=FD,  B(4)=CS,  B(5)=W,  B(6)=A
1     READ(1,100)B(1),B(2),  B(3),  B(4),  B(5), B(6)
      IF(B(1).LT.0.0000000001)  STOP
100   FORMAT(2F10.4,5F10.0)
      WRITE(2,50)  (B(I),I=1,6)
50    FORMAT( 5H1CP=      ,F10.0, 3HV=    ,F10.2,  4HFD=     ,F6.4,   4HCS=
     1 ,F10.4, 3HW=    ,F10.4, 3HA=      ,F6.4)
      WRITE(2,60)
60    FORMAT(////26H      FP     RATIO     COST        ,//)
      Z=      B(1)/(B(2)*(B(4) + B(5)/2.0))
      X=B(3)
10    CONTINUE
      F=B(1)/(X**B(6)) + (1.0-B(3)/X)*B(2) * (B(4) + X*B(5)/2.0)
      WRITE(2,100)X,Z,F
      X=X+0.05
      IF(X.LE.1.0)  GO TO 10
      GO TO 1
      END
1000000.  3000.    0.25     20.         40.        0.6
```

Figure 7.20

Listing of Program SEASON for determining Optimum

Production Period for a Seasonal Product

When plant costs relative to storage are high, it is best to have a small plant and operate over the whole year. When storage is expensive compared to the plant, then the plant should operate only during the demand period. Over the range 0.3 to 1.5, given by the ratio;

$$\frac{C_p}{V(C_s + w/2)}$$

a study should be carried out to determine the optimum operating period. The optimum can produce a saving equal to 10% of the capital costs compared with operating either the whole year, or just the whole demand period.

This treatment can, of course, be modified to determine the optimum policy for a product with a demand throughout the whole year, but with a seasonal peak.

The discussion on seasonal products is, of course, highly idealised and, before any final decision on capacity is made, many others factors must be taken into consideration. Often the raw materials cannot be obtained on a seasonal basis - they may be produced from a continuously operated plant in the same firm. There may be other advantages in flexibility or reliability in having either a large plant or large storage. This analysis here is by no means the whole story.

Table 7.8

Optimum Fraction of year for production of a Seasonal Product

Solution of equation   7.30 by program SEASON

Data $a = 0.6$ $C_p = 1,000,000$

          $V = 3,000$ $f_d = 0.25$

Warehouse costs			
C_s Capital costs per ton	w Working capital per ton	f_p^* Optimum fraction of year for production	$\dfrac{C_p}{V(\frac{W}{2} + C_s)}$ Plant capital for continuous plant Warehousing Capital for Storage of whole year's demand
50	100	1.00	3.33
100	200	1.00	1.67
150	300	1.00	1.11
200	400	0.80	0.83
300	600	0.50	0.55
400	800	0.25	0.42
500	1000	0.25	0.33

7.12 THE DESIGN "SAFETY FACTOR"

7.12a The Plant Safety Factor

Any discussion on plant capacity is incomplete without consideration of the Safety Factor to be included in the final design.

A planning study starts by choosing the best overall principle to employ; should a new process or an old be used, should the increased capacity take the form of an expansion to an existing plant, a further plant or a complete replacement plant?

In the next stage of the planning work the optimum capacity for the chosen principle is determined. Since this optimum is fairly flat, it is admissible to separate the choice of principle from the optimisation of capacity.

In the last planning stage we recognise that no plant has a definite output, but only a probability distribution of maximum capacity, and we have the problem of deciding which point on this distribution should be taken as the "plant capacity". In this final stage we are only dealing with a few percent of the capital costs, but nevertheless designers still require guidance as to the safety factors they should include in their designs. The employment of too conservative safety factors or the demand for a laboratory program to determine physical property data to improve design accuracy are examples of expenses whose levels should be chosen out of a planning study.

Clearly, if a product is only just profitable there is no point in incurring great expense to ensure that demand will be met, since the profit it brings will be small -- very small, or even negative, safety factors would be more reasonable. If however, a product makes a large profit compared with the capital required for its plant, it is ridiculous to lose sales simply because the plant is a few percent too small. Clearly, in such a case, a comparatively large safety factor is justified.

The following discussion will relate the economics of the process to the safety factors to be applied in the design -- based on locating the minimum cost. The discussion only applies to these situations where the objective can be defined economically. This is not the case when:

a)   an engineering department of a manufacturing company
     has a constraint that all its plants will meet
     rated capacity for political reasons (or, if promotion
     to Chief Engineer requires that all one's plants have
     met rated capacity!)

     or

b)   the plant is a link in a chain of activities which
     on the whole is very profitable, but which is not obvious
     from the analysis of politically fixed transfer prices.
     For example, oil refineries should, and do, have very
     high confidence levels for achieving rated output. Not
     because of the profit the refinery makes, since prices
     are usually adjusted so that they run at a loss, but
     because the parent company losses revenue at the oil well
     if the material cannot be processed.

     or

c)   if the optimum of the total system -- design plus operation
     is not the objective of the organisation. For example, an

engineering contractor may be legally required to provide
plants that will meet design capacity, even though it
may not be economic sense considering the whole system.

Because of errors in design methods and design data every item designed
has a probabilistic distribution for its maximum capacity. Figure 7.21
shows a distribution representing "plant capacity" which has an Expected
Capacity μ_p. If a planning study, such as described in Section 7.9, determines the optimum plant capacity to be C tons per year, the problem remains
to locate the point C on the probability distribution. If C is taken as the
mean value μ_p, then there is a 50% chance that the plant will be undersized.
If it is located at $\mu_p - 3\sigma_p$, there is 99% certainty that the plant will
meet capacity. Figure 7.21 shows this point diagramatically. This figure
also shows that the % safety factor normally used in general discussion
is 100 (μ_p - C)/C .

This problem was tackled by Saletan and Caselli (15) who showed,
analytically, that the optimum confidence level for the design was dependent
on the profitability of the project, the growth of demand and the uncertainty of demand.

For the simplest case of a plant required for a certain and constant
demand, taking typical figures for design accuracy, they showed in the case
of one rather profitable project, it was optimal to design to a 75%
confidence level of meeting designed capacity -- given by a 15% safety
factor. For a second, marginally profitable venture, a confidence level of
only 55% was justified, representing a safety factor of only 4%.

When more realistic cases were considered, which involved market
growth, the safety factors became even smaller because the only time the
safety factors could possibly be important was at the end of the project and
this discounting further reduced the benefit of the safety factor to
2 -5% range.

Their work concluded that safety factors to compensate for the uncertain
plant capacity should be small; at most 15% for profitable projects designed
to run immediately at full output, 10% for projects with lower growth rates
down to zero for marginally profitable projects. This is in agreement with
other workers who have studied the subject (9). The interesting point in
their work is that it showed, on purely economical grounds, that designs
should have a 30% probability of <u>not</u> meeting design capacity. From within
an engineering department one usually gets the impression that designs
should have probabilities of less than 5% of "failing" to meet design
capacity.

The details of Saletan and Caselli's method will not be given here because the method is too involved for use as a routine tool within an engineering department, and a more robust, though less elegant method will be later described in Section 7.12c.

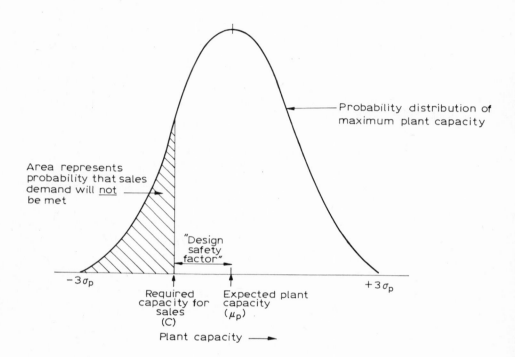

Figure 7.21.

Location of Required Capacity C on the Maximum Capacity

Distribution of a Plant

7.12b   The Plant Item Safety Factor

So far we have discussed the plant as a whole, with its mean capacity μ_p and standard deviation σ_p. In reality a plant is composed of a string of plant items, each with its own mean capacity and standard deviation, and by concatonating these the plant distributions μ_p and σ_p can be obtained. Moreover, the engineer needs to know the safety factors for each item of equipment and not the "plant safety factor".

As already explained in Chapter II Section 2.9c, the relation between the capacity distribution for the plant item and the plant itself is obtained by a simple combination of probabilities. If a plant containing 20 items of equipment was designed with no safety factors on any item, and all functions were independent, then the probability that the plant would achieve rated output would be 0.5^{20} one change in a million. Alternatively, to obtain a 50% probability of achieving rated plant capacity, each item must be designed to a confidence level of $(0.5)^{1/20}$ i.e. 96.6% or about 2 standard deviations.

A string of 20 items in a plant is not an unrealistic number and so one simple solution to the problem is to design all items with a safety factor of +2 standard deviations -- the 95% confidence level. This approach is commonly used in engineering departments. For example, some heat exchanger design methods involve using coefficients for the heat transfer correlations which are not from the mean fit of the experimental data, but from the 95% confidence limits.

This general rule however is rather rough, particularly when one considers that in a chain of processes, a simple pump can have as much effect on the plant capacity as an expensive compressor. Clearly, there should be a relation between the confidence level taken for the design of each item and the cost for providing this increased confidence.

For small items of equipment -- pumps and small heat exchangers, for example, and for proprietary equipment items with predictable performance, it is sensible to use two standard deviations on the safety factor. These then have a high enough degree of confidence to be omitted from the string of items which dictate the plant capacity distribution, leaving only the expensive items with large standard deviations to be studied further.

Consider that a plant is composed of a string of these significant plant items, each with the same standard deviation σ_i. We can calculate the general safety factor to be employed on them and also the standard deviation of the combined distribution by simply combining probabilities, using normal Tables. This data is given in Table 7.9 for a number of different string lengths. Notice that it is possible to define the safety factor for each item, and the total plant distribution in terms of the standard deviation of the equipment σ_i.

Table 7.9 shows the strong relation between the required safety factor and the number of items in the string. We have already argued that this number should be much less than the total number of items actually involved, because it is sensible to use a high safety factor for most of them. There

Table 7.9

Design Capacities of Significant Plant Items (μ_i)

to Achieve a Defined Expected Plant Capacity (μ_p)

σ_i = item capacity standard deviation

σ_p = total plant capacity standard deviation

Given: a desired mean plant capacity (μ_p) and a standard
deviation of item designs σ_i.

Equivalent No. of independent significant items in plant string (N)	Capacity to be taken for item design $\mu_i = \mu_p + q\sigma_i$	Standard deviation of resulting plant capacity $\sigma_p = p\sigma_i$
1	$\mu_p + 0.0\ \sigma_i$	$1.0\ \sigma_i$
2	$\mu_p + 0.5\ \sigma_i$	$0.9\ \sigma_i$
3	$\mu_p + 0.8\ \sigma_i$	$0.8\ \sigma_i$
5	$\mu_p + 1.1\ \sigma_i$	$0.7\ \sigma_i$
10	$\mu_p + 1.5\ \sigma_i$	$0.6\ \sigma_i$
15	$\mu_p + 1.7\ \sigma_i$	$0.6\ \sigma_i$
20	$\mu_p + 1.8\ \sigma_i$	$0.6\ \sigma_i$

are further grounds for reducing this number even below the number of
important items:

either   a)   if the items are not completely independent and so some
excess capacity on one item may relieve the load to
some degree on the neighbouring equipment. For example
a too small reactor can be compensated for by a higher
temperature, or as a second example, a poor distillation
separation can be compensated for by a higher reflux
ratio.

or   b)   if some bottlenecks can be easily removed and so they
do not justify as serious a treatment as the situation
where no compensation is possible. For example, a further
shell could easily be attached to a heat exchanger
bank, or a column packing could be changed to a higher
performance type , but a compressor or a reactor may be
incapable of performance improvement without replacement.

Without carrying out extensive plant simulations of the actual process,
the accurate evaluation of the number of units in the string cannot be made.
Since this is not generally available, the engineer must subjectively judge
an equivalent N after listing only the major equipment, and assessing

their interdependence. Recommendations in the literature suggest that 3
is a reasonable value for a typical chemical plant, though no substantiation
is given (15).

7.12c   Total Design Safety Factor - Item plus Plant

We now have to combine the safety factor for the plant items which
ensures that the final plant will have a required Expected value, with
the economic confidence level of meeting rated capacity. This has been
done analytically by Saletan and Casselli (15), but since their analytic
estimate of the degree of confidence is laborious, we will now describe
the second method which is easier to apply as a routine procedure:

> Having chosen a suitable value for N, and knowing σ_i
> from the design procedures employed, a value of σ_p can be
> obtained from Table 7.9 and a series of slightly different
> sized plants all with the same standard deviation σ_p can be
> proposed to meet a required capacity C.

For example they could be:-

1)	mean capacity = $C - \tfrac{1}{2}\sigma_p$	- equivalent to 31% confidence level
2)	mean capacity = C	- equivalent to 50% confidence level (zero plant safety factor)
3)	mean capacity = $C + \tfrac{1}{2}\sigma_p$	- equivalent to 69% confidence level
4)	mean capacity = $C + \sigma_p$	- equivalent to 84% confidence level
5)	mean capacity = $C + 2\sigma_p$	- equivalent to 97.5% confidence level.

The computed costs and fixed costs for these five alternatives
can then be calculated by some simple scaling rules, such as
Equation 7.1, and these five proposals separately evaluated with a
Monte-Carlo evaluation program. The resulting Expected NPW's can
then be plotted against mean capacity to determine the maximum
profit. The difference between the required capacity and the
optimum mean capacity is $(k^*\sigma_p)$ and this represents the "plant
safety factor".

This procedure can be summarised as follows:

    a) Given N equivalent significant equipment items in the
        plant, Table 7.9 gives the value of p and q for the equations

$$\sigma_p = p\sigma_i \tag{7.31}$$

$$\mu_i = \mu_p + q\sigma_i \tag{7.32}$$

    b) Evaluate the NPW for a number of plant capacities
        using a Monte Carlo simulation and plot the cost
        vs. capacity to give the optimum mean plant capacity $\mu_p{}^*$
        Evaluate k* from the following equations:

$$\mu_p{}^* = C + k^*\sigma_p = C + k^*p\sigma_i \tag{7.33}$$

        $k^*\sigma_p$ is the plant safety factor, and k* can be used to
        find confidence levels.

    c) The significant plant items must be designed for a
        mean capacity of $\mu_i{}^*$ where

$$\mu_i{}^* = \mu_p{}^* + q\sigma_i = C + (k^*p + q)\sigma_i \tag{7.34}$$

These "significant" items should be designed for a capacity $\mu_i{}^*$,
and no "safety factor" should be added. Non-significant items
should be designed to have a 95% confidence of achieving the
plant output by adding a safety factor in the conventional way.

Example

    Consider the design of a plant for making a liquified gas, by a
process shown on Figure 7.22. The raw materials are fed to a large reactor
and the resulting product is purified by passing through an absorption
train of four towers. One raw material is recovered from the absorption
and returned to the reactor. The purified gas is then compressed, conden-
sed, and stored in tanks prior to bulk dispatch.

    The plant can conveniently be divided into six sections, as shown
on Figure 7.22:

        1) Feed storage and pumping
        2) Reaction
        3) Absorption train
        4) Raw material recovery and recycle
        5) Product compression
        6) Product liquifaction and storage

    We can immediately discard Section 1 as not being part of the string
of units; both tanks and pumps would be sized such that no process restriction occurred. Section 4 would not be part of a string if unrecovered

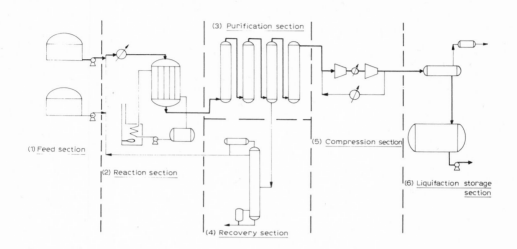

Figure 7.22
Example of Safety Factor Evaluation - Sectioned Plant Flowsheet

raw material could be disposed of in waste water, in this example we
assume that this is not the case -- if the recovery column does not remove
sufficient from the aqueous effluent, then the plant rate must be reduced.
The absorption train performs two independent functions, -- removal of
acid and drying, -- and its throughput is limited by a maximum pressure
drop restriction.

Though there are twenty-four items of equipment in the plant, it
is more reasonable to consider it as a string of six since there are six
major groups of equipment. However, these six are not completely independent; for instance there is some latitude to alter reactor conditions if
the recovery section is overloaded, or to compress if either the absorption
columns or the liquifaction condensers are undersized. It is therefore
reasonable to reduce the number in the string still further to represent
the equivalent number of independent items -- let us say 4.

With N = 4 for our process , Table 7.9 gives
p = 0.75 and q = 0.95

Let us also assume that the commercial data suggests a rising market
for four years followed by a levelling off, with some uncertainty -- as

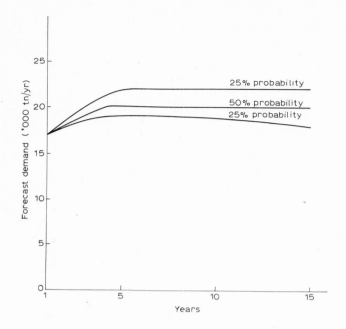

Figure 7.23

Example of Safety Factor Estimation - Market Forecast

shown by Figure 7.23.   There is one cost estimate available for a
20,000 ton/yr plant of £2,000,000, and it is considered adequate to use
the six-tenths power rule to determine the cost of other-sized plants.
Fixed costs are mainly labour and will remain substantially constant at
£80,000 per year for plants around 20,000 ± 10,000 tn/yr.

Further necessary economic data for this example are as follows:

Selling price   - £60/tn for the three forecast levels
Variable cost   - £20/tn
Tax 50%, 10 years straightline depreciation.

The demand/year and sales/demand correlation cofficients for the product
are considered to be 1.0 in both cases.

The engineers possess limited design data for this system, and this,
they claim, will limit their design accuracy to ± 20% (95% confidence limit
i.e. $2\sigma_i$).

i.e . $\sigma_i \approx 0.10\ \mu_p$

and so from Table 7.9

$$\sigma_p = 0.075\mu_p$$

We are now in a position to run a Monte-Carlo simulation with each of the five plant sizes, given in Table 7.10 to locate the optimum Expected NPW by interpolation.

Table 7.10

Example of Safety Factor Estimation - Proposed Alternative Plant Size

Plant	Capacity (μ_p)	Capital Cost $\left\{\dfrac{\mu_p}{20,000}\right\}^{0.6} \times \text{£2mm}$
1	$20,000 - 1.0\sigma_p$ = 18,300	1,896,000
2	20,000 = 20,000	2,000,000
3	$20,000 + 1.0\sigma_p$ = 21,700	2,100,000
4	$20,000 + 2.0\sigma_p$ = 23,400	2,197,000
5	$20,000 + 4.0\sigma_p$ = 26,800	2,384,000

Figure 7.24 shows the data input for the PLADE program that was used for the simulation.

Figure 7.25 is a plot of NPW, expressed as a percent of the capital involved. Curve 1 shows that the optimum point corresponds to a plant of 20,000 tons -- a zero safety factor. The conclusion from this study is therefore, that the economics and uncertainty associated with this product are such that the engineers should include no safety factor in their designs -- i.e. they should design a 20,000 tn/yr plant assuming their data is accurate. A converse of this is that there would be no benefit to this project if a data collection program was instigated to reduce the present $\pm 20\%$ design accuracy.

If the project had been more profitable, with a selling price of £90/tn and not £60/tn, then a curve as shown by curve 2 on Figure 7.25 would have been obtained. In this case, the optimum plant should be sized for 21,700 tons and the k* of equation (7.33) is 1.0. A safety factor equivalent to 8.5% ($+1\sigma_p$) would be justifiable, giving a 84% confidence level

```
HEAD
SAFETY FACTORS EXAMPLE ,SIGMA IS 15 PER CENT
TAX
50 10
DISCOUNT
15
LIFE
15 10
PROBABILITY
0.33 0.34 0.33
DEMAND
17000 18500 20000 21300 22000 22000   22000 22300 22000 22000
22000 22000 22000 22000 22000 22000   22000 22000 22000 22000
17000 18000 19000 20000 20000 20000 20000 20000 20000 20000
20000 20000 20000 20000 20000 20000 20000 20000 20000 20000
17000 17700 18400 19000 18900 16800 18700 18600 18500 18400
18300 18200 18100 18000 17900
PRICE
0.060
```

contin:

```
CORRELATIONS
1.0 1.0
TRAFZ
2.0
ALTERNATIVE
-2.0 SIG   4360 1558.68 80 0.020
-2.0 SIG  11840 1558.68 80 0.020
-2.0 SIG  14560 1558.68 80 0.020
-2.0 SIG  22040 1558.68 80 0.020
END
OPERATE
-2.0 SIG        1
END
EVALUATE
ALTERNATIVE
-1.0 SIG   7760 1738.45 80 0.020
-1.0 SIG  15240 1738.45 80 0.020
-1.0 SIG  17900 1738.45 80 0.020
-1.0 SIG  25440 1738.45 80 0.020
END
OPERATE
-1.0 SIG        1
END
EVALUATE
```

```
ALTERNATIVE
-0.5 SIG   9460 1836.19 80 0.020
-0.5 SIG  16940 1836.19 80 0.020
-0.5 SIG  19660 1836.19 80 0.020
-0.5 SIG  27140 1836.19 80 0.020
END
OPERATE
-0.5 SIG        1
END
EVALUATE
ALTERNATIVES
NO SIG  11160 2000 80 0.020
NO SIG  18640 2000 80 0.020
NO SIG  21360 2000 80 0.020
NO SIG  28840 2000 80 0.020
END
OPERATE
NO SIG        1
END
EVALUATE
ALTERNATIVE
0.5 SIG  12113.6 2100.33 80 0.020
0.5 SIG  20224.4 2100.33 80 0.020
0.5 SIG  23175.6 2100.33 80 0.020
0.5 SIG  31291.4 2100.33 80 0.020
END
OPERATE
0.5 SIG        1
END
EVALUATE
ALTERNATIVE
1.0 SIG  13057.2 2197.564 80 0.020
1.0 SIG  21838.8 2197.564 80 0.020
1.0 SIG  24991.4 2197.564 80 0.020
1.0 SIG  33742.8 2197.564 80 0.020
END
OPERATE
1.0 SIG        1
END
EVALUATE
ALTERNATIVE
2.0 SIG  14954.4 2383.926 80 0.020
2.0 SIG  24377.6 2383.926 80 0.020
2.0 SIG  28622.4 2383.926 80 0.020
2.0 SIG  36645.6 2383.926 80 0.020
END
OPERATE
2.0 SIG        1
END
EVALUATE
STOP
```

Figure 7.24

Example of PLADE to determine Design Safety Factor
- Data Input for Program

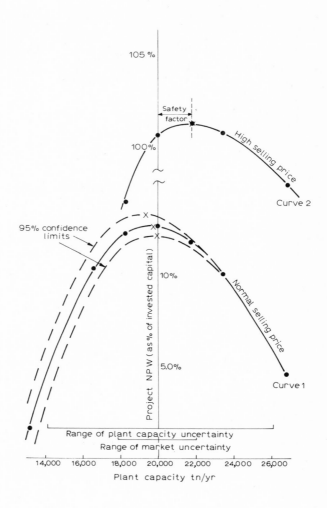

Figure 7.25

Determination of Economic Design Safety Factor

of meeting 20,000 tn/yr. Equation 7.32 shows that the significant items should be designed to have a capacity of 21,700 (1 + 0.95 x 0.1) i.e. 23,700 tns/year. The remaining items must be sized so that they have a 95% probability of handling a rate of 21,700 tn/yr.

By running a further evaluation with a plant of known fixed capacity, the corresponding optimum for a plant with a certain design can be determined. We cannot assume the optimum is 20,000 tn/yr because the market is uncertain. The difference between these two cases will give us the "value of the information" -- in this case the value of perfect design data.

In our example, this is equivalent to about 10% of the capital costs for the £90/ton case. These savings could well cover the cost of a data improvement study.

It is very difficult to generalise on the size of safety factor and the value of better design information. Work with the PLADE program has confirmed Saletan and Casselli's conclusions that economically justified safety factors are far lower than those currently used in practice. The value of better data differs greatly from case to case. If the demand is accurately known, then there is considerable benefit in having accurate design data. If the design uncertainty is of the same order as the market uncertainty then there is little advantage in having accurate design data, so resolving only one of the two uncertainties. If the market is not expected to reach a constant level but to increase beyond the proposed plant capacity, then the safety factor study is virtually converted to a plant optimum cycle study, and as Section (7.7) showed, variations of 10% in capacity about the optimum involve costs of only 0.2% of capital.

The whole treatment of safety factor here has been very general. It considers:

> "if the general level of accuracy of the design data was
> $\pm x\%$" or

> "if the general accuracy of the design method chosen was
> $\pm x\%$",

Clearly, specific data points can provide concrete, easily measured savings. For example, to measure a viscosity may enable a cheaper pump to be purchased, or to measure VLE data for a system may reduce the cost of the proposed distillation equipment. To consider safety factors rigorously requires a probabilistic study of every item design in addition to a probabilistic economic assessment. There are at present no tools available to do this, though methods are being assessed (12).

7.12d Notes on the Use of Monte-Carlo Simulation for Safety Factor Studies

Monte-Carlo simulation is a rough method of getting an approximate answer when an accurate answer is too difficult to obtain. Safety factors are dealing with variations of less than 20% on plant capacity, which represent 1 - 5% variations in the mean total cash flow in a project. The use of an approximate procedure to discriminate between values varying by

1 - 5% would be expected to, and does, lead to difficulties. Since the
standard deviation of the Expected Value produced from a Monte-Carlo
simulation is inversely proportional to the square root of the number of
iterations, it is usually impractical to obtain adequate accuracy by simply
increasing the number of iterations. It was found essential to incorporate
some accuracy improving techniques in the PLADE Monte-Carlo routine before
meaningful results could be obtained.

Firstly, the runs were stratified, so that the correct number of
interations were made at each probability level. This is described further
in Chapter V. Secondly, the random number sequence was made reproducible
by seeding the random number generator for each simulation.

Using exactly the same sequence of random numbers results in very
smooth curves, which is useful in appreciating the shape of the curve, but
dangerous in that the smoothness of the curve belies the accuracy, as their
positions change as the number of iterations increase.

It is important that any simulation program used for these studies
prints out the accuracy of the Expected value. This enables upper and
lower bounds to be drawn -- as shown in Figure 7.25 and with these bounds
two extreme curves can be drawn. The range in the optimum given by the
extreme curves indicates the accuracy that can be attributed to the located
optimum.

Figure 7.25 shows this for the safety factor example. The optimum is
20,000 ± 800. This method has an accuracy of 4%, which represents a
possible cost of ½% of capital cost through missing the optimum. This
accuracy is adequate for practical evaluations, but would not be adequate
for, for instance an academic study of safety factors.

7.13 CONCLUSIONS

The first step of a planning study is to select the best mode of
production -- should it be an expansion of an existing plant, a new plant,
a new process, or buying in product. Once the mode of production is clear,
the next stage is to determine the size of the capacity increase as a
result of an economic balance between savings due to Economies of Scale and
losses due to initial overcapacity. The third and final stage is to
consider the uncertainty in the plant capacity itself and determine the
safety factor that should be used in the plant and equipment sizing.

There are a number of instances where there is a relation between
plant capacity and warehouse capacity. Delay of expansions through

long-term storage of product, improved customer service by installing over-capacity plants, and manufacturing policy for seasonal products are areas where a balance between warehousing and manufacturing capacity is called for.

This chapter has shown that it is possible to treat these topics in a quantitative way and so arrive at a plant capacity with a confidence deserving of a decision of its magnitude.

The striking feature of the numerical work reported in this chapter is the flatness of the optimum. It is comforting to know that the losses in not being exactly at the optimum are usually small and that after the optimum has been determined there is scope to make capacity changes for qualitative reasons. Typical costs of a 20% move from the optimum plant size are equivalent to a 1-5% change in capital. Changes of ± 10% result in 0.2% which are immeasurably small. The costs in installing a plant of half the optimum size, as an insurance against a possible future event such as complete market failure, incur cost increases, which can be considered to be the insurance premium, of the order of 5-10% of the capital involved.

Design safety factors that can be economically justified are surprisingly small, except for highly profitable non-growth projects. Safety factors for marginally profitable uncertain ventures are negative. Studies of design safety factors are useful in giving detailed guidance to the engineer on the item safety factors he should employ in this design. A safety factor study can also estimate the value of providing a more accurate design, either by a more detailed design method or by obtaining more accurate design data.

7.14 BIBLIOGRAPHY FOR CHAPTER VII

1.   Berretta, J.C., and Mobasheri F. An Optimal Strategy for Capacity Expansion,
     The Engineering Economist, 17, 79, (1972)

2.   Cameron , D., Three Simple steps to Determine Optimum Plant Capacity,
     Long Range Planning 7, (Feb. 1974).

3.   Chenery, H.B., Overcapacity and the Acceleration Principle,
     Econometrica, 20, 1, (1952).

4.   Chilton, C.H.,'Six-Tenths-Factor' Applies to Complete Plant Costs,
     Chem. Eng. 112, (April, 1950).

5.   Chilton, C.H. Cost Engineering in the Process Industries,
     McGraw-Hill, New York 1960.

6.   Coleman, J.R., and York R., Optimum Plant Design for a Growing Market,
     Ind. and Eng. Chem. $\underline{56}$, 28, (1964).

7.   Erlenkotter, D., Investment for Capacity Expansion - Size, Location
     Time Phasing, Allen and Unwin, London 1967.

8.   Generoso, E.I., and Hitchcock, L.B., Optimising Plant Expansion-Two
     Cases, Ind. Eng. Chem. $\underline{60}$, 15, (1968).

9.   Kittrell, J.R., and Watson, C.C., Don't Overdesign Process Equipment,
     Chem. Eng. Prog. $\underline{62}$, 4, 79, (1966).

10. Manne, A.S., Capacity Expansion and Probabilistic Growth,
     Econometrica, $\underline{24}$,4, p.632 (1961).

11. Manne, A.S., Investment for Capacity Expansion, - Size, Location
     and Time Phasing, Allen and Unwin, London, 1967.

12   Marketos, G., The Optimal Design of Chemical Plant Considering
     Uncertainty and Changing Circumstances,
     Dissertation No. 5607. E.T.H Zurich 1975.

13. Rose. L.M. The Application of Mathematical Modelling to Process
     Development and Design,
     Applied Science Publishers, London 1974.

14. Sims. S.P., Michelson, D.L., and Fricke, A.L., Optimisation of
     Sequential Investment with Comprehensive Risk Analysis,
     A.I. Ch.E. 72nd National Meeting, A.I.Ch.E. New York, 1972.

16. Saletan, D.I., and Casselli, A.V. Optimum Design Capacity of New
     Plants, Chem. Eng. Prog. $\underline{59}$, 5, p.69 (1963).

17. Srinivasan, I.N., Investment for Capacity Enpansion-Size, Location,
     Time Phasing, Allen and Unwin. London. 1967.

18. Veinott, A.F., Investment for Capacity Expansion - Size, Location,
     Time Phasing, Allen and Unwin, London, 1967.

CHAPTER VIII

THE ACCURACY OF THE EVALUATION
AS A PREDICTION OF PROFITABILITY

Hansdampf im Schnaaggeloch,
häd alles was er will,
und was er will, das häd er nüd,
und was er häd, das will er nüd,
Hansdampf im Schnaaggeloch
häd alles, was er will.

8.1   INTRODUCTION

As has been repeatedly emphasized throughout this text, the evaluation
that precedes any decision on capital expenditure is an attempt to simulate
the future to obtain a prediction of the profit that would result from the
project over its whole lifetime.

However, it is unusual for any firm to look back to see how well past
evaluations agreed with what actually occurred. It is considered to be
"water under the bridge" and it is reasoned that events are so stocastic
that the explanation for discrepancies will be so different for every
project that no generalized conclusions can be expected.

In the extreme case this relegates the Evaluation to a formed ritual
that must be passed, by fair means or foul, before the project can proceed.

A notable exception to this appears to be Amoco Chemical Corporation,
judging from a series of papers published by Malloy et al (8,9,12). The
theme of all these papers concerns methods of more accurately predicting
the profitability of a project, and in particular, the reduction in the
over-estimation of profit.

Malloy identifies changes that occur in plant capacity, processing
costs, and selling price which are not fully thought out when the data for
the Evaluation is being estimated. Experience shows that plant capacities
grow through learning and minor debottlenecking exercises, and so one can
generally make more product than the rated capacity. This learning has

been investigated for catalytic crackers (5) but it is unlikely that any
generalizations can be made. It is really up to the design engineer to
estimate the possible, ultimate, fully-learned and debottlenecked capacity
from his knowledge of the process and the safety factors he has had to
employ.

Malloy claims that processing costs also change with time and should
not therefore be considered constant for every year in a detailed evaluation.
Variable costs--raw material usage and utilities--can be expected to decrease through learning, but labour costs can be expected to rise.

Of all these changes the most important, particularly for new products,
is the price fall--or profit margin decay. Malloy shows for four chemicals
that they displayed profit margin decays between 14 and 23% per year. It
is therefore very necessary to obtain realistic prices and price falls with
time before making a profitability evaluation. This is, in fact, a very
difficult forecasting exercise, and since the price is dictated by production costs this is, strangely enough, best done by referring back to project evaluations:

> By considering the economics of a hypothetical, modern, advantage-
> ously situated, large plant it is possible to estimate the capital
> charges--at a rate of return equal to the cost of capital--as well
> as processing and all other costs, and to arrive at a Floor Price.
> By considering learning which increases capacity and decreases pro-
> cessing costs, and inflation, which increases labour costs, it is
> possible to construct a Floor Price curve vs. time. To this can be
> added a profit margin based on existing prices, but also including
> an exponential decay at some predetermined rate. The addition of
> the profit margin plus Floor Price can then be taken as a realistic
> price forecast with which one's own proposed project--which may not
> be a modern, advantageously situated, large plant--can be evaluated.

It is common practice to neglect learning and price changes when forecasting the data for the evaluation even though experience shows that these
changes generally occur. Malloy shows that every attempt should be made to
include them in the evaluation.

Other factors which affect evaluation results are: inflation, uncertainty, strategy, correlation between uncertain variables, and estimation of

probabilities. These points are discussed separately in the following five
sections.

8.2 THE EFFECT OF INFLATION

     Over the life of a project the value of money may well decrease two or
more-fold as the results of general inflation. It is therefore important to
define the basis of the profitability calculation.

     Chapter V, Section 5.2f argues that constant value money is the most
logical basis for the actual calculation, with cash flows being converted
into current money at the end of the calculation if actual amounts of money
have to be reported. The effect of inflation, which is extremely difficult
to predict in the long term, is then restricted to reducing the value of
the tax allowance on depreciation.

     Consider Example 2 of Chapter I (Section 1.9b) evaluated ignoring
inflation and then again with an overall inflation rate of 20% per annum:
Tables 8.1 and 8.2 are the corresponding outputs from the PLADE program

TABLE 8.1		TABLE 8.2	
Evaluation of Example 2 (Chapter 1)		Evaluation of Example 2 (Chapter 1)	
assuming No Inflation		assuming 20% Inflation Rate	
Cumulative Undiscounted Cash Flow £(Current Money)		Cumulative Undiscounted Cash Flow £(Current Money)	
Year		Year	
1	-3886.87	1	-3884.25
2	-3652.93	2	-3591.36
3	-3253.07	3	-2973.22
4	-2717.77	4	-1970.58
5	-2162.77	5	- 738.41
6	-1420.47	6	1279.50
7	- 606.17	7	3938.97
8	- 262.73	8	7345.10
9	1186.83	9	11697.27
10	2052.33	10	16537.04
11	2952.33	11	23224.12
12	3847.08	12	31201.81
13	4751.95	13	40883.35
14	5880.20	14	52544.54
15	7075.21	15	74345.47

DCF Discount Rate   10.83   DCF Discount Rate   7.91
     Payback Year is 7.70   Payback Year is 5.37

and they summarise the two cases. Even with the 20% inflation rate both DCF
rates of return are reasonably close, - 10.8% compared with 7.9% respectively.
The effect on pay-back time is more marked, - 7.7 compared with 5.4 years,
but there is a large effect on the quantity of money flowing in later years.
In the last year there is a 10-fold difference in the quantity of current
money that flows. The "With Inflation" case recovers the capital cost 18-
fold compared with 1.7-fold for the "Inflation-free" study. This project
has a real yield of less than 8% yet it covers the capital spent 18-fold
for the company balance sheet! This apparent contradiction brought about
by inflation emphasizes how important it is to define a clear basis for the
study from the outset. There must be many firms who are satisfied with their
long-term growth in book-value terms, but who have actually contracted in
size in real-value terms.

The above discussion holds for situations with equal inflation on all
costs and prices. In practice, it is better to sub-divide inflation into
three distinct categories: material produced within the same country as
manufacture, materials imported, and labour costs.

The importance of relative inflation rates between different countries
is compensated in the long run by changes in exchange rates between different currencies. It is reasonable therefore not to consider specifically
different inflation rates for imported materials in the long term. This is
not necessarily true for short terms of 1 to 2 years, where it is found
to be worthwhile estimating individual country's inflation rates to make
suitable corrections -- in capital costs estimation this is the usual
procedure. Fluctuations in exchange rate in the short term do not necessarily
properly reflect the value of money.

The situation is somewhat different for the relative inflation rate
between labour and material. All countries show a positive growth in standard of living with time. This is because wage rates increase faster than
material costs as a result of productivity increases. Table 8.3 gives the
corresponding figures for England over the year 1966-1973. In Project
Evaluation one can therefore expect labour costs to rise faster than material costs. Inflation rates for material and labour will be different.

This can be a very important effect if the project has a high labour
content, for instance, when the project is a labour-saving proposal such as
rationalisation or automation of an existing labour-intensive process.

Table 8.3

Relative Inflation Rates for Engineering Materials, Wages and Composite Installed Plant Costs (6)

Year (January Figures)	Index for Material	Index for Labour	Composite Index 0.6 Labour + 0.4 Material
1966	100	100	100
1967	103.2	102.4	102.7
1968	109.6	107.0	107.6
1969	109.1	118.8	114.7
1970	124.2	130.0	126.9
1971	127.7	146.0	138.0
1972	134.6	158.3	148.7
1973	143.9	181.9	166.2

It requires no mathematics to appreciate that the optimum degree of automation should be beyond the point indicated by considering today's wage-rates, because as labour gets relatively more expensive it justifies a greater degree of automation in a long-life project.

This effect should be taken into consideration in evaluations involving significant labour costs. This can be done simply on a material constant value money basis by assuming an annual increase in labour costs equal to the difference in labour and material inflation rates, corrected for material constant value money.

Example

Consider a 5-year project costing £200,000 capital, the income of which is based on annual savings of £50,000 in wages and £20,000 in material. An assumed inflation rate (material) of 5% is expected and a wage inflation rate of 9%. What is the DCF of the project?

Table 8.4 derives the yearly cash incomes in constant value terms. The last two columns of this table can be used to evaluate the DCF in a constant value money calculation. The PLADE Program gives a DCF rate of 6.7% for this example using a 50% tax rate and a 5 year depreciation period.

A DCF evaluation ignoring all inflation rates gave a rate of 8.5% and an evaluation assuming 5% inflation, but neglecting the differential between labour and materials, gave a DCF of 5.8%.

There is unfortunately no general rule that can be given concerning

the effect of unequal inflation rates, even though firms have been known
to work on rules such as "if the major saving is labour costs a DCF Rate of
Return of 7% and not the normal 12% will be accepted."

Inflation reduces profitability because less tax rebate is allowed than
would appear from an inflation-ignored calculation, whereas payback times
and cash flow profiles are improved by the presence of inflation.

The effect of unequal inflation alone can be predicted. The usual
case of Labour-saving proposals is to involve a progressively more valuable
saving which results in an improved profitability over a single inflation-
allowed calculation.

However, no relation can be given between the totally inflation-ignored
evaluation and an unequal inflation-allowed evaluation. The profitability
may be worse because of the reduced tax allowance, or better because of the
increasing value of the saving.

The only sure way to make evaluations in an inflating environment is
to define the mean inflation rate, make annual cash flow adjustments to
cover inflation rates which are expected to deviate from the average rate,
and then carry out an evaluation.

Table 8.4

Example: Differential Materials and Labour
                          Inflation Rates

Year	Yearly Income £ Current Money		Constant Value of £1 Current Money (material based)	Yearly Income £ Current Money	
	Labour	Material		Labour	Material
1	50,000	20,000	1.00	50,000	20,000
2	54,500	21,000	0.9524	51,906	20,000
3	59,405	22,050	0.9070	53,880	20,000
4	64,751	23,152	0.8638	55,931	20,000
5	70,579	24,310	0.8227	58,065	20,000

8.3 THE IMPORTANCE OF CONSIDERING UNCERTAINTY

Of the literature describing Monte-Carlo simulations for the evaluation
of investment proposals, the majority show that Monte-Carlo simulations give
a lower profitability than a parallel deterministic analysis, (1,3,4,7,8).
Hertz (3) showed a Monte-Carlo simulation DCF of 14.6% compared with 25.2%
for the deterministic study; Malloy (8) showed 9.6% from Monte-Carlo

simulation compared with 12.4% for a deterministic analysis, and Cameron (1)
showed 26% compared with 30%.

All these examples reported were actual projects and Malloy went further by showing, from a study of 3 projects, that there was only a 26% probability of achieving the profitability given by the deterministic analysis.
26% was also the proportion of eleven actual operating plants that were
achieving their predicted probability. This agreement is probably fortui-
tous, but at least Malloy has published some results on re-appraisal of
profitabilities which shows that, in actual experience, about 75% of plants
do not meet their predicted profitability.

The reason for this lower profitability from a Monte-Carlo simulation
is two-fold. Firstly, the distributions of the data are not symmetrical and
it generally appears that there is a greater chance of pessimistic outcomes
occurring than optimistic; secondly, we have the effect of lost sales through
shortages in capacity giving a cut-off and skew distribution of income even
when the distribution of demand is symmetrical.

These conclusions are further verified by the study of four projects
described in Chapter VI. Of the four levels of analysis considered, the
difference between level 2 (mean sales plan) and Level 3, (probabilistic
plan) is the difference between probabilistic and deterministic analyses.
Table 6.16 shows that the deterministic analysis in an expanding market
predicts a profitability that is higher than the more correct probabilistic
analysis, sometimes by an amount equal to 20% of the capital costs involved.

There are therefore grounds for carrying out Monte-Carlo simulations
just to obtain a reasonable estimate for the expected profitability in
addition to the more obvious reason of obtaining an appreciation of the
distribution of profitability that is to be expected. This distribution
is an important result in the evaluation of profitability, and the remainder
of this chapter will discuss the importance of factors on both the estimate
of the mean value and the distribution.

8.4   THE IMPORTANCE OF CONSIDERING STRATEGIES

Chapter VI demonstrated the importance of careful analysis by taking
four "risky" synthetic projects and then planning and evaluating them with
different degrees of thoroughness. Two criteria were important: determining
the lowest-cost plan and accurately predicting the profitability.

Table 6.16 shows the effect on profitability of not using the most
detailed level of analysis to create the investment pattern to be evaluated.

If our planning was restricted to Level 3, -- that is a correct probability
analysis but restricted to a single plan -- then Table 6.16 shows errors
between 0 and 50% in the predicted profitability would be obtained compared
with what one would expect to happen -- that is, the management would make
different later decisions depending on the market outcomes and so would
effectively have employed a strategy approach (Level 4 of Table 6.16).

The reason for employing a strategy rather than a single plan is to
alter the single plan to minimise losses or make more gains from advantageous developments. That is to say, it shifts the whole distribution
to improve gain. The PLADE program was used to determine the distributions
for Case 3 of Table 6.16 for both Level 3 and Level 4. The results are
shown in Figure 8.1.

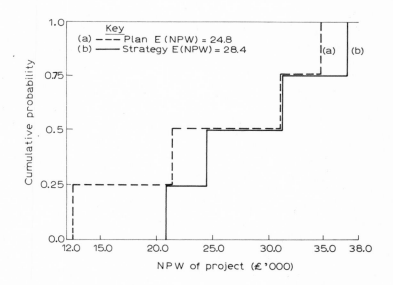

Figure 8.1

Comparison of Profitability Distributions for a Project calculated with:

(a) a fixed plan

(b) a strategy   (Case 3 of Chapter VI)

Clearly, the consideration of a strategy can have a marked effect on
both the Expected profitability and the distribution. The concept of
strategies only holds when, at some date during the life of a project, the
uncertainty in sales can be largely resolved -- in other words, when the

demand/year correlation coefficient is high. As will be shown in the next
section, when the correlation coefficient is low it is better to employ
a plan than a strategy.

8.5 THE EFFECT OF THE CORRELATION COEFFICIENTS

Chapter II, Section 2.5e described how important correlation is between
uncertain variables and particularly, how it affected the distribution of
profitability. All statistical texts emphasize the need to know whether the
variables are "dependent" or "independent" of each other. It is surprising
therefore to find so little on correlation discussed in the literature concerning investment analysis. Hertz (3) provided a relatively simple treatment. For instance to represent a dependence of price distribution on
demand level he divided the demand distribution into a number of segments
and then for each segment provided a separate price distribution. His
model then used the price distribution appropriate to the demand level generated. This method is obviously suitable for data that is already segment-
ed, i.e. if it is available in histogram form.

Gershefski and Phipps (2) provided a Monte-Carlo program which allowed
for any degree of correlation between any two variables. This was done by
taking the next value generated as a composite of a random fraction and a
fraction determined by the value of the correlated variable, -- the fraction of each being determined by the correlation coefficient. This method
is particularly suitable when continuous distributions are given for all
the variables, but complications occur when the approach is applied to
histograms or other discontinuous distributions.

Other authors seem remarkably quiet on the subject, although Pouliquen
(10) emphasizes it as being extremely important.

Chapter V, Section 5.2d(1) describes a technique that has been developed for the PLADE program. To some extent the choice of this technique
was dictated by the need to keep the discrete characteristics of the demand and price forecasts, to be compatible with the rest of the PLADE
System. The technique does provide a means of allowing for different
degrees of correlation and the results are a good simulation of the way in
which market demands and prices do change, as shown in Figure 1 of Appendix
I.

To investigate this problem of correlation in detail, the PLADE program
has been run with a number of examples. The results are described in the
following sections.

8.5a   The Demand/Year Correlation

When a plan is being evaluated and the demand/year correlation is 1.0, those evaluations concerned with the lowest demand in the first year will all have this low demand for their whole life -- so forming the low end of the profitability distribution. Similarly, at the high end there will be a considerable proportion of results that have had the highest demand every year. With zero correlation the chance of a low demand staying low for its whole life is very small, therefore the low end of the distribution will have a much lower probability. Similarly, the highest profitability will have a much smaller probability. The lower the correlation coefficient, the less risk incurred -at least so long as a single plan is involved.

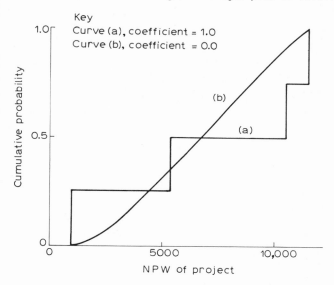

Key
Curve (a), coefficient = 1.0
Curve (b), coefficient = 0.0

Characteristics of distributions:

	(a)	(b)
Demand/year correlation coef.	1.0	0.0
Mean	7000	7000
Standard deviation	4200	3000
Characteristics of low profit end:		
95% chance the NPW will be above	1000	2100
90% chance the NPW will be above	1000	2700

Figure 8.2
Comparison of Evaluation with a Demand/Year Correlation
Coefficient of 1.0 with 0.0 (Case 1 of Chapter VI)

Figure 8.2 shows this clearly with both distributions plotted on the same curve. The data for Figure 8.2 was taken from Case 1 of Chapter VI, using a single plan determined as the best plan by the Level 3 analysis.

The importance of the Demand/Year correlation increases when a strategy rather than a plan is being employed. The principle behind using a strategy requires the demand to settle on to a particular demand level and stay there -- i.e. it relies on the demand/year correlation coefficient being 1.0. As the correlation coefficient falls from 1.0 the possibility of the decisions in the strategy being inappropriate for the later demands increases, which results in a lowering of profit. As the coefficient falls still further, the market movement is so rapid that to make decisions relying on a steady market becomes worse than making decisions good for an average market. In other words, there comes a point when it is better to stick to a single mean plan and forget about strategies.

Looking at the level of profit, one would expect the highest profit to be with a coefficient of 1.0 because this represent most knowledge of the system and hence one would expect maximum gain.

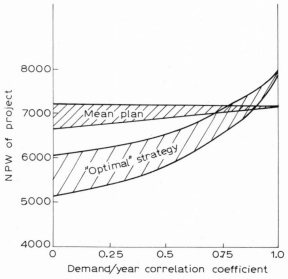

Hatched areas represent 95% confidence region
for the expected value from the Monte-Carlo simulation

Figure 8.3
Effect of the Demand/Year Correlation
Coefficient on the Evaluation of Strategies and Plans

These characteristics can be demonstrated with the optimum plant strategy from Case 1 of Chapter VI; running this example with a Monte-Carlo simulation of PLADE for a series of the different values for the demand/year correlation coefficient gave the results shown in Figure 8.3

Figure 8.3 shows that, for this project, if the demand/year correlation coefficient is expected to be less than 0.8 -- that is, if one expects changes in level about once every 8 years (see Section 8.6b), -- then there is no point in applying a strategy -- a single plan will be as profitable or even more profitable.

This crossover point will vary from case to case, since it will depend on the divergence of the levels, on the divergence of the plans making up the strategy, and on the profitability, which will govern the time required to recover the investment.

8.5b The Price/Demand Correlation

The effect of the price/demand correlation coefficient is described in Chapter II, summarised by Tables 2.2, 2.3 and 2.4. The definition of this coefficient made in Chapter V gives two extremes: that the price is bound to the demand level, or that it is completely independent. The importance of each in simulating a realistic price is defined by the value of the coefficient. Price is often the most important factor in an evaluation. It follows therefore that the value of this coefficient can be extremely important in evaluating the profitability of a project. To illustrate this by means of an example , Case 2 of Chapter VI was given four price levels and evaluated using PLADE with a zero and a 1.0 price/demand correlation coefficient, and a third run was made with a 1.0 coefficient, but with the correlation in the opposite sense. The three NPW distributions which represent high demand/high price, independent demand/price, and high demand/low price, are shown in Figure 8.4.

8.6 THE IMPORTANCE OF "SUBJECTIVE JUDGEMENTS"

8.6a Probability Estimates for Demand and Price Levels

All methods described for dealing with uncertainty rely on somebody making "subjective judgements". It can almost be guaranteed that, following any talk on investment analysis, one of the first questions will be implying that subjective estimates of probability are impossible. Though the questioner is likely to hold strong "opinions" concerning the outcome of

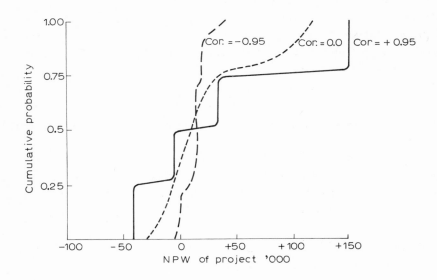

Example based on case 2 of chapter 6,
with price levels: 4 5, 42, 38 and 35.

	Project NPW '000	
	Mean	Standard deviation
+0.95 price corrl.	31	72
0.0 price corrl.	21	41
-0.95 price corrl.	10	9

Figure 8.4
The Effect of the Price/Demand Correlation Coefficient

events, he does not believe these opinions can be quantified to give, say,
"29% probability" -- and in that he is right. He would, however, agree
that his opinion could be expressed as a "nearer 1 in 100 than 99 in 100".
The general disagreement over subjective probabilities is really a question
of accuracy, but rarely is there any discussion on how accurate the subjective estimate needs to be to be useful. In fact, the accuracy need not
be very high for the following reasons:

    a)   each probability contributes only a fraction to
        the Expected Value.

b) since all probabilities must add up to unity, and an error in one changes the others, this provides some compensation on the Expected Value,

c) since the result is itself a distribution, there is no particular sense in knowing the Expected Value particularly accurately before making the investment decision.

But how inaccurate can it be?

Consider an investment with four possible market forecasts which produces a benefit/cost of 10, or 20, or 30,or 40% of the capital costs - a fairly wide range of outcomes. Assume also that the correct probability for each level is 0.25 -- the four levels have equal probability. The true Expected Value will be:

$$0.25 \times 10 + 0.25 \times 20 + 0.25 \times 30 + 0.25 \times 40 = 25$$

Now let us assume that in obtaining "Subjective Probabilities" our expert has made an error by under-estimating by half one of the extreme cases; his estimate was only 1 in 8 for this case, the other being again equal.

Hence, by normalising the set his estimates are :

$$0.125, 0.291, 0.291, 0.291$$

and the Expected result is:

$$0.125 \times 10 + 0.291 \times 20 + 0.291 \times 30 + 0.291 \times 40 = 27.4$$

The error, by a factor of two in the most sensitive level, has produced a 9.6% error in the mean value, which moves it over only 8% of its distribution range.

Now let us assume his twofold error was an overestimate:
In this case his probabilities are :

$$0.50, 0.167, 0.167, 0.167$$

and the Expected Value is given by

$$0.50 \times 10 + 0.167 \times 20 + 0.167 \times 30 + 0.167 \times 40 = 20$$

The error is 20% and the mean has moved 17% of the distribution range.

Much greater errors can be accepted for the inner levels without getting worse accuracy. For example, a three-fold error in Level 2 will result in:

$$0.083, 0.75, 0.083, 0.083$$

with a mean value of 21.7 -- an error of 13%.

All the accuracies quoted here are acceptable in an investment decision context , since they represent shifts of less than 20% within the distribution width. It is also important to know that errors in probability do not affect the maximum and minimum profitability but only the shape of the distribution and the mean value.

A sensible questioning technique for obtaining probability estimates
for sales forecasts would therefore be to present a restricted list of
chances and ask the forecaster to mark the one he feels most appropriate
for each forecast level. The planner can normalise these later, and return
for further questions only if the sum of the probabilities differs seriously
from unity.

Table 8.5 is a proposed questionnaire for the Sales Forecaster:

Table 8.5

Questionnaire for Estimation of Probability for Sales Forecast

Chance of Sales Level Occurring	Please tick Appropriate Row							
	Level 1	Level 2	Level 3	Level 4	Level 5	Level 6	Level 7	Level 8
1 in 100								
1 in 50								
1 in 25								
1 in 10								
1 in 5								
1 in 4								
1 in 2								
2 in 3								
3 in 4								
9 in 10								
24 in 25								
49 in 50								
99 in 100								

NB. Particular care should be taken with the estimates for the
     outer levels. Accuracy for the intermediate levels is less
     important.

8.6b Subjective Estimates for Correlation Coefficients

The estimation of correlation coefficients depends on the definition
of the correlation coefficients. Since literature is very sparse in this
area, we will consider only the coefficients as defined in Chapter V which
were developed for the PLADE program.

Demand/Year Correlation Coefficient

There are two ways of making a subjective estimate of this coefficient; either one can study Figure 1 of Appendix 1, which displays typical market paths for different values of the coefficient, and choose the figure most appropriate to the product in question, or alternatively, one can try to be more quantitative.

A zero correlation for all but the extreme levels means that each year there is an equal chance that demand will move up, move down, or stay on the same level next year as this year. Correlation of 1.0 means that demand stays on the same level until the end of the project. Correlation of 0.5 means the probability of demand staying on the same level as this year is :

$$0.5 \times 1.0 + 0.5 \times 0.33 = 0.66$$

That is, there is a 1 in 3 chance of a new level every year -- that is on average it stays on one level for three years before it moves off.

This means the correlation coefficients can be estimated if the forecaster is willing to estimate how often he would expect a change of level during the life of the project.

For example, if he thought three times in a 15 year life project there may be market upsets that cause a change of level, then the mean time between changes is 5 years, and so the yearly probability of leaving is 0.2. Hence, the probability of remaining on the same level is 0.8, and we can determine the correlation coefficient C_D by the equation:

$$C_D \times 1.0 + (1-C_D) \times 0.33 = 0.8 \qquad \text{hence} \quad C_D = 0.70 \qquad (8.1)$$

Table 8.6 relates an adequately spaced set of steps in the correlation coefficient to the equivalent average number of years between probability level changes.

Table 8.6
Relation between Expected Number of Years on a Fixed Level and the Value of the Demand/Year Correlation Coefficient

Average No. of Years on Same Level	Equivalent Correlation Coefficient
"will never change"	1.0
20	0.92
10	0.84
5	0.69
3	0.5
2	0.25
1.5	0

The Price/Demand Correlation Coefficient

A discussion, similar to that above for the Demand/Year correlation, can also be developed for the Price/Demand correlation coefficient. With this case, the correlation is between the price level and the demand level.

The value for the coefficient can be estimated again either by comparing with the graph, shown as Figure 1 of Appendix 1, or by the following more numerical method.

A coefficient of 1.0 means that 100% of the time the price will correspond to the demand level for that year. A value of 0.0 means that the price has an equal chance of being on any level, any year, regardless of the demand level. If there is a 70% chance of the price level corresponding to the demand level and the project has four levels, the correlation (C_S) is given by:

$$C_S \times 1.0 + (1-C_S) \times 0.25 = 0.7 \qquad (8.2)$$

therefore

$$0.75 \ C_S = 0.45$$

and $$C_S = 0.6$$

Table 8.7 gives the relationship between the probability of price being at the same level as demand, and the price/demand correlation coefficient.

The Forecaster should be asked to make a subjective judgement for the probability level and Table 8.7 then gives the appropriate correlation coefficient.

Table 8.7

Relation Between Probability of Price Level
Corresponding to Demand Level and the Price/Demand
Correlation Coefficient

| | Coefficient | | | | |
| | No. of Levels in Forecast | | | | |
Probability	2	3	4	6	8
0 - "demand and price independent"	0.0	0.0	0.0	0.0	0.0
30	0.0	0.0	0.07	0.16	0.20
40	0.0	0.10	0.20	0.28	0.31
50	0.0	0.25	0.33	0.40	0.42
60	0.20	0.40	0.47	0.52	0.54
70	0.40	0.55	0.60	0.64	0.66
80	0.60	0.70	0.73	0.76	0.77
90	0.80	0.85	0.87	0.88	0.88
100 "demand and price completely dependent	1.00	1.00	1.00	1.00	1.00

8.7 INTERPRETATION OF CUMULATIVE DISTRIBUTIONS OF PROFITABILITY

Having obtained a profitability distribution for a project, the last step is to translate it into reasons for and against accepting the project, or, in a study involving selection of an alternative, how to use it for making the selection.

Profitability distributions are generally presented as cumulative and not frequency distributions because frequency distributions have no direct scale on the Y axis, and so are more difficult to interpret than cumulative probability distributions.

The cumulative probability distributions are at first sight confusing, as they are generally nondescript shapes with the curves of competing alternatives usually crossing each other at least once. However, order can easily be put into such a diagram by drawing in the 50% cumulative probability line, (see Figure 8.5). The order in which each curve crosses this line is an initial ranking of projects based on 50% probability values. Where the curves cross this line is neither the Expected Value or the "most likely"

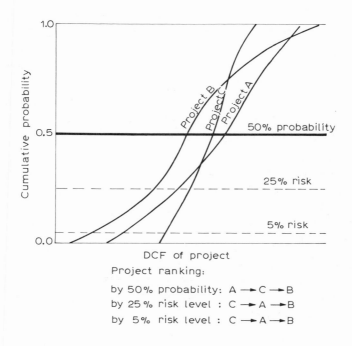

Figure 8.5

Interpretation of Cumulative Probability Profitability Distributions

but it is some guidance for giving an intial ranking of the projects in
terms of mean profitabilities.

The next step is to consider the risks associated with the various
alternatives. There are two reasonable risks to consider; firstly, the
maximum possible loss that could be expected to occur, though the chance
be small, and the second is the loss that is quite likely to occur.

The first of these corresponds to looking at the start of the curve.
Although there is always a finite chance of a total loss, -- explosion,
vast damage claims concerning the product, expropriation by revolutionaries,
etc., -- these are usually taken as normal business risks and excluded from
the quantitative risk analysis. On similar grounds it is usual to quote
not the beginning point of the curve but the risk with the 1 in 50 or 1 in
20 chance of occurring. That is, to quote the 0.02 or 0.05 level of cumulative probability as the minimum profitability level, and not the very
start of the cumulative distribution curve.

Because of the very low chance of this minimum profit condition being
realised, it is useful also to quote a low profitability that is quite
likely to be realised -- say, the 0.25 cumulative probability level.

These criteria are marked on Figure 8.5. In this figure, A has the
highest 50% probability, DCF, but has a risk of being much less profitable
than C. B has both a lower 50% profitability and a higher risk, and so can
be rejected. A choice between A and C depends on the importance one places
on the small chance of a greater loss compared with the larger chance of a
greater profit. Expected Values must also be considered because they
include the higher profitability end of the curve which, up to now, we
have neglected.

This "other end" can also be included by quoting the 75% cumulative
profitability (or the mean 50% range, i.e. the 25%/75% values), since this
represents the profitability range that is reasonably likely to occur.

It is often difficult to make a selection from cumulative probability
graphs. All the information is there but a quantitative comparison fails.
Utility theory is aimed at removing this difficulty associated with the
final interpretation of the distribution. Given a certain attitude to
risk, summarised by a Utility Function, the Certainty Equivalent for the
distribution can be evaluated and the alternative with the highest Certainty Equivalent is the most attractive project.

Chapters III and V describe Utility Functions at some length; the
PLADE program evaluates the Certainty Equivalent of its cumulative

distributions; but generally management has not yet understood the technique and so the final stage of investment analysis usually takes the form of a qualitative discussion.

8.8 CONCLUSIONS

This Chapter began by discussing various points that have an important effect on the results of an Evaluation but that are frequently neglected. Capital investment sums are so high that investment decisions should never be based on incomplete and inaccurate analyses.

When a careful detailed analysis has been made, including proper allowance for all factors, there is much more confidence in the figures given and the recommendations made concerning the decision to be taken.

As a fitting summary to this chapter, a number of interpretations of probabilistic planning studies are quoted -- since this, after all, is the final objective of any planning study.

Quotation 1.

Summary of a Project involving the Building of a Chain of Hotels (1)

"There is a 0% chance of exceeding 30% DCF return. There is a 65% chance of achieving or exceeding a 26% DCF return, and there is a 100% chance of the DCF return being greater than 22% It can be seen that the chance of the DCF occurring falls in the relatively narrow band of 22-30%. This indicates the 'good risk' characteristic of the project."

Quotation 2.

Summary of an Analysis for a Proposed Chemical Plant (8)

"The risk analysis provides a clear picture of the likelihood that profits will depart from the expected level by any given amount. If the minimum acceptable return is taken as 10% for example, the figure shows that there is only a 48% chance that return will be above the minimum acceptable level. The chance that present worth above 10% will be positive is necessarily identical. A reasonable range in which profits might fall can be taken as the 50% probability band centered on the expected result. Thus, the figure shows that there is a 50% chance of a return on an investment between 6.8 and 11.8%. In NPW terms there is a 50% chance of present worth between $-1.7 million and + $1.5 million.

The extremes of the distribution show that there is very little chance of reaching rates of return below 2% or above 16%, or of reaching present worth below -$5 million or above + $5 million."

Quotation 3.
A Highway through Africa (10)

"The probability analysis indicates that while there is a high degree of uncertainty with respect to some of the major variables there is a relatively small risk that the project will not yield a satisfactory return. The analysis showed that there is less than 5% probability that the project will earn less than a 10% return for Tanzania -- i.e. more than 95% probability that the rate of return will be over 10%. The mean returns under probability analysis are 20.1% with the Zambian traffic and 15% without."

Quotation 4.
Evaluation of a Proposal to Build a Port in Africa (10)

"We find there is a 99% probability that the rate of return will exceed 5%, a 94% chance of its exceeding 7% down to a 2% chance of exceeding a 15% rate of return. We find there is about 40% chance that the rate of return will be between 10% and 13%. The figure also shows that the probability of getting a return inferior to 12.2%, the rate of return we obtained in the conventional analysis using best estimates for each variable, is 70%."

Quotation 5
Summary of a Project to Build a Manufacturing Plant

"The expected after-tax profitabilities are 12.8%, 11.3%, 11.2% and 10.54% for the plants with annual capacities of 5,10,15 and 20 Mio. packages, respectively. There is an 85% chance that the smallest plant will offer the highest expected return compared to only a 15% probability for the plant with a capacity of 13 Mio. packages. An estimated £2.5 Mio investment corresponding to the smallest plant with a capacity of 5 Mio. packages per annum yields at 12.8% the highest expected, after tax, profitability.

The downside risk exposure of the proposed venture is acceptable. Based on our assumptions, there is only a 5% chance that the profitability will fall below 7%.

On the other hand, the manufacturing operation does not promise to be highly profitable. There is a 25% chance that we can achieve a profitability above 15% but only a 5% probability that the return on investment will exceed 18%

In accordance with the results of this analysis and in view of the uncertainties pertaining to the long-term economic outlook, implementation of the smallest plant is recommended.

Should it turn out that higher price levels can be obtained we should then bring to mind that an early plant expansion may improve the consolidated profitability of the business."

8.9 BIBLIOGRAPHY FOR CHAPTER VIII

1. Cameron, D.A., Risk Analysis and Investment Appraisal in Marketing, Long Range Planning, 5, 43,(Dec. 1972).

2. Gershefski, G., and Phipps, A.J., A Higher Order Computer Language for Risk Analysis, Decision and Risk Analysis Summer Symposium Series VI, The Engineering Economist, 131, 1972.

3. Hertz, D.B., Risk Analysis in Capital Budgeting, Harvard Business Review, 42,   95, (1964)

4. Hess, S.W., and Quigley, H.A. Analysis of Risk in Investment Using Monte-Carlo Techniques, Chem. Eng. Prog. Symp. Series No. 42, 59, 55, (1963)

5. Hirschmann, W.B., Harvard Business Review, 42, No. 1,   125 (1964).

6. Liddle, C.J. and Gerrard, A.M., Application of Computers to Capital Cost Estimation, Inst. Chem. Eng. London 1975.

7. Naylor. T.H. Computer Simulation Experiments with Models of Economic Systems, Wiley, New York, 1971 .

8. Malloy, J.B. Risk Analysis of Chemical Plants, Chem. Eng. Prog. 67, 10, 68, 1971.

9. Malloy, J.B. Projecting Chemical Prices, Chem. and Indus.942, (1974).

10. Pouliquen, L.Y. Risk Analysis in Project Appraisal, World Bank Occasional Paper, No. 11, John Hopkins Press, Baltimore, 1970.

11. Reutlinger, S. Techniques for Project Appraisal under Uncertainty, World Bank Occasional Papers, No. 10, John Hopkins Press,Baltimore, 1970.

12 Twaddle,W.W., and Malloy J.B. Evaluating and Sizing New Chemial Plants, Chem. Eng. Prog. 62, 7, 90, (1966).

CHAPTER IX

INVESTMENT DECISION
OUTSIDE THE PROCESS INDUSTRIES

> Chämifäger, schwarze Maa,
> häd es schwarzes Hämpli aa,
> nimmt de Bäse und de Lumpe,
> macht die böse Buebe z'gumpe.

9.1 INTRODUCTION

The last 8 chapters have been concerned with investment in the process industries--i.e. Chemical, Petroleum, and Food, and in the chemical industry in particular. All discussion and examples have been drawn from this field, although the problems associated with making decisions concerning capital investment are met much more widely. The rationale behind this restricted approach has been to develop and discuss the planning problems in a specific (well understood by the author!) context.

The objective of this chapter is to show how the planning problems of other industries are similar to the process industry, and show where they can be tackled by similar techniques. It is by no means an exhaustive survey, but merely a collection of examples to spark the imagination of the reader. It is basically the job of the planner to use his own imagination to isolate the points from this "planning in the process industries text" which are useful within his particular industry.

Investment planning problems in the process industries have the following characteristics

  (a) A number of alternative solutions can be proposed, which can
       be compared economically.

  (b) The cost data on which the economics is based is uncertain.

  (c) The future performance demanded from the investment is uncertain.

and (d) The final capacity of the project can be achieved as a series
       of further investments.

Two steps are involved for the making of investment decisions

  (a) The planning step, where the best principle and best

"capacity" has to be selected for the investment.

followed by

      (b) The evaluation step, where the best decision is more carefully evaluated and compared with some accept/reject economic standard.

Any planning problems which have all, or some, of these characteristics can benefit to some degree from the techniques discussed in this text.

9.2 PRODUCTION ENGINEERING

Production engineering, the mass-production of goods such as cars, electrical equipment, textiles, furniture, etc. is discussed first because it is so similar to process engineering.

A capital investment (a production line) must be made, which must be followed by fixed costs (factory and management overheads) and variable costs (raw materials, wages, and services) in order to produce large quantities of a product which will command an uncertain price and will sell in uncertain amounts.

The capital investment could take a variety of forms:

    A new factory

    A new line in an existing factory

    A revamp of an existing product line

    A well-automated line

    A partially-automated line

    The modification of an existing product line to include the new product.

Later production capacity increases could be achieved by

    (1) Building a second factory

    (2) Building a second production line

    (3) Uprating an existing product line by improving the automation

    (4) Going to shift working on the existing line

    (5) Contracting part of the work to be done to an outside firm

    (6) Removal of a different product from a multi-product line and providing separate manufacturing facilities for this product.

In fact, production engineering presents the same type of, though more difficult, planning problem than any process industry problems. Whereas it

is admissable for a chemical plant to work at part load for a number of
years while the market builds up, a car plant, because of its lower economies of scale and a more fluid market, cannot adopt such a policy. Here the
planning problem is more difficult because it is essential to consider
future expansion plans to maintain high occupancy of automated machinery
and labour. Markets are difficult to forecast, and in the consumer market
there is a high price/demand correlation.

The PLADE program which embodies much of the content of the work, has
successfully been applied to the analysis of a packaging plant which,
although it was within a process industry, had more the characteristics of
production engineering than process engineering.

9.3 CIVIL ENGINEERING

Civil Engineering projects usually take the form of extremely expensive,
long-life investment such as roads, bridges, dams, docks, irrigation projects
and the like. They rarely involve selling a product, and the concept of
price and demand is missing, although "capacity" has a definite meaning.
The projects are usually undertaken for communal or social reasons, and so
the idea of profit is also irrelevant.

In place of these missing objectives are two goals, firstly "minimum
costs"--that is total payment of capital and running costs that must be
made during the project life, and secondly "worthwhileness"--that is, that
the benefit that the community will accrue from the project will cover the
total effort necessary to carry it out.

There are two goals corresponding to two stages: "planning" to obtain
the "minimum costs" project, and "evaluation" to determine whether the
project is worthwhile.

Although community benefits from civil engineering projects are qualitative in nature--more convenience, better living conditions, more leisure
time--it is possible to attempt to express their benefits in monetary terms.
For instance, in highway engineering, standard methods have been developed
for estimating the value of saving journey time for different types of
vehicle, and there are methods for estimating the expected accident rate,
and figures for the mean cost of an accident. Hence the total benefit of
time savings and accident reduction produced by a road by-pass scheme can

be determined in money units, which can then be divided by the cost of the
project to give a ratio to be used to assess priorities, when funds are
limited.

Other examples of quantifying benefits in monetary units are airport
location studies and the justification of an underground railway system.
Cost-Benefit analysis is described by Frost (3) and referred to in Chapter 1.

The World Bank is interested in confirming that the money it lends to
developing countries is for "worthwhile" projects, and they have published
detailed studies of road and dock projects (6,7) which show how they estimate tangible benefits. Naturally, one would expect a bank to consider
only "tangible" benefits--reduced transport costs, reduced maintenance,
lower handling charges and damage costs--but an economic analysis using
only these sums, excluding social benefits, is necessary when money is
being raised from the money market.

Such calculations are full of uncertainty. There are the difficulties
in forecasting the capital cost of very large projects due to geological
and geographical features inadequately known by the engineers. The benefits
are also difficult to estimate accurately--forecasting future traffic
patterns is a difficult procedure, complicated by the autocatalytic effect
of a good road generating more traffic. If this analysis attempts to include a monetary benefit for social factors--increased leisure time, or
reduced noise--then the exact value of these constituents is open to discussion, and should therefore be treated as uncertain variables with a
distribution.

The World Bank papers (6,7) propose that a simulation program should
be written for every project. This may well be the best solution for the
extremely large projects, but for intermediate projects there is considerable scope for general purpose evaluation simulation models employed on a
routine basis.

The model could be very similar to the models for process-engineering.
For example, PLADE could be used to produce a distribution for returns on
capital for a road project by taking capital (with uncertainty), a set of
future demands (probablistic forecast traffic usage levels), price (unit
traffic benefit in time, wear and tear and fuel) and annual fixed cost
(maintenance and management costs).

Providing the data distribution for capital, price and fixed costs involves the aggregation of distributions for the individual components. Such aggregation can be achieved using the Combination of Variance Formula given in Chapter 2.

Civil Engineering projects also involve a planning stage; there are always a number of ways of achieving the same objective, and the best has to be chosen. Very often the choice is a multi-criterion problem of the type discussed in Chapter 3, Section 3.3 with "poor economics" being sub-jectively judged against "politically more acceptable" for example. When such comparisons can be reduced to a single monetary criterion, a choice can be made by selecting that alternative with the lowest total cost.

There are, however, planning situations where simple comparisons are inadequate. Take for example alternative road routeing problems, with each proposal resulting in a different saving to the user because of dif-ferent distances travelled at different degrees of convenience. Add to this a number of possible vehicle patterns, and we have a situation where the presentation of a Decision Matrix based on a total-life evaluation instead of a capital cost would be of enormous help in coming to a decision.

A second area where one is confronted with a difficult planning prob-lem is where multi-stage projects may be advantageous. It may be opportune to make a minor road improvement followed by a later additional two lane by-pass as an alternative to a single major four lane by-pass project (possibly delayed through lack of funds). On top of this add uncertainty in the traffic pattern and a high demand/year correlation and we have a classic first-decision problem. Another multi-stage investment example may be this three alternative comparison of:

> One four lane bridge
>
> <u>with</u> a two lane bridge plus the possibility of a second being built
> later if traffic patterns develop in a particular way.
>
> <u>with</u> a two lane bridge being designed to be extendable to four lanes
> at a later date.

A third example would be an irrigation scheme with somewhat unpredict-able benefit, since it is difficult to forecast the success of altering the ecology sufficiently to grow new crops. A pilot scheme, or a multi-stage project which can be modified or cut-short if pessimistic outcomes are realised, may be better approaches than that of the single large project.

These are examples of first-decision problems which could be handled by the planning stage of PLADE without modification. It is, no doubt unusual for such problems to be analysized in the depth advocated in this book, but there would appear to be adequate enough gains--particularly in the sense of completeness that a Decision Matrix conveys--for at least considering such an analysis for these problems.

9.4 ENERGY GENERATION

The generation of electrical energy is an activity which involves large capital investments and large integrated systems for which large numbers of alternatives are possible. It differs from the process industry primarily in its objective which is not to maximize profit but to minimize costs. It is generally accepted as a constraint that demand must be satisfied, and the price is so calculated that costs are covered.

Uncertainties exist, particularly with respect to the prediction of demand, but also in the cost of basic fuels. For example, when new power stations are required to meet future demand increases, oil (high raw material cost--low capital) coal (medium raw material cost, medium capital costs) gas turbines (low capital and highest running costs) or nuclear (highest capital costs, lowest running costs) are available. However there is nothing more volatile than the cost of coal or oil--the price of both are determined on political rather than purely economic grounds. Hence studies on the choice of fuel must include the uncertainty associated with these fuel prices.

Following the choice of process, further planning problems exist in sizing the generating stations. There are Economies of Scale that must not be ignored, but whether to install small plants, or medium plants catering for large areas, or large plants catering for larger areas to meet longer term future demands, are all alternatives that require careful study.

Clearly the contents of Chapters 5, 6 and 7 are appropriate for such problems. In fact a number of writers have used many of the techniques described here for energy generation planning problems. Dynamic Programming to select alternatives (5), first-decision analysis and the identification of Dominant Decisions (4) have been reported.

The problems of the gas industry are similar to those of electrical energy generating, with the gas source being natural gas (piped or liquified), oil cracking or coal pyrolysis, and similar problems of capacity

choice exist. It is interesting to note that the first reported paper in
the literature on optimal capacity was that of Chenery (2), concerned with
the capacity of a gas distribution system in the steadily growing market.

9.5 TRANSPORT

The transport industry, be it passenger or freight, air, sea, road or
rail, is concerned with making capital investments with the aim of making
a profit by charging for the service of moving things or people. Even
nationalised transport organisations work to this same objective, although
it is not hard to devise a more suitable objective when viewing a country
as a single system.

To choose the capital investments that yield the best profit it is
necessary to predict the operating costs of the transport service--which
are not too difficult--and predict the future transport demands--which is
much more difficult. Transport demands are affected by prices charged,
competitors actions, the general level of prosperity and simply by fashion
(it is presently more fashionable to travel than to fish for one's holiday).
In effect, a transport demand forecast could reasonably be represented by
a series of possible profiles, with a high price/demand correlation, each
with a probability level.

The type of capital investment can be concerned with:

Very large single investments--representing the building of a new
railway or purchase of a passenger liner;

Or the important policy decision that implies a series of future
investments based on that decision--for example, the decisions
to build up a lorry fleet, or a Concord fleet, or to initiate a
form of container rail service.

Or the purchase of a single unit to enlarge a fleet.

The choice from alternatives is an important activity. The selection
of Jumbo Jets or smaller craft, the selection of size or make of lorry, the
selection of diesel or electric train must all be made as a result of
comparative cost studies.

Described in this way there seems to be little difference between the
Transport Industry and the Process Industry. Both are concerned with
estimating profitability in the face of uncertainty, and accepting or

rejecting projects on the basis of the profitability criterion. Both are
concerned with decisions to enter markets with the implied decision to make
later investments in that field if business turns out well, but make fewer,
or no later investments if the demand proves to be pessimistic. Both are
concerned with choosing from alternatives.

9.6 RESEARCH AND DEVELOPMENT DECISIONS

Because the process industries have a strong research and development
interest, this area has already been mentioned in the main text. However,
it is included in this review of application areas because it is a particularly important area also for mechanical and electrical production
engineering, for aeronautical engineering, steel, mining, in fact any
industrial activity.

The objective of all industrial research and development is to discover a technique which will eventually, sometimes in the very long term,
benefit the organisation concerned. Decisions have to be made in research
and development departments in selecting those projects on which to concentrate effort. The decision should be made by determining the profitability
of the individual projects and then selecting the most profitable. Once
more there are the familiar ingredients--the capital being the future
research and development costs (which naturally are uncertain), the benefit
being a future, often continuing, benefit from applying the result of the
research. If the research is successful the benefits may be high, if the
research is unsuccessful they may be zero. Hence these benefits are a
distribution which could be conveniently summarised as a histogram with
each level having a subjectively judged probability.

The techniques described in the main text of this book, and the program
described in the appendix could profitably be employed to evaluate the
profitability distribution of each project, and then evaluate the Certainty
Equivalent and other Decision Criteria, in order to determine whether the
individual projects are themselves worthwhile.

Much of the research in the chemical industry is concerned with the
creation of new products and new processes. The benefit of this research is
the total profitability of the new process less the research costs yet to
be spent. When this situation exists the PLADE program is ideally suited

to provide quick answers as to how worthwhile is the research being carried
out.

The selection of worthwhile projects is only part of the research
manager's job. He probably has more potentially worthwhile projects in
hand than staff and money to carry them out. He then has a resource allocation problem, rather like the problem a company has in allocating its
funds to new investments. He should choose to keep high profits but choose
also to add stability to the department in event of some pessimistic outcomes, but have sufficient relation between the projects to build up expertise and maintain a staff flexibility. This is a portfolio selection
problem, which is beyond the scope of this book.

9.7 NON-ENGINEERING AREAS

Although the intent has been to handle only engineering investments, a
few words concerning purely business investments are worthwhile.

Any activity which follows to some degree points a-d given in the
introduction of this chapter, should benefit from risk analysis.

Many business ventures--the development of a chain of stores, or
restaurants, or hotels--fall into this category. Investments--property
rather than engineering--are made in the hope of recovering good profits
from the venture. Forecast sales, prices or hotel occupancy are subject to
uncertainty and so any realistic study must result in a distribution of the
profitability. The decision-maker should consider the maximum loss that
might be incurred and its probability as well as the Expected Value, before
making a decision (1). Since such investments are rarely made at one
instance in time, but built up over a number of years, there is the possibility of making future investment plans to hedge against losses or to take
advantage of optimistic situations.

This description applies equally to a process industry or to an owner
of a hotel chain. The same decision techniques can apply, and even similar
computer programs can carry out the analysis.

9.8 CONCLUSIONS

This last chapter has been written to indicate that the planning and evaluation techniques described in the main body of this book are applicable to a much wider area than merely the process industries.

The chapter is no more than a post-script to remind the reader that professional boundaries are largely artificial, and the overlaps are much greater than is normally appreciated.

After all, has this book been about Chemical Engineering or Production Engineering or Operations Research or Econometrics or Economics?

9.9 BIBLIOGRAPHY FOR CHAPTER IX

1. Cameron D.A., Risk Analysis and Investment Appraisal in Marketing, Long Range Planning 5, 43, (Dec. 1972).

2. Chenery, H.B., Overcapacity and the Acceleration Principle, Econometrica, 20, 1, (1952).

3. Frost, M.J., Value for Money--The Techniques of Cost Benefit Analysis, Gower Press, London, 1971.

4. Manne, A.S., Waiting for the Breeder, IIASA Report RR-74-5, Schloss Laxenburg, 2361 Laxenburg, Austria, 1974.

5. Petersen, E.R., A Dynamic Programming Model for the Expansion of Electrical Power Systems, Management Science, 20, 656, (1973).

6. Pouliquen, L.Y., Risk Analysis in Project Appraisal, World Bank Staff Occasional paper No.11, John Hopkins Press, Baltimore, 1970.

7. Reutlinger S., Techniques for Project Appraisal under Uncertainty, World Bank Staff Occasional Paper No. 10. John Hopkins Press, Baltimore, 1970.

APPENDIX 1

PLANNING AND INVESTMENT ANALYSIS

M A N U A L

USING

P L A D E

for Planning and Evaluating Extensions
to Manufacturing Capacity

MANUAL CONTENTS

2 SCOPE OF THE PROCEDURE

A method has been developed for searching for the best from a number
of presented alternative ways of investing in plant manufacturing capacity.
The method concentrates on the first decision, but takes into account later
decisions in order to enable alternatives with very different lifetimes to
be properly compared. This search gives both the best alternative and the
best time at which to install it. In addition, best estimates of the
timing and choice of later decisions are given, for although these may be
changed by later information, we are presented with useful information for
longer range planning and research guidance.

The method accepts forecast data in probabilistic form and accepts
that decisions other than the first will depend on how the market develops.
It develops these alternative decision sequences before choosing the first
decision. In this way, flexibility and risk are introduced into the
criterion for the selection of the first decision.

Following this planning stage a detailed investment evaluation is
carried out and NPW, DCF, Benefit/Cost Ratio, Equivalent Annual Value, Pay-
Back Time and annual cash flows can be obtained. A sensitivity analysis
can also be performed.

If the planner is lucky enough to have any of the data in probabilistic
form, the investment analysis can produce a distribution of the profitability criterion and an estimate of the risk as well as the Expected value.

Though the method has been developed to handle problems involving
multi-alternatives and probabilistic data, the programme can very easily
be used for simple investment evaluations. It has therefore the potential
of being a company standard procedure for all its evaluations.

3 COLLECTING THE DATA

3a The Selection of Alternatives

The planning phase requires a set of alternatives to be presented so
that the search for the best combination can be carried out. The type of
alternatives: extensions, new plant, new processes etc., are produced from
earlier studies and by this stage these possibilities must have been
collected together. However, in many cases the planner has to determine
the capacity of these alternatives. This has to be done using common sense

and intuition, since the planning procedure checks the suitability of a proposal rather than generating capacity levels. In selecting capacity levels the following points should be considered:

a) The proposed set of alternatives must be capable of being built up into sensible looking plans, capacity-wise, for the whole project life.

b) The search method does not differentiate between capacity and process, hence the chosen plan may include a particular process, not because it is the best process, but because it has been given the most suitable capacity. This can be overcome in difficult cases by providing alternative processes with identical capacity levels. If adequate capacity of the preferred process is not available, the less attractive process will be introduced simply to make up the capacity.

c) When probabilistic forecast data is used, the capacity level should enable intelligent plans for every probability level to be built up.

d) The capacity level need only be considered in discrete steps which have a practical significance, for instance, 1,000, 1,250, 1,500 or 2,000, 2,500, 3,000 tons/year.

When there is no clear choice of capacity level possible the problem can be presented as a series of sub-problems and individual comparisons made on each sub-problem. Out of this will come an understanding of the system to enable intelligent capacity levels to be presented for the final search.

3b Defining the Alternatives

There are two particular constraints that must be defined before the planning search can begin. Firstly, if the alternative is a physical extension of another plant, it can only be installed or operated if the parent plant is operating. Furthermore, if there are two mutually exclusive extension projects, one extension can only be installed if the other extension project has not already been carried out. Hence, this extension data must be communicated to the search technique.

Secondly, if some projects cannot be available until a certain time, either because they are large projects requiring extensive engineering, or new projects requiring further research and development,--they must be kept out of any search plans until they will be available. This is done by defining a "first year available" for each alternative. This gives the number of the year at the beginning of which the alternative will first be operating. Hence, 1 means it is immediately available for inclusion in the search. Most alternatives will neither be extensions nor have delayed availability.

The list of alternatives includes all the ways of manufacture, including those already existing and operating. It is therefore necessary to define whether the alternative is already operating, and this is transmitted to the programme by 0 or a 1 flag as part of the data with each alternative. This gives the starting point for the search and also enables the base-case "do not invest" for both the planning and evaluating phases to be calculated.

The remaining data to define each alternative are Capital Cost, Annual Fixed Cost, Variable Cost and Scrap Value. All these are in constant value money units, i.e. inflation corrected money. Hence, despite inflation, the fixed and variable costs remain constant every year, and can therefore be represented by single numbers. The cost data should be given in constant value money with the beginning of year 0 as the basis, except for capital, which requires year-1 money units, since that is when the money could be spent. The programme makes the necessary modifications to capital cost and tax allowance introduced by the inflation rate, which is read in as a separate data item.

The division of yearly costs between fixed and variable should be done so that the variable cost properly reflects the marginal cost of production. These data can be better obtained as the axis intersect and slope of the production/total annual cost curve than by assigning total costs of various cost centres either to "fixed" or "variable". Should changes be expected over the years to these costs for reasons other than inflation, then variable cost changes can be incorporated by modifying the yearly sales price data, and fixed costs changes by the yearly additional cash flow facility.

The scrap value is the definite value that the plant has, should the manufacture of the product under question cease. It is useful in planning studies if the plant could be employed on some other definite product. The scrap value should not be used to convey vague corrections to represent

further plant use beyond the project lifetime. This should be done by the
Extended Life Method, described in detail in Section 5.2c of Chapter 5.

If any of the capacity, capital, variable or fixed costs or scrap value
data is known probabilistically, it can be incorporated into the analysis
by the four characteristic points of the trapezoidal distribution. Chapter
2 describes some techniques for evaluating these four values. If any of
these data are summations of items whose distributions are known, then the
distributions for the sums can be obtained by analytically aggregating the
individual distributions, as described in Section 5.2c of Chapter 2.

3c Short-Fall, Penalty and Working Capital

These two data items represent less important costs which are sometimes
necessary in planning studies.

The Shortfall Penalty is a cost/unit of market not satisfied, when
demand outstrips capacity. It has been incorporated in the technique
because it is a commonly held concept in some fields--e.g. warehousing.

However, its use is not to be generally recommended because of the
difficulty in setting a sensible value for it. An ill-chosen value can
completely dominate any search, or the result of any evaluation. It should
not therefore be used to represent a vague "loss of future market" without
intensive justification. It should be used to represent definite losses--
contractual payments on failure to supply, or loss in profits from later
processes using the product as raw material within the same company. It
does not represent loss of profit on the short-fall: the economic evaluation
itself allows for this.

Company policies to meet at least 90, 95, or always 100% of the market
are better incorporated with the lower bound search constraint (Section 4),
rather than by using short-fall penalty factors.

The working capital is important in economic evaluation but not
usually so in optimum planning studies. Hence there is no provision to
incorporate working capital in the planning stage but this is incorporated
in the evaluation phase. Working capital is a cash outflow which is not
allowed against tax.

A correct risk analysis should use the "Cost of Capital" for the
discount rate, and incorporate risk by a probabilistic data input. However,
when probabilistic data is not available, risk can be treated in an

evaluation by discounting at a somewhat higher rate than the "Cost of Capital". When this is done, working capital may still be discounted at the "Cost of Capital" because it is comparatively risk-free. This approach can be incorporated in the computer evaluation by modifying the value of the working capital. An accurate evaluation can be obtained by entering:

$$\text{Working Capital} \quad \text{x} \quad \frac{\text{"Cost of Capital"}}{\text{Discount Rate}}$$

in place of working capital. The working capital is automatically re-covered at the end of the project. The short-fall penalty can be probabilistic, again by expressing it as the four points of a trapezoidal distribution. Working Capital is correlated completely with demand. If a probabilistic demand is given, working capital for each demand probability level must be supplied.

3d Market Forecast Information

Any form of sales volume/year relationship can be incorporated because the data is read in as yearly sales volume and sales price figures. The sales volume can be increasing or decreasing, it may have a maximum or a minimum. The sales price can likewise have any relationship with respect to time.

The units of sales price are again constant value money. Hence, if the only factor influencing price is inflation, the price data per year will be constant. Price changes will be brought about only by price changes in real terms--a reduction in profit level after a new product has been established, for instance. The yearly price data can also be used to allow for yearly changes in variable cost. If some non-inflation cost changes are to be expected in variable cost--say an above average cost increase in some raw materials is to be expected--then if this cost change is subtracted from the yearly price data, the effect has been adequately represented for the study.

The basis for the price calculation must be carefully chosen. The price is that figure which gives annual cash inflow before tax as:

(Price-Variable cost) x quantity sold - annual fixed cost.

"Price" is a mean net realisation per unit sold, and it must have the same

basis as the variable and fixed costs. The basis could be "ex plant,"
"ex works," or "at customer". Each basis gives a different price, variable
cost and fixed cost because of works overheads, packaging and selling
expenses.

The data can also be probabilistic. When this is so, triple data sets,
sales demand/price/working capital per year, must be given for every
probability level, up to a total of 8 levels. Every attempt should be made
to obtain probabilistic sales demand data from the sales forecasters, and
4 levels are recommended as a good compromise between length of computation
and accuracy.

The provision of probabilistic sales data enables decisions to be made
which have a considered fall-back position, and the risks involved to be
quantified. Section 6.3 of Chapter 6 shows how one can calculate the
savings attributable to having probabilistic data available which enable a
proper risk analysis to be carried out. The planning technique can be
equally well applied to deterministic or probabilistic data, but its
advantages are greatest when the data is probabilistic. It is very wrong
for a sales forecaster to obtain an uncertain forecast and the reduce this
to a single one for a planner to handle. It is far better for him to spend
his time assessing the probabilities of his uncertainies.

It is also necessary to supply the programme with the degree of both
sales volume/price and sales volume/year correlation. The sales volume/
price refers to the likelihood that any one sales volume level will be
associated with any one price level. If the high sales level were the
result of the low price level, one would have complete correlation. The
correlation coefficient would be 1.0. If the cause for the price
variations and the sales variation were completely independent the coefficient would be 0.0. The coefficient is the fraction of the variation
due to the sales, and (1 - coefficient) is that fraction due to random
effects. For intermediate cases the coefficient will be between 0 and 1,
the actual figure being a judgement made with the aid of the forecasters.

The sales volume/year correlation coefficient is a very important
data item which again must be estimated subjectively. If the next year's
sales level is entirely dependent on this year's--as would be the case if
the only factor was the suitability of the product for various fractions of
the market--the sales volume/year coefficient would be 1.0. A zero coefficient implies that the next year's demand level is equally likely to be

the same as, or above, or below, this year's.

A rough guide as to the meaning of the correlation coefficient can be obtained from Figures 1a to 1c which show demand and price projections with various values for the coefficients.

The coefficients can be estimated by subjectively offering answers to the two following questions:

a) What is the probability that the price level will be associated with the demand level?
Completely dependent (100% probability), completely independent (0%), or three intermediate levels, (25, 50 or 75%) would be an adequate answer. Table 1 converts this probability to the sales/demand correlation coefficient.

b) How many years, on average, would you expect demand to stay on the same level?
(or: How many demand level switches do you expect in the project life?)
Every two years would represent a very volatile market situation. Once in five to ten years may be a normal situation, and one in twenty may represent a monopolistic position. Table 2 converts this estimate of number of years into the demand/year correlation

Table 1

Relations between Probability of Price Level Corresponding to Demand Correlation Coefficient.

Probability	Value of Correlation Coefficient Number of Probability Levels				
	2	3	4	6	8
0	0	0	0	0	0
20	0	0	0	0.04	0.08
30	0	0	0.07	0.16	0.20
40	0	0.10	0.20	0.28	0.31
50	0	0.25	0.33	0.40	0.42
60	0.20	0.40	0.47	0.52	0.54
70	0.40	0.55	0.60	0.64	0.66
80	0.60	0.70	0.73	0.76	0.77
90	0.80	0.85	0.87	0.88	0.88
100	1.00	1.00	1.00	1.00	1.00

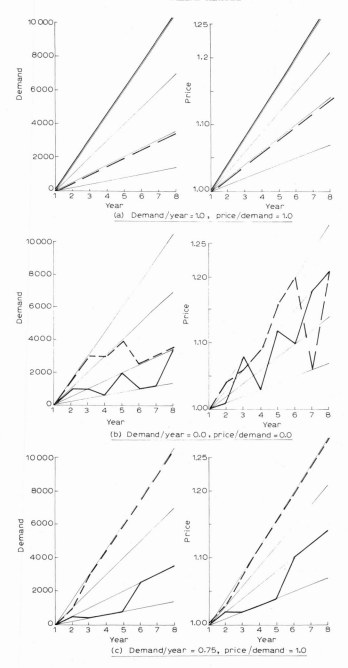

(a) Demand/year = 1.0 , price/demand = 1.0

(b) Demand/year = 0.0 , price/demand = 0.0

(c) Demand/year = 0.75 , price/demand = 1.0

Figs. 1a-1c, Forecasting Correlation Coefficients

Table 2

Relation between Expected Average Number of Years on a Fixed Level, and the Value of the Demand/Year Correlation Coefficient.

Average No. of Years on same level.	Demand/Year Correlation Coefficient
∞	1.0
20	0.92
10	0.84
5	0.69
3	0.50
2	0.25
≤ 1.5	0.

3e Multi-Product Plants

If the plant produces more than one product, either by campaigns or by means of an adjustable product mix, the demand and price for each product (up to a maximum of 10) can be entered for each year, for each probability level. The variable cost and plant capacity with respect to each product must also be given. The program converts the product mixed into a pseudo-tonnage and a pseudo-price for the first product and then treats the situation as a single product case.

The treatment should not be employed for multi-product plants when the plant has only one main product and a series of by-products. Unavoidable by-products should be allowed for in the variable cost by crediting the income as a negative raw material cost.

3f Investment Information

For the purpose of the computer programme, the project life is defined as the length of time for which forecast data has been supplied. After this time the plant can be either scrapped with the project benefitting by the scrap values, or run for a further defined number of years with an identical yearly cash flow to the last year of operation. This extended life has been found to be the preferred method in the planning phase with an extension period of 5 or 10 years. This also provides a realistic way of handling the end of the project for the investment phase. However this is contrary to normal practice and will result in a better profitability than scrapping at the end with a zero scrap value. If objections are raised, the evaluation phase can revert to the more usual scrapping procedure.

Care must be taken with the extended life procedure because it simply takes the last year's cash flow, excluding capital and working capital and extends these into the future. Hence, no plants can be shut down in the last year if the extended life option is employed. The recovery of working capital is done in the last year of the forecast data. The error introduced by recovering it at the end of the forecast data and not at the end of the extended life period is generally negligible. For the calculation of tax payments, straight-line depreciation is assumed over a definite number of years. No differences are made between "buildings" or "machinery" for depreciation purposes. Provision is made for reading in extra Alternative data which could be used for this, but the programming of this into the tax routine has been left to the user. Instructions are given in the Appendix for revising the individual taxation subroutines to enable users to program in their own specific tax regulations. Because the calculation is carried out in constant value money terms, when inflation rates are positive the real allowed depreciation sum reduces with time, giving correspondingly less tax relief. For this to be calculated, it is necessary to provide the programme with an estimate of the expected future inflation rate. The single figure is an estimate of the average inflation over the whole life of the project. This can be zero when inflation is being excluded from the evaluation, or the programme is being manipulated to carry out a current money calculation. The inflation rate can be given a probability distribution as the 4 points of a trapezoidal distribution for inclusion in the Monte-Carlo simulation.

The discount rate used for both the planning phase and the Net Present Worth evaluation phase should be the true cost of capital to the firm after correcting for inflation. If it is the firm's own capital, it should be equal to the firm's past achieved performance if this has been considered satisfactory, or alternatively, the average rate for that particular industry. It should not be artificially increased to "allow for risk" if probabilistic data is available and a proper risk analysis is to be carried out.

The procedure also uses the concept of Certainty Equivalent and Utility

to enable choices to be made in risk situations. As data it is necessary
to provide one figure--the minimum permitted profitability that the firm
can stand. This represents the loss that would so shake the firm that it
could not be contemplated even if the probability was less than 1%. The
planning stage uses NPW as the evaluation criterion and so this minimum
profitability figure will be a negative value corresponding to the NPW of
the highest acceptable loss. The evaluation phase can also be run as a
DCF evaluation, in which case, the minimum profit represents the lowest DCF
rate of return bearable by the firm.

4 THE PLANNING PHASE

Having assembled data together, the first stage of the procedure is
the planning phase in which the optimum timing and choice of alternatives
is determined. If the sales data is deterministic, or if it is probabilistic
but there is zero sales/year correlation, the single plan (mean plan) that
maximizes the Expected NPW can be planned. This is done by the OPLEX subroutine using the Dynamic Programming algorithm described in Chapter 5.

If the sales/year correlation is 1.0, the sales will always stay on
the same probability level, so the optimum plan for each probability level
can be determined by repeating the DP search deterministically with each
forecast level in turn. However, because the operating levels are usually
not known until the first decision has been operating, the programme has
been arranged to repeat the searches for a number of enforced first decisions.
Each of the searches with the same enforced first decision forms a strategy
for that enforced first decision.

If the sales/year correlation is between 0.01 and 0.99, reality lies
somewhere between the two extreme cases described above, and the final
choice has to be left to a subjective assessment using both extremes as
guides. Therefore, in this case, both mean plans and strategies are produced by the programme to enable this assessment to be made.

The management of these searches is done by subroutine EXEC so that
the PLADE programme carries out this sequence of searches automatically.
The OPLEX subroutine is the DP Search routine.

The NPW's of the various searches are displayed as a Decision Matrix,
with possible actions--the possible first decisions--down the RHS and the
possible states of nature--the probability levels--across the top.

Figure 10 is an example of the Decision Matrix output.

This Decision Matrix should be used to decide as to the best first decision to take. This can be done off-line, so enabling judgement as to risk or other non-quantitative or non-economic factors to be brought into the choice. To help quantify risk the Certainty Equivalent is also calculated wherever strategies are evaluated. This is effectively a weighted average which weighs losses more heavily than gains, with a loss equal to the maximum acceptable loss having infinite weighting.

Having chosen the preferred first decision, the conditional plan--in terms of years of building and scrapping alternatives--that make up the chosen strategy can be found from the part of the computer output that summarizes each search. (See Fig. 9.)

The planning phase is now over and the selected plan or strategy can be evaluated by the second stage of the procedure.

The inspection of the Decision Matrix is a very important stage in the procedure, and it should never be omitted. However, if the Certainty Equivalent is accepted as the decision criterion and it is not overridden by other judgements, and if the sales/year correlation is near 1.0, the progress to the later stage of the procedure can be done automatically. Since these conditions hold much of the time, the programme has been arranged to select the first decision based on the best Certainty Equivalent or, in default, the best Expected Value, and move on to the next stage without the need for manually transferring the optimum strategy data from the output back into the programme for the evaluation phase. This facility must be used with great care, and the Decision Matrix must always be inspected to ensure that the choice of strategy made by the programme is sensible.

There are two further data items associated with the planning phase which require some explanation.

As well as enforcing a certain alternative to be the first decision, it is possible to enforce the year in which that decision is taken. In some planning studies it is interesting to know the optimum year for the first decision so that this year can be used as the start-up date for the new investment. In other planning studies, it is necessary to enforce the year of the first decision as well as enforcing the alternative. In particular, when strategies are being developed for different sales probability levels, following an enforced first decision, the time of this decision must also be enforced if it is necessary that the start-up data of the

first decision be the same for all probability levels. This situation
usually exists when there is no way of determining which level will occur
before the start-up date. A second situation where the year of the first
decision must be fixed is in the "If we go-ahead and install the plant
next year ...." type of planning activity.

The DP search also requires capacity constraints to be provided to
reduce the length of the search and concentrate in sensible areas. The
lower limit prevents any plan being accepted if in any year its capacity
is less than a fraction of the lowest forecast for that year. Fractions
between 0.5 and 1.0 are suitable. In deterministic studies this can also
be used to enforce management policy--for example "90% (or even 100%) of
the demand will always be satisfied." Because this could cause discarding
of attractive plans if very late decisions in the plans do not meet the
constraint, for the last 30% of the project life this constraint is
automatically relaxed by the programme. The upper constraint is the factor
by which the capacity must not exceed the maximum sales forecast (regardless of year). This constraint is very important in reducing the search
to be within practical limits. An ill-chosen upper bound can result in
sensible plans being discarded in the same way that a too high lower band
can. This occurs when insufficient alternatives have been provided and
the upper bound is too near 1.0. If a good initial plan has to make up
capacity towards the end of a project by adding a large plant because there
is no other alternative left, and if this then lifts the capacity above
the upper constraint, then the whole plan is lost. The most satisfactory
solution is to provide further alternatives, but the high upper constraint
is useful as it prevents the plan from being entirely lost. Hence there
are disadvantages in having this factor too close to 1.0. Factors between
1.3 and 2.0 are a good compromise between reduction of computation and loss
of over-capacity plans.

5   THE EVALUATION PHASE

Having produced a plan of strategy, the next stage is the evaluation
stage. If the project is very straightforward, a single alternative for
instance, and the plan is obvious by inspection, the planning phase can be
omitted, and the project taken directly to the evaluation phase.

If plans are conditional, i.e. if a strategy is presented, or if any data is probabilistic, the evaluation should take the form of a Monte-Carlo Simulation. If there is no probabilistic data, the evaluation should be carried out without the Monte-Carlo Simulation.

These evaluations are carried out by three basic subroutines. INVEST does general supervision, FLOW evaluates NPW and other criteria, and MONTE provides random data and the general management of Monte-Carlo simulations.

Every effort should be made to obtain probabilistic data for all the input information. A simulation is probably the surest way of doing anything new, be it landing on the moon, designing a chemical reactor, or justifying an investment. To use probabilistic data is a serious attempt to simulate all possible paths the project could take. The resulting distribution shows what the chances are of the misfortunes combining and also what the profit level will be when this does occur.

The programme itself determines whether it should carry out a Monte-Carlo simulation or a single evaluation. If a Monte Carlo simulation is required, it simulates a sufficient number of times for the Expected value of the resulting NPW or DCF to be within a requested tolerance. The resulting distribution is expressed graphically and in tabular form, and summarised by the Expected value and standard deviation and as a Certainty Equivalent. The Certainty Equivalent is the single value which summarises the complete distribution after weighting for risk.

This information is suitable for presentation to management to enable them to make the final accept or reject decision. Figure 14 shows the significant Monte-Carlo output. This, and the Decision Matrix (Figure 10) are the two important results from this investment procedure.

If there is no probabilistic data, a deterministic calculation in place of the Monte-Carlo is carried out. A series of deterministic calculations are also carried out for the special case of probabilistic forecast data with both correlation coefficients being greater than 0.99. The advantages of this calculation mode is that it requires much less computer time than the Monte-Carlo simulation.

For all evaluation calculation modes, a sensitivity analysis can be carried out. The sensitivity analysis is a rough attempt at a Risk Analysis when probabilistic data is not available. It is very much better than nothing, but it is inferior to the Monte-Carlo simulation because:

a) the setting of the +ve and -ve limits of the limits are
    arbitrary, whereas a simulation uses estimated distributions

b) the combined effects of changes in more than one variable at
    a time cannot be determined from a sensitivity analysis.

The sensitivity analysis does find a use in deciding in what areas to
concentrate effort. For example, the first step in a study could be a
sensitivity run with a deterministic evaluation. The results would immediately point out what factors have the greatest effect on profitability.
This may result in re-directing the research or engineering effort into the
more profitable areas.

The sensitivity analysis can also be used with Monte-Carlo simulations.
Though it seems a mixture of concepts, there are possible occasions where
one has a distribution, for example capital cost, with a rider that there
is a chance that the whole distribution may be shifted by 10%. A sensitivity
analysis may then be of use in investigating the effect of this displacement.

6   THE PROGRAMME

The computer programme is obviously the centre of this procedure for
decision taking. It has been written in ANSI Standard FORTRAN IV, and
every care has beeen taken to ensure that it is transferrable between
machines. The listing is given as Manual Appendix II, and Manual Appendix I
gives the logic flow charts of the more important parts of the programme.

Appendix I also describes how the tax subroutine can be altered to
meet individual situations. The programme has made use of the high integer
size of the CDC machines to pack data. On machines with maximum allowable
integers less than 10^{10} (equivalent to 36 bit word size) some variables
must be made double-precision. This has not been done for the standard
version because it increases the DP search time. Instructions are given
in Appendix I as to how this change to double precision can be done. On
machines with maximum integer size below 10^{6} (word size below 24 bits),
further changes are necessary. These are again detailed in Appendix I.

6a   The Data Input

Numerical input data is free-format and Keywords are used throughout.
Hence, repeat runs can be made by repeating only the Keywords and data to
be changed. It is this principle that also enables consecutive runs to be

made--OPLEX, MONTE, FLOW for example.

Table 3 defines the input data required for the programme. Names and Keywords are not free-format. Keywords must be started in Column 1 and all alternative names must be within their designated field, columns 1-9. Following data should not encroach on this field.

The mode of operation can best be explained by referring to a simpli-fied, logic flow chart of the programme presented as Figure 2. This diagram shows that after every computation the programme returns to read a further Keyword. For instance, on returning from carrying out the instruc-tion PLAN, it has completed a DP search and put the best strategy into OPERATE. Hence, a following Keyword EVALUATE will carry out an evaluation. If the sales/year correlation coefficient and the sales/price correlation are both greater than 0.99, a single evaluation run will be made. If these coefficients have other values a MONTE-CARLO simulation will be performed.

Hence, the sequence of calculation recommended as the procedure for decision-making can be obtained in a single computer run by completing the data input for the problem with the following instructions:

PLAN

EVALUATE

STOP

Further Keywords can be entered before STOP. A sensitivity analysis could be called for or a different strategy employed through Keyword OPERATE or a complete new study with the exclusion of one of the alterna-tives, and so on.

There is no need to begin with a DP search if strategy data is given as input. There is no need to carry out an evaluation with the strategy chosen by the DP search. An EVALUATE following PLAN can use any strategy data given with the OPERATE keyword between PLAN and EVALUATE. Later data simply overwrites earlier data but all the data remain available to the programme until it is overwritten.

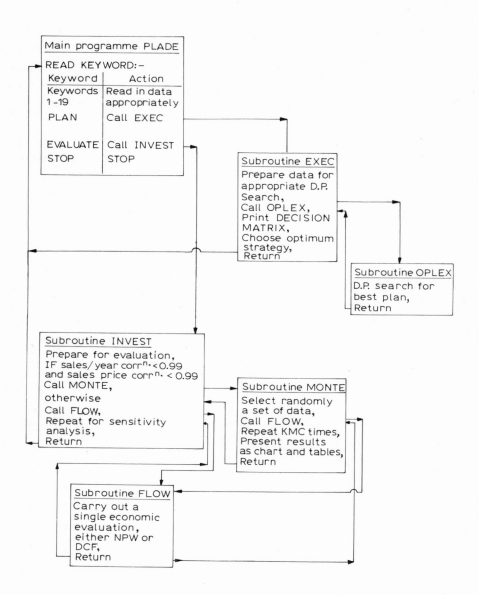

Figure 2: PLADE Programme Structure

6b Scaling of Numerical Values of Input Data

Because in practice the application of this planning procedure is
limited by the computation time required by the DP search, integer arithmetic has been used with the search to make use of the greater computational
speed of integer over fixed or floating point arithmetic in some machines.
Hence, errors are introduced due to round off error which can reach a
magnitude of around 10. There is also an upper limit for the size of
integer numbers, depending on the machine. This program has been written
for a range of -500,000 to +500,000 which is suitable for machines with
word lengths 20 bits or greater. Hence, the working range for NPW in the
search is ±500,000 with a sensitivity of 10. This necessitates occasional
scaling of the money units so the NPW will at all times be less than
500,000 and differences of less than 10 are insignificant. An adequate
rule of thumb for this scaling is that the capital cost should be within
the range of 250,000 to 1,000. When this is not so, the total cash unit
for the whole program input must be changed accordingly, £,000 for
instance.

Similarly, the demand is handled as an integer in the DP search and
so this again must be in units not greater than the allowed maximum
integer size and not so small that one unit represents a significant error.
Demands between 500,000 and 100 are recommended for 24 bit and greater
word size machines. There are no other restrictions on money units or
capacity units as long as they are consistent.

6c Estimating Computation Time and Problem Size

There are two restrictions to the size of planning problems that can be
tackled. Firstly, the program limitations that allow only 256 states to be
handled in any one year, and secondly, the practical limitation that the
computation time should not be excessive.

It is therefore useful to be able to estimate the maximum No. of expected
states and the expected computer time before the computation is made. The
most significant variables for these estimates are the number of alternatives
presented for the search and the number that can be operated simultaneously
without violating the upper or lower capacity boundaries. Figures 3 a-g
show charts which give guidance as to the maximum number of expected states.

Consider a problem with 9 Alternatives and end states consisting of
3,4 or 5 plants; then, the maximum number of intermediary states in any year
may reach 80 + 130 + 130 = 340 (from Figures 3b,c and d). This is greater
than the 256 the program can handle, and so if no other constraints apply,

Table 3

Input Data for PLADE Program

Notes:

1. Names must be accurately located within the field designated.
2. Numerical data is free format, separated by blank or comma, decimal point optional. All data read as real.
3. Money and quantity units may have to be scaled to get NPW within the ±500,000 range (See section 6b).
4. Each Keyword has a number. This can be punched in columns 77-80, as an aid to ensure that Keywords are submitted in order between Calculation Execution Cards (EX).

No. Cols. 77-80 (optional)	KEYWORD (Cols. 1-3 suffice)	Required by:		Following Data Cards A = No. of Alphanumerics N = No. of numeric data items	Default:	
		Plan	Evaluate		Allowed	Value Assumed when omitted.
1	HEAd	yes	yes	72A	Title of Study	yes	
2	DIScount	yes	yes	1N	NPW calculation requested, %Discount Rate as Data Item	yes	15
3	DCF	yes	yes	-	Initiates DCF evaluation in place of NPW (no following data card required)	yes	NPW assumed
4	TAX	yes	yes	2N	% tax rate and years to depreciate (straight line).	yes	50, 10
5	UTIlity	yes	yes	1N	Initiates Certainty Equivalent Evaluation Minimum acceptable profit (NPW or DCF, corresponding to Keywords 2 and 3)	yes	Utility evaluation omitted
6	LIFe	yes	yes	2N	Years of forecast data available (normal project life), and years extended life beyond forecast data (when zero, scrapping all at end of forecast)	yes	15, 5

#	Keyword			Format	Description		Notes
7	PRObability	yes	yes	xN	A fractional probability for each probability level; must correspond in number and order to the forecast data input. Maximum 8 levels.	yes	1.00 (1 level)
8	DEMand	yes	yes	10N	First 10 years sales volume forecast for 1st probability.	yes	requires Multi Keyword to be used
				(10N)	Second card if more than 10 years forecast. Maximum of 20 years. Only 10 years per card.		
				=	Following 1 or 2 cards with data for remaining probability levels. Two cards must be supplied per probability level when more than 10 years.		
				=			
				10N	If the demand/year correlation is not 1.0 or 0.0 the demand level must be sections of a		
				(10N)	cumulative distribution and submitted in order. If the price/demand correlation is not 1.0 and the prices are different for each		
				=	probability level, all probability levels should have equal probabilities.		
				=			
9	WORKing	no	yes	10N	Incremental yearly working capital cash flow. Outflow positive.	yes	no working capital
				(9N)	No entry allowed for last year of project. The Programme automatically recovers working capital in last year of project. Two cards		
				=	must be entered if the life is greater than 10 years.		
				=			
				10N	One set of data is required for every probability level. The working capital levels are alway paired with the demand level corresponding to the read-in order. Working capital is ignored in the planning stage.		

TABLE 3   continuation 1

						Format	Description	Notes
10	PRIce	yes	yes	yes	yes	10N (10N) = = = 10N (10N)	The first ten years sales price on the first card for 1st probability level. A second card is necessary if the forecast life is more than 10 years. (Zeros or blanks before the end of the forecast period denote equal price to last given figure.) Following 1 or 2 cards for corresponding following forecast levels. Blank cards denote equal price/year profile to previous level. Two blank cards per level must be supplied if there are more than 10 years. The order must correspond to the probability level order given with the DEMand Keyword.	requires Multi Keyword to be used.
11	MULti	yes	yes			1N 10N 10N 9A, 10N 9A, 10N 9A, 10N =	This Keyword replaces DEMand and PRIce and enables data for a multi product plant to be entered. Number of products (up to a max. of 10) Plant effectiveness factors. The plant capacity for each product alone. Individual variable costs for each product. The variable cost for the 1st product must agree with the mean value in the Alternative data. Pairs of cards. Each card has the first 9 columns free for identification. First card gives annual demand for each product, for year 1, for 1st probability level. Second card gives corresponding sales prices. Second pair given data for year 2 Pairs are continued to end of the project life. The whole set is then repeated for the next probability level, this continues until all probability levels have been entered.	yes

	Keyword			Format	Description		Default
12	CORrelations	yes	yes	2N	Demand/year, and price/demand correlation coefficients. Omit this Keyword if there is only one probability level presented.	yes	0, 1.0 (i.e. No Monte-Carlo, but full matrix evaluation.)
13	TRApezoidal	no	yes	1N	Initiates 4-fold read-in of data with Key-words 14 to 16 inclusive, to enable 4 points of trapezoidal distribution to be read in. This card initiates a Monte Carlo simulation if it has not already been initiated by the correlation data. One data item: the allowable error for the expected value.	yes	Deterministic data read in
14	INFlation	no	yes	1N or 4N	Average % inflation rate/year for whole project life. If Trapezoidal card is entered, 4 values must be given on the same card. They can be the same value if no distribution is being taken into account.	yes	0.0
15	PENalty	yes	yes	1N or 4N	Penalty per unit capacity short of demand. If keyword TRApezoidal has been entered, the card must contain the 4 data points of the trapezoidal distribution.	yes	0.0
16	ALTernatives	yes	yes	9A,7N or 4(9A,7N)	Each alternative must be represented by 1 card. Each card must contain: a) - name in first 9 columns, no other data must be before Col. 10. b) - capacity (based on first product if multiproduct) c) - capital cost d) - annual fixed cost e) - variable cost/unit produced (based on first product if multiproduct) f) - scrap value	no	

For row 16, the Format column entries for items b) through f) are each shown as ` = " ` (same as above).

TABLE 3 continuation 2

No.	Keyword			Card format	Description	Example
16	ALTernatives (continued)	yes	no	9A,7N or 4(9A,7N)	g) - 1 = already operating 0 = not yet built h) - year, the beginning of which is the earliest time that the alternative is available for production. The OPERATE Keyword states when the plant will be in production.	
				" = "	If the TRApezoidal Keyword has been entered, each card must be entered 4 times, each time with the different point of the trapezoidal distribution for items b) to f) inclusive. For data without Uncertainty the four values will be the same.	
				" = "	Following the last alternative must be the card END. (Maximum number of alternatives is 10.)	
				END		
17	BOUndaries	yes	no	2N	Lower and upper capacity boundary factors for planning search.	yes 0.5 1.75
18	EXTensions	yes	no	9A,9A,2A	Alternative, (name, 9A) is an extension of Alternative (name, 9A) which Excludes (EX,2A) all other extensions of this plant.	
				" = "	Each extension relationship takes a new card. After all such relationships a card with END must be supplied.	
				" = "	This data is not free format. Names must begin in Cols.1 and 10). EX must be in Col. 19-20 to denote exclusion of alternative extensions.	
				" = "	Leave blank if extension not exclusive.	
				END		

				Description		
19	OPErate	no	yes	9A,16N	Each card contains the Alternative's name followed by year production started (Jan.1st) and the year it shut down (Jan.1st). If a strategy rather than a single plan is available, pairs of start and shut-down years should be given, one for each forecasted probability level, in the order of the sales demand forecast input.	no	for EVAluation alone.
				"		yes	for PLAN
				"	Alternatives operating in the final year and Alternatives already existing should have a zero as the appropriate data item.	no	for EVAluation following a Plan.
				"	Alternatives not operating, or existing and never shut down, need not be entered.		
				END	Finish data with END card.		
20	SENsitivity	no	yes	5N	The ±levels to be studied in a sensitivity Analysis. The five figures correspond to % changes in: (a) Price (b) Demand (c) Capital (d) Fixed Costs (e) Variable costs A zero denotes no sensitivity analysis for that particular item is required.	yes	0,0,0,0,0
21	ADDitional	no	yes	10N (10N)	Yearly additional <u>costs</u> associated with the project. Second card if forecast more than 10 years. These costs are considered as modifications to the annual fixed costs, they are allowed against tax.	yes	0
22	CHEck	yes	yes	1N	Controls intermediate printout = 0 no intermediate printout = 1 extended printout = 2 debugging printout	yes	0

TABLE 3   continuation 3

					Card	Description
23	TRApz	yes		yes	1N	TRApz can be used in this position to initiate a Monte-Carlo Simulation, to a required accuracy, without having to feed data in quadruple for items 14-16. The allowable error for the expected value.
	NOTrapz	yes		yes	1N	Cancels the TRApz Keyword, i.e. M.C. Simulations not performed. Useful for sensitivity studies about the means value. Can be entered at any point in the input deck.
	PLAn	yes	no	yes	1N 9A	This is the Calculation Execution Card which initiates the planning phase, and produces a decision matrix. Planning phase not carried out.
	EX				1N	The first data card defines the enforced year of first decision. (Zero if no enforcement required). The following cards, with a name on each, list the alternatives to
					9A	be enforced as first decisions: The program
					=	carries out a search with every enforced alternative named in the order given. The
					=	list of names must be followed by an END card.
					=	Any number of enforced first decisions can be requested up to 10. If the sales/year
					=	correlation is greater than 0.01, the number of searches carried out will be approxim-
					=	ately: (No. of enforced 1st decisions) x No. of probability levels.
					=	If there are no alternatives named between the enforced year and END card, the search
					9A	will produce the unconstrained optimum plan. The naming of any Alternative as enforced 1st
					END	decision supresses the overall optimum search

EX	EVAluate	no	yes	This is a Calculation Execution Card which initiates the evaluation phase of the program.	yes	Evaluation phase not carried out.
				A Monte Carlo simulation will be carried out, unless both sales/year correlation is less than 0.01 and sales/price correlation is greater than 0.99.		
LAST	STOp	yes	yes	Denotes end of data for problem	yes	ends with a failure message.

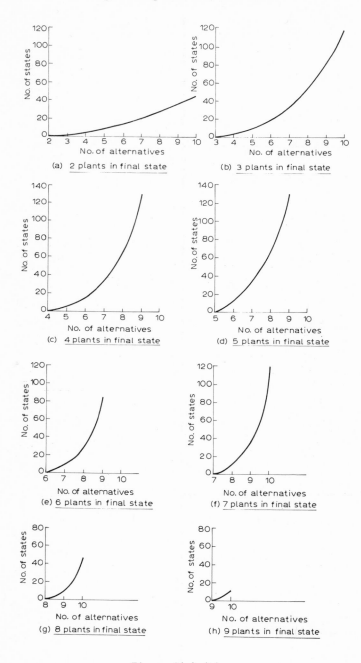

Figure 3(a)-(h)
Estimation of Number of States

the problem is too big for the program. However, there are generally
sufficient other constraints -- for instance, it is unlikely that all combinations of the nine can be used in all 3,4 and 5 states and still remain
within the capacity limits -- to allow considerable overstepping of the
figure 256. Should the problem be found, during the computation, to be
too big, an error message is printed out. An unconstrained maximum size
problems with 10 alternatives could produce over 1,000 yearly states, --
4 times the memory requirement provided by the program.

Since each state requires an NPW evaluation this estimate of the number
of states can also be used to estimate computer time. For a 15-year project
life, computer times of the order of 1 second per state are required. Hence
for our 9 Alternatives example, one DP search may be expected to take 340
seconds, if no other constraints apply.

The multiple searches that occur when the sales/year correlation is
greater than 0.01 and series of repeat runs with enforced alternatives as
first decisions increase the time proportionately. The number of searches
when a full Decision Matrix is required is calculated as:

(Nc. of probability levels + 1) x (No. of enforced first decisions)

In practice, the computation time will be somewhat less than directly proportional to the number of searches because:

a)   some parts of the program are used only once, not once
     per search,

b)   the enforced first decision is a further constraint which
     reduces the magnitude of the search.

Experience has shown that the calculated time for a Design Matrix search
is really only one quarter to one tenth of that calculated by the above
method because of the effect of the constraints. These times are clearly
for rough guidance only; accurate time cannot be predicted. Further computer
times for specific problems are given with the examples which are presented
at the end of this Appendix.

When the planning phase is being following by the evaluation phase, the
computer time for the evaluation must be added to that calculated above.

A non-Monte-Carlo NPW evaluation takes about 0.4 seconds. A DCF
evaluation requires about 4 iterations of the NPW, giving a time of less than
2 seconds. Hence, the maximum time possible, a 5-fold sensitivity analysis
using DCF, is less than 20 seconds.

On the other hand, a simple NPW Monte-Carlo simulation can take between
20 and 100 seconds, depending on the accuracy requested. Hence a 5-fold

Sensitivity DCF in the Monte-Carlo mode could produce a bill for 4,000 seconds computation! All times here correspond to the use of a CDC 6400/6500 computer system. Other computers have other "run times", and it is necessary to multiply all the times here by the factor that approximately relates the speed of the machine in question with the CDC 6400 machine speed.

6d Techniques for Reducing Computer Time

The planning program in particular can easily result in long computation times, but by careful consideration of the constraints, many of the sub-optimal paths can be excluded, so producing times reduced by factors of 5 or even 10. The following list gives methods of handling the constraints to reduce computer time:

1) Capacity Boundary constraints are the most useful parameters for reducing the search time. Care must be taken that they are not so narrow that they limit sensible combinations of Alternatives, but they can be narrowed down so that non-sensible combinations are not evaluated. For example, it doubled the computer time to open the bounds sufficiently to evaluate the one extreme case -- install a large plant for pessimistic market -- in Example 3, Section 9c, and in fact this did not yield any useful information.

An error message is printed out when the bounds are too restrictive.

2) By defining any identical plants as being a chain of extensions to one plant, a marked reduction in computer time is produced, (see Example 3, Section 9c).

This technique cannot be used when there is a decline in the market because, by definition within the program, an extension cannot be scrapped without scrapping the parent plant.

3) The number of first decision in the Decision Matrix should be restricted to those Alternatives that could be sensible first decisions. Clearly, the computer time is proportional to the number of first decisions investigated; the Decision Matrix should not necessarily include all Alternatives.

4) The number of probability levels should be limited. To use three levels will often be adequate. To divide forecasts into six levels will double the computation costs and may not yield any more useful information.

5) If a Monte-Carlo simulation is to be performed, the requested accuracy of the Expected Value should be chosen with care. A too accurate request can multiply computer time four-fold.

6e Error Messages

The program has the following Error messages incorporated:

a) "Zero allowed states found, -- constraints too restrictive"

The constraints in the problem are: upper and lower bound, first year available for production, enforced first decision year, enforced first decision and extension structure. Great case must be taken in setting these up so that conflicts do not occur. Any conflict where one constraint cuts off everything that is allowed by another constraint, results in the above message. This can very easily occur for a number of reasons:

> Too close upper and lower bounds can result in no combination of alternatives meeting both boundary constraints.

> an enforced first decision and year may disagree with the first available data for the alternative to be in the prediction

> enforced first decision timing may be at variance with the lower boundary constraints,

to detail just a few possibilities.

Following this error message the particular search must stop, and -500,000 is put in the corresponding NPW.

b) "Number of states generated greater than 256, --
    the allowed number. Best state may be lost".

The program has room for 256 combinations of alternatives. This corresponds to every unconstrained combination when 8 Alternatives are presented. When 9 Alternatives are presented the upper boundary constraint has to be such that the maximum number of installed alternatives is 4, and 10 alternatives must have no more than 3 installed at one time.

When the number of combinations becomes greater than 256, the later ones are simply forgotten and the above error message is printed out. The computation continues because the results obtained are numerically correct. However, it cannot claim with certainty that the best plan has been found. It may be that the best plan was amongst those discarded. The results are still valuable, the best out of a large number of alternatives will be printed out.

c) "Some NPW outside range, values set on limits --
    best state may be lost".

Since the NPW is stored as integer numbers there is a maximum allowable value. If it is above +500,000 the above error message is printed and the value is set to the +500,000 limit. Hence, if the optimum plan oversteps these limits the resulting NPW will be wrong and any number of plans will have been set equal, so the best plan might be lost. However, if the optimum plan has not gone through this route, the results are valid. This can be seen from the optimum plan yearly NPW print-out.

In the Decision-Matrix and other result print-outs, -500,000 denotes that the value has not been evaluated. This could either be because the calculation was not requested, or because the search was not successful. The value of -499,999 is used in Certainty Equivalent print-outs to denote a successful evaluation with a result of minus infinity.

7. THE ANALYSIS OF THE RESULTS

7a The Planning Phase

The most important result is the Decision Matrix. This should be carefully studied, taking due note of the risks associated with each possible decision, expressed either by the range of outcomes, or by the Certainty Equivalent. Non-quantitative factors should be introduced into the study at this point in order to choose one, or more first decisions to go on to the evaluation stage. If the evaluation stage has been called automatically, it is important that the optimum case that has been chosen is in agreement with that which would be chosen from a study of the Decision Matrix.

It is important to know the contents of the total plan before coming to a final decision. This can be obtained from the summary print-out of every search, which gives a complete plan, both Alternatives and timing, that results in that case's NPW. This enables the corresponding later incorporated Alternatives and also the occurrence of short-falls to be considered before the decision is finally made.

If it is necessary to know how much better one plan is than another, it is necessary to have an extended print-out with CHECK = 1. This produces the best NPW for every possible end-state. From this can be answered the type of question; "how much would it cost us to choose Alternative A rather than Alternative B?" This question is frequently asked for technical or commercial reasons.

This evaluation of end-states must be used with care. If B is very much better than A, the best way of having A as the end-state may be to

install B until the last year, and only then scrap B and install A! Hence,
this end-state data does not always answer the question. If this is suspected, the study should be repeated exluding B from the search, or including
A as an enforced first decision.

Having chosen the first decision, or a number of possible first
decisions, the next stage is the evaluation stage.

7b The Evaluation Phase

The evaluation phase produces much better information for the go/no-go
investment decision than the NPW from the search, for the following reasons:

 a) Probabilistic Distributions for all the major
    variables can be included.

 b) The right degree of sales/year and sales/price
    correlation can be incorporated

 c) Working capital can be included in the analysis

 d) Economic criteria other than NPW can be evaluated.

The most important output from this phase is the distribution of
profitability -- this can be either NPW or DCF. The distribution is
shown both graphically and in tabular form. The decision-maker can use
this information to decide whether the Expected Profitability is acceptable,
and if the risk of low profit or loss is also acceptably small.

The DCF and Benefits/Cost Ratio indicate how efficiently the project
is using capital. The NPW or the Equivalent Annual Value shows the effect
of the project on the company's annual balance sheet and the undiscounted
cash flow profile indicates the strain imposed by the project on the
company's liquidity. These are three different but important aspects to
consider before making the go/no-go decision. This information is printed
out in the evaluation following the Monte-Carlo analysis.

Sensitivity analysis results can be useful if:

 a) the data for the Monte-Carlo simulation are not to hand

 b) despite probabilistic data there is ground for suspecting
    a systematic error in addition to the reasons for the
    proposed distribution. This for instance could be an
    inflationary effect on the capital costs or a market
    development feature such as the introduction of legislation
    restricting a product's use.

 c) it is demanded by a company's existing rules for presentation
    of requests for capital.

8. PROGRAM/MAN INTERACTIONS

Planning is a trial-and-error activity. Like the Design activity from architecture to engineering, it involves the sequence:

- a) understand the problem
- b) propose alternatives
- c) select a proposal
- d) test the proposal
- e) if the test is satisfactory - finish
     otherwise modify proposal in light of test, and return to test (d).

It is unlikely that one computer run can be expected to complete a study. The computer program only enables a large number of proposals to be systematically checked, but different alternatives must be tried which necessitate more computer runs to produce the complete study. This is particularly true for the selection of plant capacity levels.

The second reason for the need for a number of runs is the difficulty in obtaining satisfactory runs on the first attempt. The program accurately answers the questions put to it, but it is not easy to pose sensible consistent questions to the program at the first attempt. Planning is a difficult activity, and when confronted with probabilistic data, enforced first decisions, and a relentlessly methodical search procedure, unexpected answers can be obtained because of inadequately defined questions. Such results contribute to the study by demanding a deeper understanding but they also increase the number of computer runs required and the need for easy man/computer interaction.

There are also other good grounds for wanting program/man interactions during the computation. The Decision Matrix is a part of the procedure where the calculation should pause. The planner consider; "should there be more enforced first decisions," "should the calculation proceed to the evaluation phase". If so, which first decision should be evaluated? These points are good grounds for real-time computer use. Again one can compare planning with design, where present moves, in architecture, chemical and civil engineering are to the development of on-line computer methods to aid the recycling steps of the design procedure.

The program PLADE is capable of operating on 16 bit word machines and with its Keyword structure it is well suited for real-time use.

9. EXAMPLES

This manual concludes with a number of examples of the use of the planning procedure and the program. They are intended to indicate the range of problems that can be tackled.

9a Example 1: Deterministic, Single Plan Investment Analysis

In order to demonstrate the use of the program as a straightforward investment analysis tool, the problem given as the example in Chapter I, Section 1.9b is evaluated by the program. Chapter 1 contains all the necessary data and problem description. The problem is deterministic, but there is the difficulty introduced by an existing plant; the investment analysis has to consider the difference above a base-case. Working capital is included, and a single investment in year 1 is being analysed. The plant is assumed to be scrapped at the end of the forecast period; the scrap values are as given in Chapter I. In addition to an NPW evaluation, a DCF result is also requested.

Figure 5 shows the input data required by the program, and Figure 4 gives the output.

9b Example 2: Probabilistic Investment Analysis Involving a Strategy

This example has been reported as Case 1 in Chapter VI, but now we assume we are in the final stages of a study where the best strategy has been selected, and we now require a comprehensive probabilistic evaluation. In this example, this strategy has been obtained by looking through the planning output, but it could equally well have been a strategy derived from a completely different source.

The strategy is defined by data given with the OPERATE Keyword. The remaining data is based on that given in Chapter VI but we are assuming that at the last stage of the calculation the majority of the data is defined probabilistically, including some working capital and inflation. It is assumed that after the project lifetime the plants are scrapped and have no scrap value. A distribution of NPW is required, with a 95% confidence limit on the Expected Value of £10,000.

Notice that the money units for the program have been selected to be £100.
Figure 6 gives the input for the problem and Figure 7 is the program output excluding the summary of the input data. Note that after the Monte-Carlo results, the program makes a final run through to calculate the mean values of further evaluation criteria. The computer time required by this example was 30 seconds.

```
EXAMPLE 1
NPW CALCULATION REQUESTED;  DISCOUNT RATE  = 15.00
PROJECT LENGTH          15 YEARS
SCRAPPING ALL AT END
DEPRECIATED OVER        10 YEARS
TAX RATE =              25.00
SHORTFALL PENALTY PER TON      0.00
ADDITIONAL YEARLY COSTS BY YEAR:

    0.    0.    0.    0.    0.    0.    0.    0.    0.    0.
    0.    0.    0.    0.
```

```
        UNIT   CAPACITY   CAPITAL   FIXED COST   VARIABLE COST   SCRAP


1 EXISTING    20000.0    3000.0     100.0           .040          0.0
2 NEWPLANT    26000.0    4000.0      80.0           .035        400.0
```

```
OPERATING STRATEGY
         EXISTING  START/END PAIRS, BY PROBABILITY

1 EXISTING   1        0 16
2 NEWPLANT   0        1 16
```

```
YR DEMAND PRICE WORK CAP/ DEMAND PRICE WORK CAP/DEMAND PRICE WORK CAPL
   NO    1
   PROB  1.0000
 119500.    .099    0.
 222600.    .096    0.
 326600.    .093    0.
 430400.    .091    0.
 533800.    .090  100.
 636800.    .088    0.
 739200.    .087    0.
 841200.    .086    0.
 942800.    .086    0.
1044000.    .086  100.
1145000.    .086    0.
1245400.    .035    0.
1345700.    .085    0.
1445800.    .085    0.
1546000.    .085    0.
```

Figure 4

CUMULATIVE UNDISCOUNTED CASH FLOW (CURRENT MONEY)

   YEAR   PRCB1   PROB2   PROB3

      1   -3886.88
      2   -3652.93
      3   -3253.07
      4   -2717.77
      5   -2162.77
      6   -1420.47
      7   -606.17
      8   262.73
      9   1186.83
     10   2052.33
     11   2952.33
     12   3847.08
     13   4751.95
     14   5660.20
     15   7075.20

                    MEAN NPW IS   -1282.1
                FOR DISCOUNT RATE   15.00
               MEAN PAYBACK YEAR IS   7.70
       EQUIVALENT ANNUAL VALUE IS   -219.26   BENEFITS COST RATIO IS -.28

MEAN DISCOUNTED CAPITAL INVESTMENT IS   4600.0

EXAMPLE 1
DCF CALCULATION REQUESTED

CUMULATIVE UNDISCOUNTED CASH FLOW (CURRENT MONEY)

   YEAR   PRCB1   PROB2   PROB3

      1   -3886.88
      2   -3652.93
      3   -3253.07
      4   -2717.77
      5   -2162.77
      6   -1420.47
      7   -606.17
      8   262.73
      9   1186.83
     10   2052.33
     11   2952.33
     12   3847.08
     13   4751.95
     14   5660.20
     15   7075.20

                    MEAN NPW IS   -.0
                FOR DISCOUNT RATE   10.83
               MEAN PAYBACK YEAR IS   7.70

MEAN DISCOUNTED CAPITAL INVESTMENT IS   4433.3

Example 1: Data Output

```
HEAD
EXAMPLE 1                                                              1
TAX
25.0 10                                                               4
LIFE
15 0                                                                  6
DEMAND
19500 22600 26600 30400 33800 36800 39200 42800 44000                8
45000 45400 45700 45800 46000
WORKING                                                               9
0 0 0 0 100 0 0 0 0 100
0 0 0 0 0
PRICE                                                                10
.099 .096 .093 .091 .090 .088 .087 .086 .086 .086
.086 .085 .085 .085 .085
ALTERNATIVES
EXISTING 20000 3000 100 0.040 0 1 1                                  16
NEWPLANT 26000 4000 80 0.035 400 0 1
END
OPERATE
NEWPLANT 1                                                           19
END
EVALUATE                                                             EX
DCF                                                                   3
EVALUATE                                                             EX
STOP                                                                LAST
```

Figure 5

Example 1: Data Input

```
HEAD
EXAMPLE 2
DISCOUNT
10.0
TAX
25.0 10
LIFE
15
PROBABILITY
0.25      0.25      0.25      0.25
DEMAND
   10200     10800     11400     12000     12600 13300   14000 14800    15500    16400
   17200     18200     19200     20200     20400
   10200     10700     11200     11700     12300 12800   13400 14000    14600    15200
   15300     16600     17200     18000     18700
   10200     10400     10700     11000     11400 11700   12000 12300    12700    13100
   13400     13800     14200     14600     15000
   10100     10300     10500     10700     10900 11100   11300 11500    11700    11900
   12200     12400     12700     12900     13200
WORKING
120 0 0 0 0 120 0 0 0 0
120 0 0 0
100 0 0 0 0 100 0 0 0 0
100 0 0 0
100 0 0 0 0 100 0 0 0 0
100 0 0 0
 80 0 0 0 0 80 0 0 0 0
 80 0 0 0
PRICE
1.2
1.2
1.2
1.2
1.2
1.2
1.2
1.2
CORRELATIONS
0.75 1.0
TRAPIZOIDAL
100
INFLATION
4.0  6.0  7.0  8.0
ALTERNATIVES
EXISTING 10000      0 1800  0.40 0 1
EXISTING 10000      0 2000  0.40 0 1
EXISTING 10000      0 2000  0.40 0 1
EXISTING 10000      0 2200  0.40 0 1
SM.EXTN    900    540   95  0.40 0 0
SM.EXTN   1000    600  100  0.40 0 0
SM.EXTN   1000    600  100  0.40 0 0
SM.EXTN   1100    660  110  0.40 0 0
LGE.EXTN  1800   2000  135  0.40 0 0
LGE.EXTN  2000   2200  150  0.40 0 0
LGE.EXTN  2000   2200  150  0.40 0 0
LGE.EXTN  2200   2300  165  0.40 0 0
2500TNPLT 2500   4500  550 00.40 0 0
2500TNPLT 2500   5000  600 00.40 0 0
2500TNPLT 2500   5000  600 00.40 0 0
2500TNPLT 2500   5500  650 00.40 0 0
5000TNPLT 5000   7900  900 0.22 0 0
5000TNPLT 5000   8000 1000 0.24 0 0
5000TNPLT 5000   9000 1000 0.26 0 0
5000TNPLT 5000   9500 1100 0.30 0 0
NEWPROC  15500  27000 1500 0.22 0 0
NEWPROC  16000  27600 1600 0.24 0 0
NEWPROC  16000  28000 1600 0.26 0 0
NEWPROC  16500  29000 1700 0.30 0 0
END
```

contin:

```
OPERATE
EXISTING     0   3    0   3    0 26    0 26
SM.EXTN      1   3    1   3    1 26    1 26
2500TNPLT    0   0   14  26    5 26    6 26
5000TNPLT   10  26    0   0    0  0    0  0
NEWPROC      3  26    3  26    0  0    0  0
END
EVALUATE
STOP
```

Figure 6

Example 2: Data Input

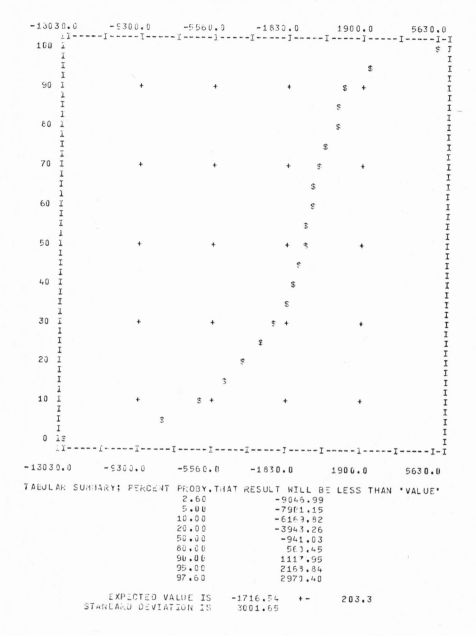

TABULAR SUMMARY; PERCENT PROBY.THAT RESULT WILL BE LESS THAN 'VALUE'

2.60	-9046.99
5.00	-7901.15
10.00	-6163.82
20.00	-3943.26
50.00	-941.03
80.00	563.45
90.00	1117.95
95.00	2163.84
97.60	2970.40

EXPECTED VALUE IS   -1716.54   +-   203.3
STANDARD DEVIATION IS   3001.65

Figure 7

EXAMPLE 2
FINAL RUN THROUGH EVALUATION TO PRODUCE CASH-FLOW,EAV AND B/C
    USING FIXED PLANS AND MEAN DATA

CUMULATIVE UNDISCOUNTED CASH FLOW (CURRENT MONEY)

    YEAR   PROB1   PROB2   PROB3

     1   -665.63   -644.74   -644.74   -687.22
     2 -31687.52 -31733.95   -445.05   -555.20
     3 -27539.26 -27729.42   -18.32   -272.20
     4 -23289.60 -23729.95   -5697.25   180.06
     5 -18205.98 -13952.85   -5133.41   -5932.95
     6 -12332.93 -13559.36   -4428.51   -5713.84
     7   -5262.46   -7076.14   -3281.99   -9189.23
     8   3242.36   452.21   -1740.56   -4444.47
     9   -1773.16   9133.08   278.92   -3456.02
    10   8573.98   19033.18   2852.92   -2195.97
    11   20385.07   30220.46   5717.86   -687.67
    12   34514.28   42502.24   9082.01   1319.03
    13   50344.25   43874.95   12646.97   3833.80
    14   68785.66   53805.61   16425.22   6776.35
    15   90454.87   77133.39   21044.55   10929.02
                 MEAN PAYBACK YEAR IS   8.37
EQUIVALENT ANNUAL VALUE IS   -120.16   BENEFITS COST RATIO IS -.06

MEAN DISCOUNTED CAPITAL INVESTMENT IS   16547.0

Example 2: Program Output (excluding
           Summary of Input Data)

9c Example 3: Selection of Capacity in a Probabilistic Forecast Study

This example is taken from Case 2 of Chapter VI. Chapter II gives the probabilistic forecasts, the arguments resulting in the selection of 6 alternatives with 4 different capacities, and the data for the problem. In order to select the best initial plant capacity a planning study has to be requested, and since a complete survey of the first decision is required, each different capacity must, in turn, be enforced as a first decision. The plant must be operating in year 1, hence the enforced year of the first decision is 1, but further extensions to capacity need only be made when it is economic to do so.

Following this search, we accept a choice based on Maximum Certainty Equivalent, and so we can accept the program's choice of optimum strategy and move directly into the evaluation phase.

An evaluation of DCF is required and because capital costs are subject to some uncertainty, a Sensitivity Analysis at the ± 20% level is required on capital cost.

To reduce computer time for the search the two sets of identical plants were declared as extensions of each other. This defines that one must be installed before the other, which considerably reduces the number of combinations to be considered. Since they are identical, the order in which they are introduced has no physical meaning.

Figure 9 gives the data input for this problem and Figures 8, 10, 11 show examples of parts of the output.

Figure 8 shows an example of the results of one search, showing the plan involved, the growth of capacity and NPW, and final NPW.

Figure 10 shows the Decision Matrix which has been fully discussed in Chapter VI, and Figure 11 gives the output for the evaluation phase, including the effect of the ± 20% change in capital cost.

No Monte Carlo simulation occurred in this case because the demand/year and price/demand correlation coefficients were both 1.0 and no trapezoidal data has been entered.

The computer time for this example was 90 seconds.

9d Example 4: Selection of the Process, Given Full Probabilistic Data

Here, Case 1 of Chapter VII is used to illustrate the choice of process faced with probabilistic data for both the forecast and the plants.

The question raised in Case 1 is; is it economic to replace existing plants by a new process, or just cover new market developments with a new process, or simply cover the new market by plant expansion and plus a small plant using the old process?

The data given in Chapter VI is used here except that a distribution is given to the plant cost data; ± 5% for the data covering the existing plants, and ± 15% for the new process data.

Each of the new plants needs to be considered as an enforced first decision to obtain a Decision Matrix. The company's policy is always to meet demand for this product, therefore the lower boundary must be 1.0.

Certainty Equivalent is the agreed criteria for making this selection, after which we can move on directly to the evaluation phase by calling EVALUATE. A Monte Carlo simulation for the chosen strategies will automatically be done.

Figure 12 gives the input data for this example. Figure 13 shows the resulting Decision Matrix, and Figure 14 the Monte Carlo evaluation output. The computation time for this example was 280 seconds.


```
                    PROBABILITY LEVEL    1
            FIRST DECISION RESTRICTED TO    4      PILOT PLT


                      OPTIMAL POLICY

 YEAR      DECISION   CAPY      WORTH      DEMAND            STATE
           BLD SCFF                       1 LEVEL


                                              INITIAL   00000  0
    1        4    0    1200     -97175     1400          00010  0
    2        0    0    1200     -83677     1680          00010  0
    3        0    0    1200     -71940     2100          00010  0
    4        1    4    3750    -179417     2700          10000  0
    5        0    0    3750    -149173     3181          10000  0
    6        0    0    3750    -119454     3470          10000  0
    7        0    0    3750     -91965     3630          10000  0
    8        0    0    3750     -66987     3750          10000  0
    9        0    0    3750     -45267     3830          10000  0
   10        0    0    3750     -26380     3900          10000  0
   11        0    0    3750      -9957     3940          10000  0
   12        0    0    3750       4323     3970          10000  0
   13        0    0    3750      16741     3980          10000  0
   14        0    0    3750      27539     3995          10000  0
   15        0    0    3750      84053     4000          10000  0
```

FIRST DECISION SHOULD OPERATE IN QUARTER   1

Figure 8
Example 3: Program Output, Optimum
Policy Plan

```
HEAD                                                              1
EXAMPLE 3
TAX                                                               4
25 10
UTILITY                                                           5
-200000
LIFE                                                              6
 15 10
PROBABILITY                                                       7
.25   .25   .25   .25
DEMAND                                                            8
1400. 1680. 2100. 2700. 3181. 3470. 3630. 3750. 3830. 3900.
3940. 3970. 3980. 3990. 4000.
1200. 1400. 1600. 1810. 2080. 2340. 2580. 2760. 2780. 2984.
2970. 2980. 2990. 3000. 3000.
1040. 1160. 1300. 1470. 1620. 1790. 1890. 1960. 1980. 2000.
2000. 2000. 2000. 2000. 2000.
 920.  960. 1020. 1060. 1110. 1150. 1190. 1200. 1210. 1240.
1260. 1270. 1270. 1280. 1290.
PRICE                                                            10
40.0
40.0
40.0
40.0
40.0
40.0
40.0
40.0
CORRELATIONS                                                     12
1.0 1.0
ALTERNATIVES                                                     16
LGE.PLT    3750. 200000. 35000. 3.5 0. 0 1
MED.PLT    2750. 165000. 32500. 3.5 0. 0 1
SM.PLT     2000. 143000. 27500. 3.5 0. 0 1
PILOT PLT 1200. 110000. 20000. 3.5 0. 0 1
2ND SM     2000. 143000. 27500. 3.5 0. 0 1
2ND PILOT 1200. 110000. 20000. 3.5 0. 0 1
END
BOUNDARIES                                                       17
0.5      3.
EXTENSIONS                                                       18
2ND SM    SM.PLT
2ND PILOTPILOT PLT
END
PLAN                                                             EX
1
LGE.PLT
MED.PLT
SM.PLT
PILOT PLT
END
DCF                                                              3
SENSITIVITY                                                      20
0 0 20 0 0
EVALUATE                                                         EX
STOP                                                           LAST
```

Figure 9
Example 3: Input Data

EXAMPLE 3

FIXED DECISION YEAR   1

DECISION MATRIX

INITIAL DECISION	NO YEARLY CORRELATION NPW	I MEAN	WITH TOTAL CORRELATION CERT EQUIV	BY PROBABILITY 1	2	3	4
1 LGE.PLT	-500000	-32370	-86303.	131780	7540	-91190	-177710
2 MED.PLT	-500000	-8890	-32410.	93250	44050	-43130	-129730
3 SM.FLT	-500000	7860	-1833.	78330	32440	3650	-82950
4 PILOT FLT	-500000	20560	17480.	84050	16440	-1940	-16280

BASED ON THE ABOVE FIRST DECISIONS,
THE EXPECTED VALUE OF COMPLETE FORECAST INFORMATION IS   20230.

Figure 10
Example 3: Program Output, Decision Matrix

CUMULATIVE UNDISCOUNTED CASH FLOW (CURRENT MONEY)

YEAR	PRCB1	PROB2	PRCB3	
1	-99400.00	-69400.00	-93780.00	-97065.00
2	-65400.00	-69800.00	-7427.00	-93075.00
3	-24200.00	-49200.00	-53675.00	-67362.50
4	-192787.50	-27600.00	-33075.00	-55595.00
5	-124207.63	-7000.00	-12475.00	-32458.75
6	-47716.38	-151400.00	8125.00	-13227.50
7	33154.88	-99272.50	23772.00	7093.75
8	117311.13	-40491.25	49325.00	27693.75
9	201467.38	17290.00	69925.00	48293.75
10	285623.63	75071.25	90525.00	68893.75
11	367025.88	130102.50	103375.00	86743.75
12	448436.13	185133.75	125225.00	104593.75
13	529842.38	240165.00	144075.00	122448.75
14	606248.63	295196.25	161925.00	142293.75
15	682654.88	350227.50	179775.00	158148.75

DISCOUNT RATE 16.54
MEAN PAYBACK YEAR IS   7.16

MEAN DISCOUNTED CAPITAL INVESTMENT IS   184196.0

CERTAINTY EQUIVALENT EQUALS   -1875.91

DISTRIBUTION OF NPW

CASE	DIS	PBK	1	2	3
NORMAL	16.54	7.2	46973.	-4772.	-14190.	-28016.
+20. O/O CAPITAL INCR	14.19	7.9	56865.	-8059.	-17098.	-31731.
-20. O/O CAPITAL INCR	19.75	6.3	37089.	-1996.	-11130.	-23966.

Figure 11
Example 3: Program Output, Evaluation

```
HEAD
EXAMPLE 4
DISCOUNT
10.0
TAX
25.0 10
UTILITY
-27800
LIFE
15 10
PROBABILITY
0.25      0.25      0.25      0.25
DEMAND
     10200     10800     11400     12000     12600  13300   14000  14800   15500   16400
     17200     18200     19200     20200     20400
     10200     10700     11200     11700     12300  12800   13400  14000   14600   15200
     15800     16600     17200     18000     18700
     10200     10400     10700     11000     11400  11700   12000  12300   12700   13100
     13400     13800     14200     14600     15000
     10100     10300     10500     10700     10900  11100   11300  11500   11700   11900
     12200     12400     12700     12900     13200
PRICE
1.2
1.2
1.2
1.2
1.2
1.2
1.2
1.2
CORRELATIONS
0.75 1.0
TRAPIZOIDAL
100
ALTERNATIVES
EXISTING  10000  20000  1900  0.40  0  1
EXISTING  10000  20000  1900  0.40  0  1
EXISTING  10000  20000  1900  0.40  0  1
EXISTING  10000  20000  1800  0.40  0  1
SM.EXTN    1000    570    95  0.40  0  0
SM.EXTN    1000    600   100  0.40  0  0
SM.EXTN    1000    600   100  0.40  0  0
SM.EXTN    1000    630   105  0.40  0  0
LGE.EXTN   2000   2400   140  0.40  0  0
LGE.EXTN   2000   2500   150  0.40  0  0
LGE.EXTN   2000   2500   150  0.40  0  0
LGE.EXTN   2000   2600   160  0.40  0  0
2500TNPLT  2500   4700   570  0.40  0  0
2500TNPLT  2500   5000   600  0.40  0  0
2500TNPLT  2500   5000   600  0.40  0  0
2500TNPLT  2500   5200   630  0.40  0  0
5000TNPLT  5000   7300   910  0.22  0  0
5000TNPLT  5000   8500  1000  0.25  0  0
5000TNPLT  5000   8500  1000  0.25  0  0
5000TNPLT  5000   9700  1100  0.28  0  0
NEWPROC   16000  23000  1500  0.22  0  0
NEWPROC   16000  27800  1500  0.25  0  0
NEWPROC   16000  27800  1600  0.25  0  0
NEWPROC   16000  33000  1700  0.28  0  0
END
BOUNDARIES
1.0 2.0
EXTENSIONS
SM.EXTN   EXISTING EX
LGE.EXTN  EXISTING EX
END
```

contin:

```
PLAN
0
SM.EXTN
LGE.EXTN
2500TNPLT
5000TNPLT
NEWPROC
END
EVALUATE
STOP
```

Figure 12

Example 4: Input Data

EXAMPLE 4   DECISION MATRIX

INITIAL DECISION	NO YEARLY CORRELATION		I------- WITH TOTAL CORRELATION ------- BY PROBABILITY			
	NPW	I MEAN	I CERT EQUIV	1	2	3	4
2 SM.EXTN	7110	7200	6963.	12240	9910	4970	1700
3 LGE.EXTN	7000	7630	7423.	12650	10030	4590	3240
4 2500TNPLT	2950	4460	4200.	9440	7300	1980	-880
5 5000TNPLT	4630	5470	5081.	11370	8940	3070	-1490
6 NEWPROC	2840	3440	2805.	10660	7800	370	-5060

BASED ON THE ABOVE FIRST DECISIONS,
THE EXPECTED VALUE OF COMPLETE FORECAST INFORMATION IS   90.

Figure 13

Example 4: Computer Output, Decision Matrix

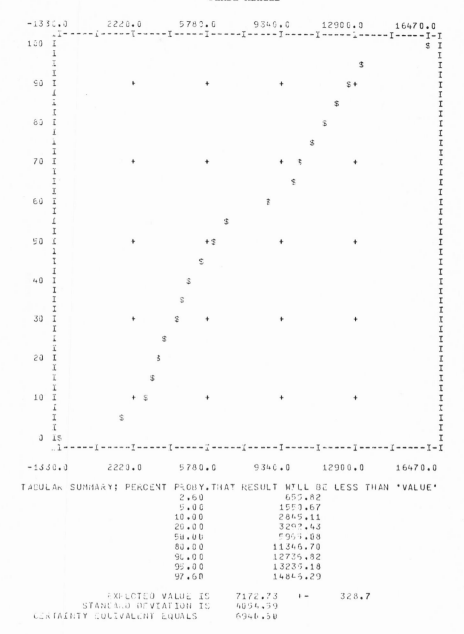

TABULAR SUMMARY; PERCENT PROBY.THAT RESULT WILL BE LESS THAN 'VALUE'

2.60	655.82
5.00	1550.67
10.00	2845.11
20.00	3292.43
50.00	5965.08
80.00	11346.70
90.00	12736.82
95.00	13235.18
97.60	14845.29

EXPECTED VALUE IS	7172.73	+-	328.7
STANDARD DEVIATION IS	4054.59		
CERTAINTY EQUIVALENT EQUALS	6940.50		

Figure 14

Example 4: Summary of Monte Carlo Evaluation

MANUAL APPENDIX 1

THE PLADE PROGRAM DESCRIPTION

A1.1 INTRODUCTION

The object of this Appendix is to describe the logical sequences that make up the PLADE PROGRAM, so that, together with Manual Appendix 2 -- the PLADE PROGRAM Listing, -- would-be users have sufficient information to make any modifications they wish to meet their particular needs. There are three particular areas where changes are likely.

Firstly, the Program presented utilizes integers up to 3.1×10^{10} which corresponds to 36 bit word machines, since this gives a compact and efficient Program. This size of integer is too big for 24 bit word machine and less, and the use of some double precision must therefore be introduced when the Program is used on some machines. Instructions for this are given in the last section of this Appendix-Section A1.12.

Secondly, straight line depreciation is assumed for the tax calculations. Application to specific projects may well require the use of specific depreciation regulations. An individual firm may therefore require a number of different alternatives to be available. Methods of programming alternatives is described under the section on subroutines TAXD and TAXF (Section A1.10).

Thirdly, the Certainty Equivalent is calculated from one data point, the minimum tolerable profit, as described in Section 5.2d(3) of Chapter V. This is a simple method, and should a better correlation be available, from an interview-based utility function for instance, it should replace that already in subroutine UTIL.

A1.2 THE GENERAL STRUCTURE

The Program has basically three parts. A main program which reads in and prepares the data for the following two sections; the planning search, together with the search sequence management, and the evaluation which may take the form of a Monte-Carlo simulation or of a single restricted probabilistic run. Fig.1 of the manual shows this structure diagramatically.

The Program is so written that after executing every request, it returns to the keyword, which enables either new data to be tried, or allows a transfer from the planning to the evaluation phase to be made with data obtained from the searches.

The Planning and Evaluation parts of the Program call on further subroutines, some of which are specific to planning or evaluation, and some are common to both. Fig. A1.1 gives the complete subroutine hierarchical structure, with the name of each subroutine and a short description of its duty.

The following pages describe the important subroutines, under the subroutine name. The less important subroutines with obvious functions are not included; they are covered adequately by the comment statements in the Program listing. The diagrams associated with these descriptions show the logical steps for the particular functions involved. They are not limited to instructions found within the particular subroutine involved, and they are not intended as subroutine flow diagrams. More detailed information can be obtained from comment statements in the listing, given as Manual Appendix 2.

A1.3 SUBROUTINE EXEC

The sequence of Dynamic Programming searches carried out by the OPLEX subroutine must be directed, and this is done by the EXEC subroutine.

Firstly, a base-case is necessary to determine the NPW corresponding to no investment. This is done by a D.P. search with the constraint that no building is allowed. The NPW so obtained is subtracted from the later outputs, to provide the difference in NPW due to investment.

Secondly, the searches themselves depend upon the degree of demand/year correlation:-

(a) If there is complete demand/year correlation,
then the conditional plans have to be found.
This is done by repeating the search for
every probability level in a deterministic fashion.

(b) If the demand/year correlation is zero, then a
probabilistic single search by OPLEX is appropriate.

(c) If the demand/year correlation is between zero and
one, then both the above is carried out.

Thirdly, if any enforced year of first decision or alternative enforced as first decision is required, this is handled by EXEC.

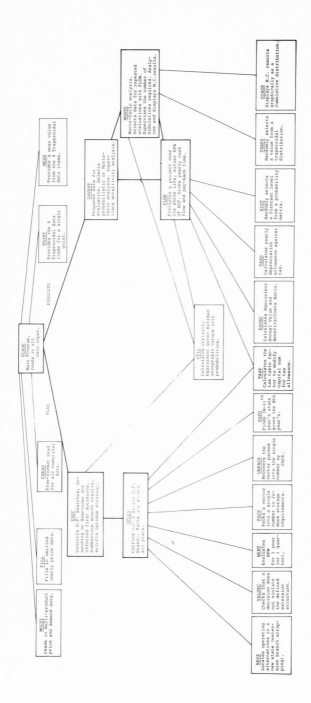

Figure A1.1: Subroutine Hierarchy

EXEC collects all the resulting NPW's and appropriate base cases, and organises them as the Decision Matrix.

It calls on UTIL to obtain Certainty Equivalent and chooses the preferred strategy as the one with the best Certainty Equivalent or the best Expected NPW, depending on the correlation. This chosen strategy, (or plan when the sales/year correlation is zero or the data is deterministic) is returned to the main Program to be used by the evaluation phase if it is not over-written by the OPERATE keyword.

Fig. A1.2 shows the Flowplan for the management of the DP searches.

A1.4 SUBROUTINE OPLEX

This subroutine contains the Dynamic Programming search routine. Starting with the existing state, the search moves forward in time creating new states from old ones, by making every build decision for alternatives not installed, and every scrap decision for alternatives installed, and also by doing nothing. Each year, the total set of existing states is subjected to this complete series of build and scrap decisions, to create a set of new states for the next year.

These new states are then checked against constraints. This is done partly be subroutine VALDEC and partly within the OPLEX subroutine itself. The checking within OPLEX concerns the capacity constraints, and year and first decision restrictions, and VALDEC checks that the extension logic is not violated.

All decisions within the constraints are then evaluated by determining the NPW for that one year, from the old state to the new. This one-year NPW is added to the NPW of the old state to give the NPW of the new state. If this NPW is better than the previously restored NPW for that new state it is remembered, as is the decision leading to it.

The NPW's are calculated by subroutine WERT. By repeating the call for each sales forecast probability level, an Expected NPW can be obtained, and the D.P. search is evaluating the Expected Value correctly.

The results from the one-year search are then a set of new states, each with an optimum NPW to that point, plus the decision which lead to that optimum NPW.

This search is repeated for every year, resulting in a set of end states with NPW's, from which the best can be chosen. At the end of the search, the sequence of decisions which lead to the optimum NPW can be

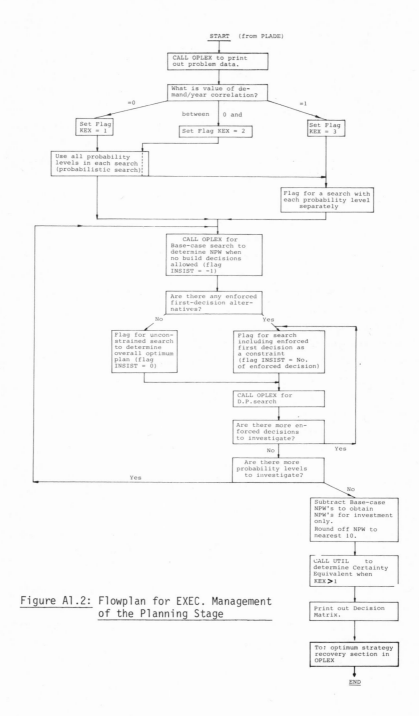

Figure A1.2: Flowplan for EXEC. Management
            of the Planning Stage

collected by recovering the stored data, year by year. This is the
optimum plan, and the unfolding is done by subroutine PATH.

After the search, OPLEX determines which quarter year gives the optimum
NPW for the start-up of the first decision. This is done by evaluating
the NPW for each quarter of the year of the first decision, with and without
the first decision. The first quarter where the NPW with the decision is
greater than without it is taken as the optimum quarter. Lower capacity
constraints still apply in the quarter's search.

Integer Packing

Much of the programming in OPLEX is concerned with the efficient
handling of the large quantity of data which is generated and must be
stored. Each state consists of 10 zero or one digits to define it ; using
the convention that 0 denotes alternative not operating, and 1 = operating.
Each state should also hold its optimum NPW and the decision that gave
this NPW and there is advantage also in storing the previous state on the
optimum path. For 256 yearly states for 20 years for 10 alternatives, the
storage requirement is therefore

$$256 \times 20 \times (10 + 10 + 2 + 1) = 117{,}760 \text{ locations.}$$

This requirement was reduced by using a packing routine, since most
of the numbers involved were small.

Firstly, the 10 zero or 1 state digits were packed into one variable
using the subroutine PACK. Secondly, a single variable was used to store
the decision, plus the last state on the optimum path, plus the NPW of the
state, again, by using the PACK subroutine. In this way the storage
requirement was reduced tenfold to

$$IRS(20,257), \quad IRR(20,256) \quad \text{i.e. } 10{,}260 \text{ locations.}$$

The OPLEX subroutine also stores all the plans obtained in the
searches, so that once the optimum strategy has been selected, it can locate
the yearly, operating data corresponding to this strategy. Again to have
10 alternatives, which may be operating or not, for each of 20 years for
a total of 10 strategies each with 8 probabilities requires a storage of:

$$10 \times 20 \times 10 \times 8 = 16{,}000 \text{ locations}$$

This was again brought within practical limits by packing. Up to 5 build
and 5 scrap decisions were allowed for in one strategy, and these were
stored as decision and year of decision. The 5 build decisions and their

corresponding years and the 5 scrap decisions and their years were then
packed into 4 variables. Hence for 10 strategies and 8 probabilities the
following matrices suffice:

    IAE(10,8), IYE(10,8), IAS(10,8), IYS(10,8) = 320 locations

The optimum strategy is located by the EXEC subroutine. The last
action of the planning phase is to return to the OPLEX subroutine to unpack
the optimum strategy into "operate" data for the evaluation program to
handle.

Fig. A1.3 gives the Flow diagram for the Dynamic Programming Search
procedure, Fig. A1.4 shows tests made to discard any state that violates
constraints, and Fig. A1.5 is a logic diagram for strategy storage and
location of optimum strategy.

These three program functions are performed within the OPLEX subroutine.

A1.5 SUBROUTINE WERT

This subroutine calculates the NPW after tax of one year's operation,
for use in the Dynamic Programming Search of subroutine OPLEX. The NPW
includes capital and scrap cash flows associated with the decision taking
effect that year. The result is not the cash flows in that year. For
example, capital is included the year the plan starts up, although it must
have been spent earlier, and the tax allowance depreciation is all credited
in that year by multiplying the capital sum with a factor, which is obtained
from subroutine TAXF.

When capacity is greater than demand, WERT firstly has to decide which
plants to operate fully, and which to reduce load, so as to minimise its
operating costs. Following this, a normal discounted cash flow after tax
calculation is made. Although capital is spent 2 years earlier than the
remaining cash flows, the Factor from subroutine TAXF makes allowance for
this, so that all sums in the NPW calculation can be discounted with a
single multiplication. Working capital is not included.

WERT is also used to evaluate the NPW obtained in a single quarter
year, to enable subroutine OPLEX to determine the optimum quarter in which
to make the first decision.

Fig. A1.6 shows the flow diagram for the evaluation of NPWs for the
D.P. Search.

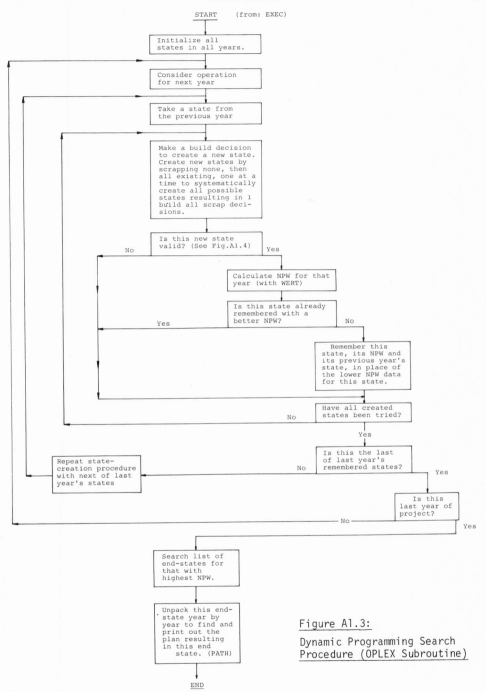

START   (from: EXEC)

Initialize all
states in all years.

Consider operation
for next year

Take a state from
the previous year

Make a build decision
to create a new state.
Create new states by
scrapping none, then
all existing, one at a
time to systematically
create all possible
states resulting in 1
build all scrap deci-
sions.

Is this new state
valid? (See Fig.A1.4)   No   Yes

Calculate NPW for that
year (with WERT)

Is this state already
remembered with a
better NPW?   Yes   No

Remember this
state, its NPW and
its previous year's
state, in place of
the lower NPW data
for this state.

Have all created
states been tried?   No   Yes

Is this the last
of last year's
remembered states?   No   Yes

Repeat state-
creation procedure
with next of last
year's states

Is this
last year of
project?   No   Yes

Search list of
end-states for
that with
highest NPW.

Unpack this end-
state year by
year to find and
print out the
plan resulting
in this end
state. (PATH)

END

Figure A1.3:

Dynamic Programming Search
Procedure (OPLEX Subroutine)

to: plan Packing Section (Fig. A1.5)
and return to EXEC.

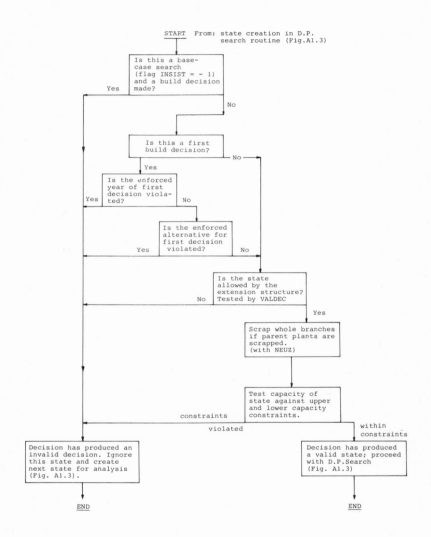

Figure A1.4: Test for Valid Decision
(OPLEX Subroutine)

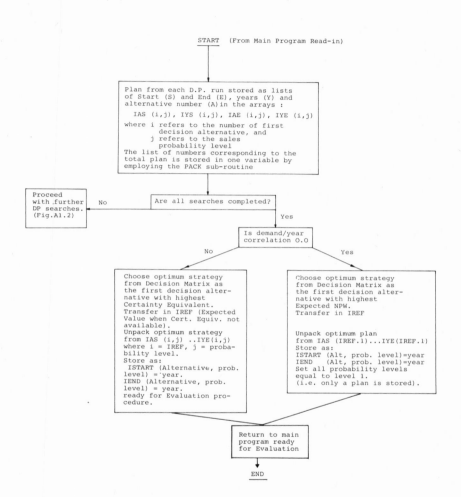

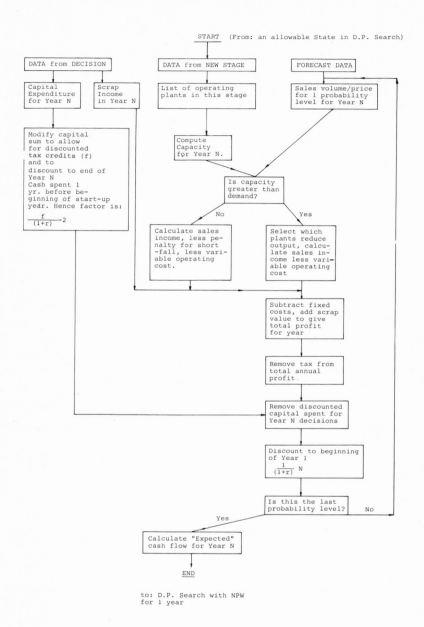

START   (From: an allowable State in D.P. Search)

DATA from DECISION

Capital Expenditure for Year N | Scrap Income in Year N

DATA from NEW STAGE

List of operating plants in this stage

FORECAST DATA

Sales volume/price for 1 probability level for Year N

Modify capital sum to allow for discounted tax credits (f) and to discount to end of Year N Cash spent 1 yr. before be-ginning of start-up yedr. Hence factor is:
$$\frac{f}{(1+r)} -2$$

Compute Capacity for Year N.

Is capacity greater than demand?

No

Calculate sales income, less pe-nalty for short -fall, less vari-able operating cost.

Yes

Select which plants reduce output, calcu-late sales in-come less vari-able operating cost

Subtract fixed costs, add scrap value to give total profit for year

Remove tax from total annual profit.

Remove discounted capital spent for Year N decisions

Discount to beginning of Year 1
$$\frac{1}{(1+r)} N$$

Is this the last probability level?   No

Yes

Calculate "Expected" cash flow for Year N

END

to: D.P. Search with NPW
for 1 year

Figure A1.6: Procedure for Calculating Cash Flow for Year N for D.P. Search (WERT Subroutine)

A1.6 SUBROUTINES VALDEC AND NEUZ

The extension structure defined by the input data effectively limits
the number of allowable states. The following decisions are illogical:

      (a) build an extension before the parent plant
           is built.

      (b) scrap a parent plant without scrapping its
           extension.

      (c) build a second extension onto a parent plant
           designated as having only exclusive extensions.

      (d) scrapping extensions without scrapping the parent
           plant. (This is based on practical experience --
           one never closes down only the newest parts of
           an extended plant).

The program logic that ensures these conditions are held is in three
parts. In the extension read-in which is in PLADE, the extension structure
is coded into three arrays; $KDAT(I,2)$, $KDAT(I,3)$, and $KTDAT(I)$. In
subroutine VALDEC, every decision is checked using these 3 arrays to ensure
that the conditions a), c) and d) are not violated. Violation terminates
further analysis of that state. VALDEC also returns a code to say whether
condition b) is violated.

In the third section, subroutine NEUZ uses this flag concerning
condition b) to scrap any extension structure branches where scrapping is
logically inferred by the single scrap decision under review.

A1.7 SUBROUTINE INVEST

The INVEST subroutine is the entry into the evaluation phase. It
first prints out the problem data in an organised way, then determines
whether the data requires a Monte-Carlo simulation or a single probabilistic run to be carried out, and finally calls appopriately either subroutine MONTE or subroutine FLOW. Table A1.1 summarises the conditions
required for a Monte-Carlo simulation to be chosen.

Table A1.1

Choice of Single NPW Evaluation or Monte Carlo

Analysis for Evaluation Phase

Single determination if: -   No trapezoidal data as input
(can be probabilistic)   and only one forecast proba-
                                     bility level supplied.

              or if: -   No trapezoidal data as input
                                     and both correlation coefficients
                                     are 1.0.

Monte-Carlo simulation if: -   Trapezoidal data given as input.

              or if: -   Trapezoidal data is not given
                                     as input, but more than one
                                     sales level is given, and
                                     correlation coefficients are
                                     both not 1.0.

The INVEST subroutine also manages the sensitivity calculations. These
are available either to the deterministic evaluation, or the Monte Carlo
simulation. Fig. A1.7 is the flow diagram for the overall management of
the evaluation phase.

A1.8 SUBROUTINE FLOW

This subroutine takes a complete set of Project evaluation data and
determines the total NPW over the Project life. It can accept up to 8
probabilistic forecast levels and repeats the evaluation for each of these
probability levels using the corresponding plan, in a single subroutine
call.

Features taken into consideration in the evaluation are:

(a) the operation of existing plants is taken as the
     base-case, and all results are expressed as a
     difference from this case.

(b) when capacity is greater than demand, the highest
     variable cost plant outputs are reduced first.
     This also applies to the base-case evaluation when
     more than one plant exist.

(c) extended life or total scrapping are optional
     methods of ending the project.

(d) inflation -- i.e. its effect on capital and tax
     allowance on capital. The calculation is based on
     constant value money, but the cumulative yearly cash flow
     profile and pay-back are expressed in current money terms.

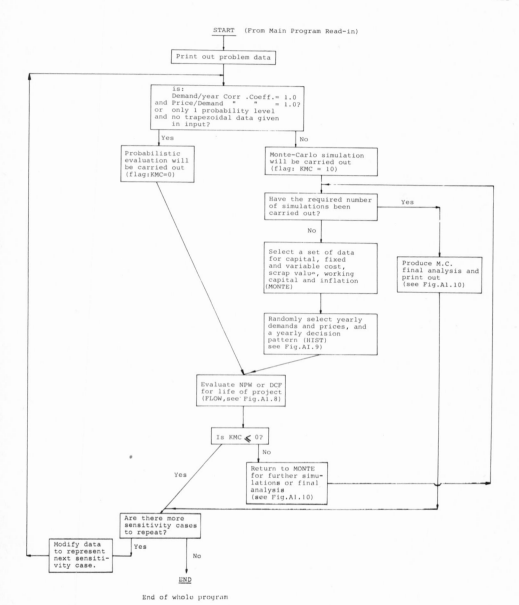

<u>Figure A1.7</u>: Management of Evaluation Phase
          (INVEST,FLOW,MONTE Subroutines)

(e) working capital is included as a factor after
     tax has been removed. In the final year, all the
     working capital is automatically recovered.

The subroutine can perform a DCF calculation by repreating the NPW
calculation until the discount rate for a zero Expected NPW is found.
The Secant method is used, after transformation to improve linearity, to
obtain this convergence.

The routine calculates Benefits/Cost Ratio and Equivalent Annual Value
using the equations given in Section 1.4c of Chapter I by calling up
subroutine EAVBC, and determines the Certainty Equivalent of any NPW
distribution by calling subroutine UTIL.

It also evaluates the mean pay-back time and provides an undiscounted
yearly cash-flow profile for each probability.

Tax allowance on depreciation is done by multiplying the capital cost
by a factor, obtained from subroutine TAXF, assuming straight line depreciation.

Subroutine FLOW is utilised by the Monte Carlo simulation by setting
the number of probability levels to 1 and then entering the required
number of times from MONTE with different, randomly generated data. Since
the simulation phase also requires randomly generated yearly data, this has
to be done within the FLOW subroutine. A part of the FLOW subroutine is
therefore the random demand and price generation and strategy selection
which is used only when FLOW has been entered from MONTE.

The demand and price generations are determined from probability
matrices, determined from the values of the correlation coefficients, in the
HIST subroutine. The strategy selection is made according to the generated
demand level. Fig. A1.8 shows the flow diagram for the evaluation, and
Fig. A1.9 shows the strategy, demand, and price yearly selection flow
diagrams for the Monte Carlo simulation.

A1.9 SUBROUTINE MONTE

The subroutine MONTE deals specifically with the Monte Carlo simulation. From the trapezoidal data for capacity, capital, variable and fixed
costs and scrap for each alternative, and penalty and inflation rate, the
subroutine chooses randomly a value for each to make up a set of data for a
single simulation by FLOW.

After carrying our 10 simulations an approximate estimate of the
variance is made, and this is used together with the requested accuracy for

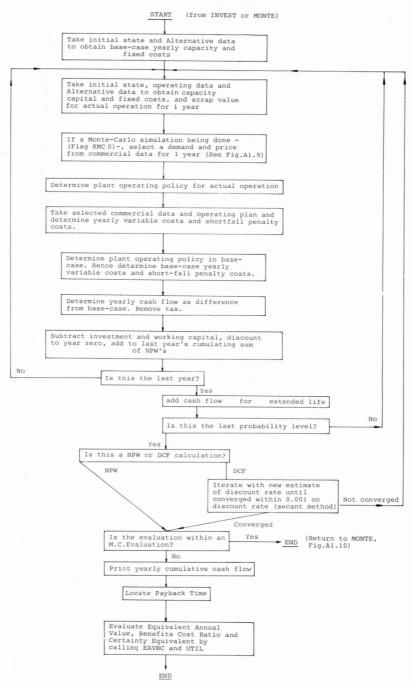

START   (from INVEST or MONTE)

Take initial state and Alternative data
to obtain base-case yearly capacity and
fixed costs

Take initial state, operating data and
Alternative data to obtain capacity
capital and fixed costs, and scrap value
for actual operation for 1 year

If a Monte-Carlo simulation being done -
(Flag KMC 0)-, select a demand and price
from commercial data for 1 year (See Fig.A1.9)

Determine plant operating policy for actual operation

Take selected commercial data and operating plan and
determine yearly variable costs and shortfall penalty
costs.

Determine plant operating policy in base-
case. Hence determine base-case yearly
variable costs and short-fall penalty costs.

Determine yearly cash flow as difference
from base-case. Remove tax.

Subtract investment and working capital, discount
to year zero, add to last year's cumulating sum
of NPW's

No

Is this the last year?

Yes

add cash flow   for   extended life

Is this the last probability level?   No

Yes

Is this a NPW or DCF calculation?

NPW   DCF

Iterate with new estimate
of discount rate until
converged within 0.001 on   Not converged
discount rate (secant method)

Converged

Is the evaluation within an   Yes
M.C.Evaluation?   ▶ END   (Return to MONTE,
                                              Fig.A1.10)
No

Print yearly cumulative cash flow

Locate Payback Time

Evaluate Equivalent Annual
Value, Benefits Cost Ratio and
Certainty Equivalent by
calling EAVBC and UTIL

END

Return to INVEST

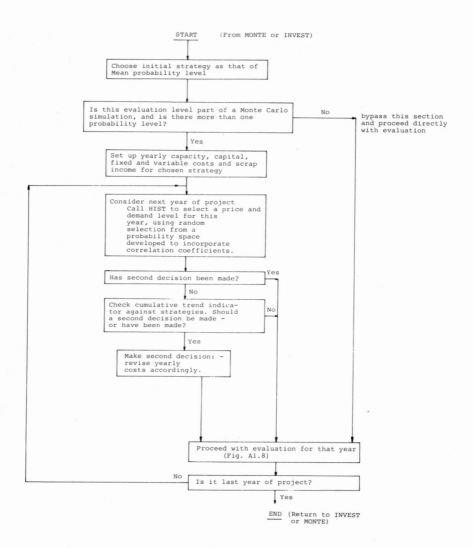

START   (From MONTE or INVEST)

Choose initial strategy as that of
Mean probability level

Is this evaluation level part of a Monte Carlo   No   bypass this section
simulation, and is there more than one   and proceed directly
probability level?   with evaluation

Yes

Set up yearly capacity, capital,
fixed and variable costs and scrap
income for chosen strategy

Consider next year of project
   Call HIST to select a price and
   demand level for this
   year, using random
   selection from a
   probability space
   developed to incorporate
   correlation coefficients.

Has second decision been made?   Yes

No

Check cumulative trend indica-   No
tor against strategies. Should
a second decision be made -
or have been made?

Yes

Make second decision: -
revise yearly
costs accordingly.

Proceed with evaluation for that year
(Fig. A1.8)

No   Is it last year of project?

Yes

END (Return to INVEST
     or MONTE)

Figure A1.9: Strategy, Demand and Price: Yearly Selection for
            MONTE-CARLO Simulation   (FLOW Subroutine)

the Expected value, to determine the number of simulations required to attain this accuracy. The simulation then proceeds for that number of iterations -- or to a maximum of 500, whichever is the smaller.

The MONTE subroutine then arranges all the results by magnitude, prints a table of results for various confidence levels, calls subroutine DIAGR to print out a cumulative probability graph of the results, and calculates the mean value, variance and Certainty Equivalent of the distribution.

The subroutine handles results as either NPW or DCF, as requested by the keyword in the data-input. The DCF will have about a 4-fold increase in computation time because each DCF result is an iteration to find the zero NPW.

After the completion of the Monte Carlo simulation, the MONTE sub-routine returns to FLOW in a non-Monte-Carlo simulation mode to produce mean pay-back, yearly cash-flows, Equivalent Annual Value and Benefit/Cost ratio. Control is then returned to INVEST.

Fig. A1.10 gives the logic Flow diagram for the MONTE subroutine.

A1.10 SUBROUTINE TAXF AND TAXD

These two subroutines are associated with the tax allowance on capital. Since the evaluations are done in constant value money terms, these sub-routines also contain the only references to inflation (apart from a return to current-value money for the cumulative cash-flow and pay-back calcul-ations).

TAXF determines the factor by which the capital sum is multiplied in order to:

(a) credit all future tax allowance by effectively reducing the capital sum (after allowing for inflation and discounting of tax allowances)

(b) make a one-year inflation correction, since capital is quoted in money terms for year -1 but the basis for the constant value money calculation is the beginning of year 0.

(c) discount by -2 years. Since there is a two-year difference between capital spent, and money received from any analysis concerning operation in a particular year, all capital sums are increased by two-year discounting, to enable a single discount factor to be used for all monies in subroutines WERT and FLOW.

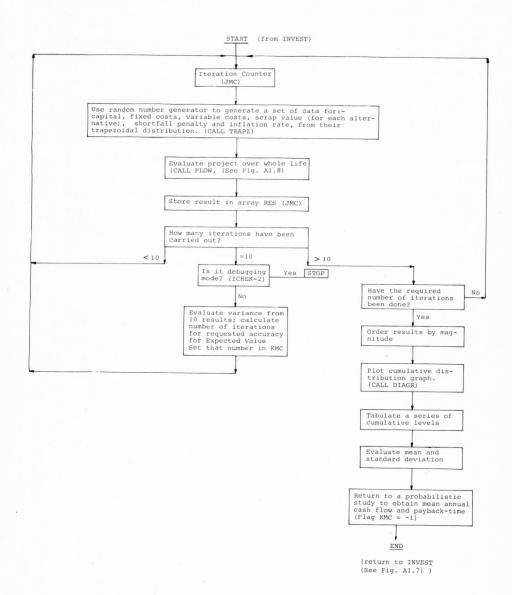

Figure A1.10: Logic Flowdiagram for
MONTE-CARLO SIMULATION Section

The TAXF subroutine has been written in these three distinct steps to aid understanding, and enable alternative methods of tax allowance to be easily programmed in.

TAXD determines the yearly sum that can be allowed against taxation as depreciation. This is quoted in current money units, and is used specifically for the cumulative cash-flow and pay-back calculations.

The subroutines given in the standard version of PLADE assume straight-line depreciation, for a number of years that can be given as a data item.

These two Tax subroutines have to be altered when a depreciation method other than straight-line depreciation is required. The subroutines can either be replaced by other appropriate routines -- sufficient description is intended in the comment statements to enable alternative routines to be easily written, or the existing routines can be extended to enable any appropriate method to be selected by a flag given in the data. To this end, a flag KTAX has been supplied as the third data item after the TAX keyword. This flag is available to TAXF and TAXD, and so can be used to direct the computation to any desired depreciation method.

The assumptions made in the existing codings are given below. It is necessary to define exactly the assumed timing of all cash flows to avoid confusion.

Consider an evaluation in Year n with plant X starting up in this year (i.e. January 1st, Year n).

(a) Capital for plant X is spent as one sum on January 1st, year (n-1). Capital is estimated for year -1. Therefore to convert to constant value money -- base year 0 -- it must be multiplied by (1+inflation rate).

(b) First tax allowance on capital for plant X is received on December 31st, year n.

(c) Scrap for any plant that is shut down in year n is received on December 31st, year n. The plant is also assumed operable until the end of the year.

(d) All other cash flows -- income from sales, fixed, variable costs, working capital, -- are assumed to flow on December 31st, year n.

A1.11 SUBROUTINE UTIL

The UTIL subroutine with the standard PLADE program calculates the Certainty Equivalent for either a set of data, each with a probability,

or a set of data of equal probabilities. Should the data be incomplete,
it returns a Certainty Equivalent equal to the lowest allowed NPW (LW i.e
-500,000 in the standard version). If it has adequate data, but some lie
below the lowest tolerable profit level, then it returns a Certainty Equivalent of (LW+1, i.e. -499999 in the standard version).

The Utility Function used is that described in Chapter V which requires
one data item. Should a better function be available, the user should
program in his better equation. To aid this, five data items (in vector
XMIN) can be read in following the UTILITY keyword, and these used for the
constants of a more sophisticated Utility Function.

A1.12 NOTES ON MODIFICATIONS NECESSARY TO RUN ON SMALLER WORD-LENGTH MACHINES

The program was developed on a CDC 6400/6500 system which has a 60
bit word length, allowing integers up to 1.5×10^{17} to be used.

To reduce program size and computation time, high integer values were
used in four places:

1) The NPW is handled as an integer number for the DP search.
    (Maximum size : 2LW, i.e. 1×10^6 in the Standard Version)

2) Data defining the operating state is packed into a single
    integer variable in OPLEX.
    (Maximum size : $2^{10} = 1024$)

3) Four data items to define each state are packed into a
    single integer variable in OPLEX.
    (Maximum size : $11 \times 11 \times 257 \times 1,000,001 = 3.11 \times 10^{10}$)

4) Five data items concerning plans -- alternative number and
    year -- are packed into single integers in the OPLEX sub-
    routine.
    (Maximum size : $21^5 = 4.08 \times 10^6$)

Whenever the program is used on a different machine, the maximum
allowed integer size must be checked to ensure that the four items above
can be handled. If the word length is less than 36 bits, the appropriate
action described below must be taken.

If items 3) or 4) cannot be met, the packing routine should be
converted to double precision.   It has been so arranged that this can be
conveniently done as follows:

a) Remove subroutine PACK and UNPACK from the deck.
b) Rename subroutines DPACK and DUNPACK as PACK and UNPACK
    according to the comment cards in the subroutines.

c) Declare all integers associated with the packed values as double precision by removing the c from the comment card No. OPLE 22.

d) Declare the packed values in Subroutine PATH as double precision by removing c from the comment card No. PATH 9.

This will result in all packing which concerns items 2), 3) and 4) to be double precision, and therefore within the capacity of all machines.

If the machine is less than 22 bits per word, item 1) will also be beyond the limits of the integer size. This can be dealt with by reducing the value of LW in the Block data to be within the permissible range. For example, for a 16 bit word machine, a value of LW of 16,000 would suffice.

This restricts the range of the money units that can be used by the program which reduces flexibility and requires more careful scaling of the cash units. With a value of LW of 16,000, it is inadvisable to use capital costs greater than 10,000.

Use on small machines may also involve difficulties with the length of the names that can be stored in one word. Three leters per word (A3 format) has been employed. This is not available on the smallest machines.

The use of double precision brings two serious disadvantages; firstly it increases the program size by 30%, because the major storage arrays are for the DP search and they take double the storage, and secondly the computer times were found, in practice, to be double on the CDC machine. This doubling of computer time must be due to the less efficient handling in double precision, which is a machine   specific feature.

It may be possible to remove these disadvantages by tailor-making the program to the machine and the type of problem. If 256 remembered states within the DP search are unnecessary, and 10^6 for NPW too accurate, and a five-stage plan unnecessary, appropriate changes can be made which may result in the problem still requiring only single precision. For such changes the following cards would have to be altered:

```
OPLE    3
OPLE   26
OPLE   34
BLK    28
OPLE  393
OPLE  414
```

If the problem size cannot be reduced to be within single precision, the creation for a further array for NPW in the DP search, instead of packing it with the other items under 3) would obviate the necessity for

double precision, and probably actually decrease computer times at the
expense of a few weeks' programming effort.

    If the program is to be used as a routine tool on a small word-length
machine, it would definitely be worth some weeks' effort by a computer
specialist to make the program and packing arrangement match the problem
size and machine characteristics.

MANUAL APPENDIX 2

THE PLADE PROGRAM LISTING

```
      PROGRAM PLADE (INPUT,OUTPUT,TAPE1=INPUT,TAPE2=OUTPUT)          PLAD    1
C  PLANNING AND DECISION INVESTMENT ANALYSIS PROGRAM                 PLAD    2
C  L.M.ROSE, C.H.D.WALTER,F.MARCOZ                                   PLAD    3
C  SYSTEMS ENGINEERING GROUP, TCL, ETH, ZURICH.                     PLAD    4
C    VERSION 1  1/3/75                                              PLAD    5
C  TO RUN THE PLANNING PART OF THIS PROGRAM ON 24 BIT MACHINES ,THE  PLAD    6
C    FOLLOWING CHANGES MUST BE MADE:                                 PLAD    7
C         REMOVE 'C' FROM DOUBLE PRECISION CARDS IN OPLEX AND PATH   PLAD    8
C         REMOVE SUBROUTINES PACK AND UNPACK FROM DECK               PLAD    9
C         REMOVE TITLE CARD FROM SUBROUTINE DPAK,AND 'C' FROM SECOND CARDPLAD  10
C         DO LIKEWISE FOR SUBROUTINE DUNPAK                          PLAD   11
C  MAIN PROGRAM READS IN,PREPARES DATA FOR OPLEX AND FLOW            PLAD   12
      DIMENSION ROUT(20),NOUT(3),KEY(30)                            PLAD   13
      DIMENSION  IISW(3,2) ,IDATA(30)                               PLAD   14
      COMMON/EX /INS(10),KEX                                        PLAD   15
      COMMON/INV /SENS(5)                                           PLAD   16
      COMMON/JL01/PN,KDCF,KS,E(8),S(5),CAPL(20),PKB                 PLAD   17
      COMMON/TRAPS/PCAPY(10,4),PCAPL(10,4),PSOP(10,4),PVOP(10,4),   PLAD   18
     *PSCRAP(10,4),PFINE(4)          ,PENF(4),MC                    PLAD   19
      COMMON/RDAT1/FINE,DR,UVOP(10),NT,NUT,NO,TX ,ENF,KTAX,ALT(10,3) PLAD   20
      COMMON/RDAT2/UFOP(10),UCAPY(10),UCAPL(10),USCFAP(10),ADDN(20)  PLAD   21
      COMMON/SALES/PROB(8),JT,TONS(9,20),SEL(3,20),WORK(8,20)        PLAD   22
      COMMON/TIME/MIEND,IST(10),ISTART(10,8),IEND(10,8)              PLAD   23
      COMMON/NAMES/ISW(24),INA(10,3)                                 PLAD   24
      COMMON/UTIL/IU,XMIN(5)                                         PLAD   25
      COMMON/RWR/IOUT,IN                                             PLAD   26
      COMMON/SIM/CORT,CORS,KMC,ERR,IYEAR,JMC                         PLAD   27
      COMMON/CUNT/LN,ICHEK                                           PLAD   28
      COMMON/CPLE/BELOW,ITRACE,UPR,INSYR,KDMD(9,20)                  PLAD   29
      COMMON/MDAT1/KDAT(10,9),NR, KTDAT(10)                          PLAD   30
      COMMON/STRAT/IFIR(8), ISEC(8) ,NPLAN                           PLAD   31
      COMMON/REF/IREF                                                PLAD   32
      DATA KEY(1), KEY(2), KEY(3), KEY(4), KEY(5), KEY(6), KEY(7),   PLAD   33
     1KEY(8), KEY(9), KEY(10), KEY(11), KEY(12), KEY(13), KEY(14),   PLAD   34
     2KEY(15), KEY(16), KEY(17), KEY(18), KEY(19), KEY(20), KEY(21), PLAD   35
     3KEY(22), KEY(23), KEY(24), KEY(25) ,KEY(26)                    PLAD   36
     4                  /3HHEA, 3HOTS,3HLIF,3HTAX,3HCOP,3HDCF        PLAD   37
     1, 3HPRC,3HDEM,3HPRI,3HPEN,3HALT,3HNOR,3HOPT,3HSEN,3HCHE        PLAD   38
     2, 3HECU,3HTRA,3HEXT,3HUTI,3HPLA,3HMUL,3HINF,3HEVA,3HADD,3HSTO,3HNOTPLAD 39
     1/                                                              PLAD   40
      DATA KEND,IEX/3HEND,2HEX/                                      PLAD   41
C                                                                    PLAD   42
C                                                                    PLAD   43
C              OPLEX                                     INVEST       PLAD   43
C                                                                    PLAD   44
C              ICHEK      : VALUE FOR OUTPUT CONTROL     ICHEK        PLAD   45
C              IDEPR      : NO OF YEARS FOR DEPRECIATION  NO          PLAD   46
C     IST1/ISW(I)I)       : STATE1                       IISW         PLAD   47
C              LIFE       : LENGHT OF THE PROJECT        NT           PLAD   48
C              MET        : EXTENDED LIFE (YEARS)        MIEND        PLAD   49
C     NR/  N              : NUMBER OF UNITS              NUT          PLAD   50
C              NKV        : NUMBER OF PROBABILITY        JT           PLAD   51
C              PROB       : PROBABILITY LEVEL            PROB         PLAD   52
C              PEN        : PENALTY PER TON SHORTFALL    FINE         PLAD   53
C     RRATE/ RATE         : DISCOUNT RATE                DR           PLAD   54
C              TAX        : TAX                          TX           PLAD   55
```

```
C                    XMIN      : VALUE FOR UTILITY SUBROUTINE              PLAD  56
C              RDMD/KDMD        : SALES DEMAND                     TONS     PLAD  57
C                    PREIS      : SALES PRICE PER TON              SEL      PLAD  58
C        CORT              :DEMAND/YEAR CORRELATION(TONS)      CORT         PLAD  59
C        CCRS              :PRICE/DEMAND CORRELATION (SEL)     CCRS         PLAD  60
        DO 1 I=1,100                                                       PLAD  61
1       KTDAT(I)=0                                                         PLAD  62
        DO 2 I=1,10                                                        PLAD  63
        IST(I)=0                                                           PLAD  64
        INS(I)=0                                                           PLAD  65
        DO 2 II=1,9                                                        PLAD  66
2       KDAT(I,II)=J                                                       PLAD  67
        DO3 J=1,20                                                         PLAD  68
        ADDN(I)=0.0                                                        PLAD  69
        DO 3 II=1,8                                                        PLAD  70
3       WORK (II,I) =0.0                                                   PLAD  71
500     READ(IN,110) IWK,(IDATA(I),I=1,26)                                 PLAD  72
110     FORMAT( 26A3,1A2)                                                  PLAD  73
        IF(IWK.NE.KEY(1))WRITE(IOUT,111)  IWK, (IDATA(I),I=1,26)           PLAD  74
111     FORMAT(1X,26A3,1A2)                                                PLAD  75
        DO 498 I=1,26                                                      PLAD  76
498     IF(IWK.EQ.KEY(I)) GO TO 499                                        PLAD  77
        WRITE(IOUT,497)IWK                                                 PLAD  78
497     FORMAT( 19H ILLEGAL KEYWORD     ,A3,15(1H*))                       PLAD  79
        GO TO 500                                                          PLAD  80
499     GO TO (501,502,503,504,505,506,507,508,509,510,511,512,513,514,    PLAD  81
      1515,516,517,518,519,520,521,522,523,524,525,528),I                  PLAD  82
C---HEAD                                                                   PLAD  83
501     READ(IN,110)  (ISW(I),I=1,24)                                      PLAD  84
        WRITE(IOUT,6) ( ISW(I),I=1,24)                                     PLAD  85
6       FORMAT( 1H1, 16X, 5HPLADE   ,/34H PLANNING AND INVESTMENT PROGRAM  PLAD  86
      1   ,/15H ETH  ZURICH     ,///1H , 24A3              )               PLAD  87
        WRITE(IOUT,100)                                                    PLAD  88
100     FORMAT(/25H      COPY OF INPUT DATA    //)                         PLAD  89
        GO TO 500                                                          PLAD  90
C---DISCOUNT RATE FOR  NPW-CALCULATION                                     PLAD  91
502     CALL FREAD(ROUT,NOUT,0)                                            PLAD  92
        DR=ROUT(1)                                                         PLAD  93
        KDCF =0                                                            PLAD  94
        DR=DR/100.                                                         PLAD  95
        GO TO 500                                                          PLAD  96
C---DCF                                                                    PLAD  97
506     KDCF=1                                                             PLAD  98
        GO TO 500                                                          PLAD  99
C     COMMERCIAL DATA                                                      PLAD 100
C---TAX                                                                    PLAD 101
504     CALL FREAD(ROUT,NOUT,0)                                            PLAD 102
        TX=ROUT(1)                                                         PLAD 103
        ND=IFIX(ROUT(2))                                                   PLAD 104
        KTAX=IFIX(ROUT(3))                                                 PLAD 105
        IF(ND.EQ.0)ND=10                                                   PLAD 106
        TX=TX/100.                                                         PLAD 107
        GO TO 500                                                          PLAD 108
C---UTILITY                                                                PLAD 109
519     CALL FREAD(ROUT,NOUT,0)                                            PLAD 110
        DO 480 J=1,5                                                       PLAD 111
480     XMIN(J)=ROUT(J)                                                    PLAD 112
        IU=1                                                               PLAD 113
        GO TO 500                                                          PLAD 114
C---LIFE                                                                   PLAD 115
503     CALL FREAD(ROUT,NOUT,0)                                            PLAD 116
        NT=IFIX(ROUT(1))                                                   PLAD 117
        MIEND=IFIX(ROUT(2))                                                PLAD 118
        GO TO 500                                                          PLAD 119
C---PROBABILITY   (READ PLUS NORMALISE)                                    PLAD 120
507     CALL FREAD(ROUT,NOUT,0)                                            PLAD 121
        JT=1                                                               PLAD 122
495     IF(RCLT(JT).LT.0.000001)  GO TO 496                                PLAD 123
```

```
         PROB(JT) = ROUT(JT)                                           PLAD 124
         JT=JT+1                                                       PLAD 125
         GO TO 495                                                     PLAD 126
496      JT=JT-1                                                       PLAD 127
         X=0.0                                                         PLAD 128
         DO 9 J=1,JT                                                   PLAD 129
9        X=X+PROB(J)                                                   PLAD 130
         DO 10 J=1,JT                                                  PLAD 131
10       PROB(J)=PROB(J)/X                                             PLAD 132
         GO TO 500                                                     PLAD 133
C---MULTIPRODUCT COMMERCIAL INPUT                                      PLAD 134
521      CALL MULTI  (NT)                                              PLAD 135
         DO 540 N=1,NT                                                 PLAD 136
         DO 540 J=1,JT                                                 PLAD 137
540      KDMD( J,N) =IFIX(TONS(J,N) )                                  PLAD 138
         GO TO 500                                                     PLAD 139
C---DEMAND                                                             PLAD 140
508      DO 30 J=1,JT                                                  PLAD 141
         CALL FREAD(ROUT,NOUT,0)                                       PLAD 142
         DO 25 N=1,10                                                  PLAD 143
         KDMD(J,N)=IFIX(ROUT(N))                                       PLAD 144
25       TONS(J,N)=ROUT(N)                                             PLAD 145
         IF(NT.LE.10)GOTO 30                                           PLAD 146
         CALL FREAD(ROUT,NOUT,0)                                       PLAD 147
         DO 26 N=11,NT                                                 PLAD 148
         KDMD(J,N)=IFIX(ROUT(N-10))                                    PLAD 149
26       TONS(J,N)=ROUT(N-10)                                          PLAD 150
30       CONTINUE                                                      PLAD 151
         GO TO 500                                                     PLAD 152
C---PRICE                                                              PLAD 153
509      DO 31 J=1,JT                                                  PLAD 154
         CALL FREAD(ROUT,NOUT,0)                                       PLAD 155
         IF(J.NE.1.AND.ROUT(1).EQ.0.)CALL FILL(J,NT,SEL)              PLAD 156
         IF(ROUT(1).EQ.0.).AND.J.NE.1.AND.NT.GT.10) CALL FREAD(ROUT,NOUT,0)PLAD 157
         IF(ROUT(1).EQ.0.AND.J.NE.1)    GO TO 31                       PLAD 158
         SEL(J,1)=ROUT(1)                                              PLAD 159
         DO 32 N=2,10                                                  PLAD 160
         SEL(J,N)=ROUT(N)                                              PLAD 161
32       IF(SEL(J,N).EQ.0.)SEL(J,N)=SEL(J,N-1)                         PLAD 162
         IF(NT.LE.10)GOTO 31                                           PLAD 163
         CALL FREAD(ROUT,NOUT,0)                                       PLAD 164
         DO 33 N=11,NT                                                 PLAD 165
         SEL(J,N)=ROUT(N-10)                                           PLAD 166
33       IF(SEL(J,N).EQ.0.)SEL(J,N)=SEL(J,N-1)                         PLAD 167
31       CONTINUE                                                      PLAD 168
         GO TO 500                                                     PLAD 169
C---CORRELATION                                                        PLAD 170
505      CALL FREAD(ROUT,NOUT,0)                                       PLAD 171
         CORT=ROUT(1)                                                  PLAD 172
         CORS=ROUT(2)                                                  PLAD 173
         GO TO 500                                                     PLAD 174
C---TRAPEZOIDAL                                                        PLAD 175
517      MC=1                                                          PLAD 176
         CALL FREAD(ROUT,NOUT,0)                                       PLAD 177
         ERR=ROUT(1)                                                   PLAD 178
         GOTO 500                                                      PLAD 179
C CANCEL TRAPS                                                         PLAD 180
528      MC=0                                                          PLAD 181
         GO TO 500                                                     PLAD 182
C---INFLATION                                                          PLAD 183
522      CALL FREAD(ROUT,NOUT,0)                                       PLAD 184
         IF(MC.GT.0)  GO TO 600                                        PLAD 185
         ENF=ROUT(1)/100.                                              PLAD 186
         CALL POINT(PENF(1),PENF(2),PENF(3),PENF(4),ENF)              PLAD 187
         GOTO 500                                                      PLAD 188
600      DO 605 I=1,4                                                  PLAD 189
605      PENF(I)=ROUT(I)/100.                                          PLAD 190
         CALL MEAN(PENF(1),PENF(2),PENF(3),PENF(4),ENF)               PLAD 191
```

```
        GOTO 500                                                        PLAD 192
C---PENALTY                                                             PLAD 193
510     IF( MC.GT.J) GO TO 581                                          PLAD 194
        CALL FREAD(ROUT,NOUT,0)                                         PLAD 195
        FINE=ROUT(1)                                                    PLAD 196
        CALL POINT(PFINE(1),PFINE(2),PFINE(3),PFINE(4),FINE)            PLAD 197
        GO TO 500                                                       PLAD 198
C       581 FOR MONTE CARLO                                             PLAD 199
581     CALL FREAD(ROUT,NOUT,0)                                         PLAD 200
        DO 582  J=1,4                                                   PLAD 201
582     PFINE(J)=ROUT(J)                                                PLAD 202
        CALL MEAN(PFINE(1),PFINE(2),PFINE(3),PFINE(4),FINE)             PLAD 203
        GO TO 500                                                       PLAD 204
C---ALTERNATIVES                                                        PLAD 205
511     IF( MC.GT.0) GO TO 583                                          PLAD 206
        J=0                                                             PLAD 207
531     J=J+1                                                           PLAD 208
        CALL FREAD(ROUT,NOUT,1)                                         PLAD 209
        IF(NCUT(1).EQ. KEND)GOTO 533                                    PLAD 210
        IF(J.LE.10)  GO TO 526                                          PLAD 211
        WRITE(IOUT,532)                                                 PLAD 212
        GO TO 525                                                       PLAD 213
526     DO 527 JJ=1,3                                                   PLAD 214
C  ALT(10,3) IS NOT USED,IT IS AVAILABLE FOR OTHER ALTERNATIVE DATA     PLAD 215
C  E.G. PROPERTY/BUILDING/MACHINARY SPLIT.IT IS TRANSFERRED TO FLOW     PLAD 216
C  AND INVEST.                                                          PLAD 217
        ALT(J,JJ)=ROUT(JJ+7)                                            PLAD 218
527     INA(J,JJ)=NOUT(JJ)                                              PLAD 219
        UCAPY(J)=ROUT(1)                                                PLAD 220
        UCAPL(J)=ROUT(2)                                                PLAD 221
        UFOP(J)=ROUT(3)                                                 PLAD 222
        UVOP(J)=ROUT(4)                                                 PLAD 223
        USCRAP(J)=ROUT(5)                                               PLAD 224
        IST(J)=IFIX(ROUT(6))                                           PLAD 225
        KDAT(J,9)=IFIX(ROUT(7))                                         PLAD 226
        KDAT (J,2) =0                                                   PLAD 227
        KDAT (J,3) =0                                                   PLAD 228
C   POINT    PREPARES DISTRIBUTION FOR A POINT VALUE FOR M C  PURPOSES  PLAD 229
        CALL POINT (PCAPY(J,1),PCAPY(J,2),PCAPY(J,3),PCAPY(J,4), UCAPY(J))PLAD 230
        CALL POINT(PCAPL(J,1),PCAPL(J,2),PCAPL(J,3),PCAPL(J,4),UCAPL(J)) PLAD 231
        CALL POINT(PVOP(J,1),PVOP(J,2),PVOP(J,3),PVOP(J,4),UVOP(J))     PLAD 232
        CALL POINT(PFOP(J,1),PFOP(J,2),PFOP(J,3),PFOP(J,4),UFOP(J))     PLAD 233
        CALL POINT (PSCRAP(J,1),PSCRAP(J,2), PSCRAP(J,3), PSCRAP(J,4), USCPLAD 234
       *RAP(J))                                                         PLAD 235
        GO TO 531                                                       PLAD 236
533     NUT=J-1                                                         PLAD 237
        NR=NUT                                                          PLAD 238
        GO TO 588                                                       PLAD 239
C       583   FOR MONTE CARLO  INPUT                                    PLAD 240
583     DO 585  N=1,11                                                  PLAD 241
C       FOR MONTE CARLO                                                 PLAD 242
        DO 584 J=1,4                                                    PLAD 243
        CALL FREAD(ROUT,NOUT,1)                                         PLAD 244
        IF(NCUT(1).EQ. KEND)GOTO 587                                    PLAD 245
        IF(N.LE.10)  GO TO 535                                          PLAD 246
        WRITE(IOUT,532)                                                 PLAD 247
        GO TO 525                                                       PLAD 248
535     DO 534  JJ=1,3                                                  PLAD 249
534     INA(N,JJ)=NOUT(JJ)                                              PLAD 250
        PCAPY(N,J)=ROUT(1)                                              PLAD 251
        PCAPL(N,J)=ROUT(2)                                              PLAD 252
        PFOP(N,J)=ROUT(3)                                               PLAD 253
        PVOP(N,J)=ROUT(4)                                               PLAD 254
        PSCRAP(N,J)=ROUT(5)                                             PLAD 255
C       FOR MEAN VALUE OF THE FOUR DATA READ IN                        PLAD 256
        IF(J.GT.1) GO TO 584                                            PLAD 257
        IST(N)=IFIX(ROUT(6))                                           PLAD 258
        KDAT(N,9)=IFIX(ROUT(7))                                         PLAD 259
        KDAT (N,2) =0                                                   PLAD 260
```

```
        KDAT (N,3) =0                                           PLAD 261
584     CONTINUE                                                PLAD 262
        J=N                                                     PLAD 263
C  MEAN  PRODUCES A POINT VALUE FROM A DISTRIBUTION FOR EVALUATION  PLAD 264
        CALL   MEAN (PCAPY(J,1),PCAPY(J,2),PCAPY(J,3),PCAPY(J,4),  UCAPY(J))PLAD 265
        CALL   MEAN(PCAPL(J,1),PCAPL(J,2),PCAPL(J,3),PCAPL(J,4),UCAPL(J))  PLAD 266
        CALL   MEAN(PFOP(J,1),FFOP(J,2),FFOP(J,3),FFOP(J,4),UFOP(J))  PLAD 267
        CALL   MEAN(PVOP(J,1),PVOP(J,2),PVOP(J,3),PVOP(J,4),UVOP(J))  PLAD 268
        CALL   MEAN (PSCRAP(J,1),PSCRAP(J,2), PSCRAP(J,3), PSCRAP(J,4), USCPLAD 269
     *RAP(J))                                                   PLAD 270
585     CONTINUE                                                PLAD 271
587     NUT=N-1                                                 PLAD 272
        NR=NUT                                                  PLAD 273
588     DO 630 I=1,NUT                                          PLAD 274
        KDAT(1,6)= IFIX(UCAPY(I))                               PLAD 275
        KDAT(1,4)= IFIX(UCAPL(I))                               PLAD 276
        KDAT(I,7)= IFIX(UFOP (I))                               PLAD 277
630     KDAT(1,5)= IFIX(USCRAP(I))                              PLAD 278
        GO TO 500                                               PLAD 279
C---WORKING CAPITAL                                             PLAD 280
512     DO 686 J=1,JT                                           PLAD 281
        CALL FREAD(ROUT,NOUT,0)                                 PLAD 282
        DO 90 N=1,10                                            PLAD 283
90      WORK (J,N) = ROUT(N)                                    PLAD 284
        IF(NT.LE.11)   GO TO 686                                PLAD 285
        CALL FREAD(ROUT,NOUT,0)                                 PLAD 286
        DO 91 N=11,NT                                           PLAD 287
91      WORK (J,N) = ROUT (N-10)                                PLAD 288
686     CONTINUE                                                PLAD 289
        GO TO 500                                               PLAD 290
C---BOUNDARIES                                                  PLAD 291
516     CALL FREAD(ROUT,NOUT,0)                                 PLAD 292
        BELCH=ROUT(1)                                           PLAD 293
        UPR=ROUT(2)                                             PLAD 294
        GOTO 500                                                PLAD 295
C---EXTENSIONS                                                  PLAD 296
518     M=0                                                     PLAD 297
        N=NUT                                                   PLAD 298
        IZ=0                                                    PLAD 299
        DO 2001 I=1,61                                          PLAD 300
2001    KTDAT(I)=0                                              PLAD 301
        DO 2030 I=1,NUT                                         PLAD 302
2030    KDAT(I,3)=0                                             PLAD 303
C  MAX NO OF EXTENSION PAIRS =30                                PLAD 304
        JJ=0                                                    PLAD 305
2021    READ(IN,1060)(IISW(1,I),I=1,3),(IISW(2,I),I=1,3),IIS    PLAD 306
        WRITE(IOUT,1059)(IISW(1,I),I=1,3),(IISW(2,I),I=1,3),IIS PLAD 307
1059    FORMAT(1X,6A3,A2   )                                    PLAD 308
        JJ=JJ+1                                                 PLAD 309
1060    FORMAT(6A3,A2)                                          PLAD 310
        IF(IISW(1,1).EQ. KEND)GOTO 2034                         PLAD 311
        DO 2029 I=1,2                                           PLAD 312
        DO 2025 J=1,N                                           PLAD 313
2025    IF(IISW(I,1).EQ.INA(J,1).AND.IISW(I,2).EQ.INA(J,2).AND.IISW(I,3)  PLAD 314
     2.EQ.INA(J,3))GO TO 2027                                   PLAD 315
        WRITE(IOUT,1061)                                        PLAD 316
1061    FORMAT(31H EXTENSION NAME NOT RECOGNISED        )       PLAD 317
        GOTO 525                                                PLAD 318
2027    IF(I.EQ.2.AND.IIS.EQ. IEX)J=-J                          PLAD 319
        JJJ=2*JJ-I+1                                            PLAD 320
2029    KTDAT(JJJ)=J                                            PLAD 321
        IF(JJ.EQ.30 )  GO TO 2022                               PLAD 322
        GOTO 2021                                               PLAD 323
2022    WRITE(IOUT,532)                                         PLAD 324
        GO TO 525                                               PLAD 325
C  EXTENSION ORGANISATION   (NETWORK THEORY )                   PLAD 326
2034    DO 2040 K=1,61,2                                        PLAD 327
        IF(KTDAT(K ).EQ.0)GOTO 2070                             PLAD 328
        I=KTDAT(K+1 )                                           PLAD 329
```

```
         IF(KDAT(I,3).EQ.0)GOTO 2035          PLAD 330
         Y7=202)                               PLAD 331
 2035    KDAT(I,3)=KTDAT(K)                    PLAD 332
 2040    CONTINUE                              PLAD 333
 2070    KR=0                                  PLAD 334
         IF(IZ.NE.0)GO TO 56                   PLAD 335
         M=M+(K-1)/2                           PLAD 336
         KV=0                                  PLAD 337
C   LOOK FOR NETWORK JUNCTIONS,COUNT THEM      PLAD 338
         DO 2120 I=1,M                         PLAD 339
         KDAT(I,2)=0                           PLAD 340
         IF(KDAT(I,3).EQ.0)GOTO 2090           PLAD 341
         KDAT(I,2)=-1                          PLAD 342
C   NO OF 'CNE ORIGIONAL JUNCTIONS             PLAD 343
         KV=KV+1                               PLAD 344
         GOTO 2120                             PLAD 345
 2090    DO 2100 J=1,N                         PLAD 346
         IF(IABS(KDAT(J,3)).EQ.I)GOTO 2110     PLAD 347
 2100    CONTINUE                              PLAD 348
         GOTO 2120                             PLAD 349
 2110    KDAT(I,2)=1                           PLAD 350
C  NO OF ORIGIONAL JUNCTIONS                   PLAD 351
         KR=KR+1                               PLAD 352
 2120    CONTINUE                              PLAD 353
C   CHECK FOR LOOPS                            PLAD 354
         IF(M.EQ.KV)GOTO 2130                  PLAD 355
         KV=KV+KR                              PLAD 356
         IZ=2120                               PLAD 357
         GO TO 56                              PLAD 358
 2130    WRITE(IOUT,1013)KR                    PLAD 359
         KV=1                                  PLAD 360
         DO 2220 J=1,M                         PLAD 361
         IF(KDAT(J,2).EQ.0)GOTO 2220           PLAD 362
C  KEEP INITIAL BLOCK                          PLAD 363
         KR=KV                                 PLAD 364
         KTDAT(KR)=0                           PLAD 365
         DO 2150 I=1,N                         PLAD 366
         IF(I.EQ.J)GOTO 2150                   PLAD 367
         IF(IABS(KDAT(I,3)).NE.J)GOTO 2150     PLAD 368
C   SEGMENT J,I  FOUND                         PLAD 369
         KV=KV+1                               PLAD 370
         KTDAT(KV)=I                           PLAD 371
 2150    CONTINUE                              PLAD 372
         KTDAT(KR)=KV-KR                       PLAD 373
         KV=KV+1                               PLAD 374
         IF(KDAT(J,2).EQ.1)GOTO 2170           PLAD 375
         I=J                                   PLAD 376
         KF=0                                  PLAD 377
 2160    I=IABS(KDAT(I,3))                     PLAD 378
         KF=KF+1                               PLAD 379
         IF(KF.GE.N)GOTO 2165                  PLAD 380
         IF(KDAT(I,3).NE.0)GOTO 2160           PLAD 381
         IZ=2160                               PLAD 382
         IF(KV.LE.0.OR.KV.GT.100)WRITE(IOUT,1043)  PLAD 383
         KTDAT(KV)=I                           PLAD 384
         GOTO 2180                             PLAD 385
 2165    GO TO 56                              PLAD 386
 2170    KTDAT(KV)=J                           PLAD 387
 2180    IF(KTDAT(KR).LE.1)GOTO 2210           PLAD 388
         DO 2190 I=1,N                         PLAD 389
         IF(KDAT(I,3).NE.-J)GOTO 2190          PLAD 390
         KTDAT(KR)=-IABS(KTDAT(KR))            PLAD 391
         KDAT(I,3)=J                           PLAD 392
 2190    CONTINUE                              PLAD 393
C  WHOLE COLUMN KDAT( ,3) NOW +VE OR 0         PLAD 394
 2210    KDAT(J,2)=KR                          PLAD 395
         KV=KV+1                               PLAD 396
 2220    CONTINUE                              PLAD 397
```

```
C   BRANCH DATA HOW IN KTDAY AND   KDAT(I,J),I=1,M  ,  J=2,3              PLAD 398
        GO TO 510                                                        PLAD 399
96      WRITE(IOUT,1042) IZ                                              PLAD 400
        GO TO 525                                                        PLAD 401
1013    FORMAT(1X,I2,23H EXTENDABLE UNITS FOUND)                         PLAD 402
1042    FORMAT(//////37H ERROR IN EXTENSION STRUCTURE,LABEL   ,I5/////)  PLAD 403
1043    FORMAT(//////32H TOO COMPLEX EXTENSION STRUCTURE//////)          PLAD 404
C---OPERATE                                                              PLAD 405
513     JX=0                                                             PLAD 406
        IREF=1                                                           PLAD 407
        DO 1050 J=1,10                                                   PLAD 408
        DO 1050 I=1,8                                                    PLAD 409
        ISTART(J,I) =0                                                   PLAD 410
1050    IEND(J,I) =0                                                     PLAD 411
        J=0                                                              PLAD 412
594     J=J+1                                                            PLAD 413
        CALL FREAD(ROUT,NOUT,1)                                          PLAD 414
        IF(NOUT(1).EQ. KEND)GOTO 598                                     PLAD 415
        IF(J.LE.10 )  GO TO 300                                          PLAD 416
        WRITE(IOUT,532)                                                  PLAD 417
        GO TO 525                                                        PLAD 418
300     DO 595  I=1,NUT                                                  PLAD 419
595     IF(NOUT(1).EQ.INA(I,1).AND.NOUT(2).EQ.INA(I,2).AND.NOUT(3).      PLAD 420
       *EQ.INA(I,3))GOTO 597                                             PLAD 421
        WRITE(IOUT,596)                                                  PLAD 422
596     FORMAT(//22H NAME   NOT RECOGNISED       //)                    PLAD 423
        GOTO 525                                                         PLAD 424
597     DO 593 JJ=1,JT                                                   PLAD 425
        JJJ=JJ*2                                                         PLAD 426
        ISTART(I,JJ)=IFIX(ROUT(JJJ -1))                                  PLAD 427
        IEND(I,JJ)=IFIX(ROUT(JJJ ))                                      PLAD 428
593     IF(JJ.GT.1)    JX=JX + ISTART(I,JJ) + IEND(I,JJ)                 PLAD 429
        GO TO 594                                                        PLAD 430
598     IF(JX.GT.0)    GO TO 500                                         PLAD 431
C   TO FILL IN A STRATEGY WHEN IT IS NOT GIVEN                           PLAD 432
        IREF=0                                                           PLAD 433
        DO 599 J=1,NUT                                                   PLAD 434
        DO 599 JJ=2,JT                                                   PLAD 435
        ISTART(J,JJ) = ISTART(J,1)                                       PLAD 436
599     IEND(J,JJ) = IEND(J,1)                                           PLAD 437
        GO TO 500                                                        PLAD 438
C---SENSITIVITY                                                          PLAD 439
514     CALL FREAD(ROUT,NOUT,0)                                          PLAD 440
        DO 94 I=1,5                                                      PLAD 441
94      SENS(I)=ROUT(I)                                                  PLAD 442
        GO TO 500                                                        PLAD 443
C   ADDITIONAL YEARLY DATA                                              PLAD 444
524     CALL FREAD(ROUT,NOUT,0)                                          PLAD 445
        DO 601 N=1,10                                                    PLAD 446
601     ADDN(N)=ROUT(N)                                                  PLAD 447
        IF(NT.LE.10)  GO TO 603                                          PLAD 448
        CALL FREAD(ROUT,NOUT,0)                                          PLAD 449
        DO 602 N=11,NT                                                   PLAD 450
602     ADDN(N)=ROUT(N-10)                                               PLAD 451
603     GO TO 500                                                        PLAD 452
C---CHECK                                                                PLAD 453
C   IF ICHEK  = 0  : SMALL OUTPUT                                        PLAD 454
C   IF ICHEK  = 1   EXTENDED PRINT-OUT                                   PLAD 455
C   IF ICHEK = 2  : DEBUGGING PRINTOUT                                   PLAD 456
515     CALL FREAD(ROUT,NOUT,0)                                          PLAD 457
        ICHEK=IFIX(ROUT(1))                                             PLAD 458
        GOTO 500                                                         PLAD 459
C---PLAN                                                                 PLAD 460
520     CALL FREAD(ROUT,NOUT,0)                                          PLAD 461
C   READS IN  INSYR  AND  ARRAY INS( )                                  PLAD 462
        INSYR=IFIX(ROUT(1))                                             PLAD 463
        KEX=1                                                            PLAD 464
        IF(CCRT.GT.0.01) KEX=2                                           PLAD 465
```

```
        IF(CCRT.GT.0.9999)    KEX=3                                 PLAD 466
        DO 600 I=1,10                                               PLAD 467
606     INS(I)=0                                                    PLAD 468
        I=0                                                         PLAD 469
620     I=I+1                                                       PLAD 470
        READ(IN,1060)(IISW(1,II),II=1,3)                           PLAD 471
        WRITE(IOUT,111)(IISW(1,II), II=1,3)                        PLAD 472
        IF(IISW(1,1).EQ. KEND)GOTO 616                             PLAD 473
        IF(I.LE.10)  GO TO 612                                     PLAD 474
        WRITE(IOUT,532)                                             PLAD 475
        GO TO 525                                                   PLAD 476
612     DO 615 IO=1,NUT                                            PLAD 477
615     IF(IISW(1,1).EQ.INA(IO,1).AND.IISW(1,2).EQ.INA(IO,2).AND.  PLAD 478
       2IISW(1,3).EQ.INA(IO,3))INS(I)=IO                           PLAD 479
        IF(INS(I).EQ.0)WRITE(IOUT,596)                             PLAD 480
        GO TO 620                                                   PLAD 481
616     CALL EXEC                                                   PLAD 482
        GO TC 500                                                   PLAD 483
C---EVALUATE                                                        PLAD 484
523     IF(MC.EQ.1)    KMC=10                                       PLAD 485
        CALL INVEST                                                 PLAD 486
        GOTO 500                                                    PLAD 487
C---STOP                                                            PLAD 488
525     STOP                                                        PLAD 489
532     FORMAT(///19H END CARD MISSING      ///)                   PLAD 490
        END                                                         PLAD 491


        SUBROUTINE FILL(J,NT,SEL)                                   FILL  1
C  TO FILL IN PRICES WHEN THEY ARE ALL THE SAME                    FILL  2
        DIMENSION SEL(8,20)                                         FILL  3
        DO 10 N=1,NT                                                FILL  4
10      SEL(J,N)=SEL(J-1,N)                                        FILL  5
        RETURN                                                      FILL  6
        END                                                         FILL  7


        SUBROUTINE MULTI (NT)                                       MULT  1
        DIMENSION A(10),VC(10),TN(10)        ,ROUT(20),NOUT(3)     MULT  2
        COMMON/SALES/ PROB(3),JT,TONS(9,20),SEL(8,20)             MULT  3
        COMMON/RWR/IOUT                                            MULT  4
C    SUBROUTINE FOR REPRESENTING MULTI-PRODUCT INFORMATION AS A SINGLE  MULT  5
C       PSEUDO PRODUCT                                            MULT  6
C    JP=NO OF PRODUCTS,TN,PR,VC =INDIVIDUAL DEMAND,PRICES VAR.COSTS.   MULT  7
C    A=PLANT EFFECTIVENESS FACTORS,A(1)=1.0                       MULT  8
        CALL FREAD(ROJT,NOUT,0)                                   MULT  9
        JP=IFIX(ROUT(1)  )                                         MULT 10
        WRITE(IOUT,5)                                              MULT 11
5       FORMAT(21H MULTI-PRODUCT DATA,    //31H  PLANT EFFECTIVENESS FACTOMULT 12
       *RS:        )                                               MULT 13
        CALL FREAD(ROUT,NOUT,0)                                    MULT 14
        DO 10 I=1,JP                                               MULT 15
10      A(I)=ROUT(I)                                               MULT 16
        WRITE(IOUT,12)                                             MULT 17
12      FORMAT( 30H INDIVIDUAL VARIABLE COSTS:         )          MULT 18
        CALL FREAD (ROUT,NOUT,0)                                   MULT 19
        DO 15 I=1,JP                                               MULT 20
15      VC(I)=ROUT(I)                                              MULT 21
        WRITE(IOUT,18)                                             MULT 22
18      FORMAT( /51H DEMAND/PRICE  FOR EACH PRODUCT,PER YEAR,PER FRCBY ) MULT 23
        DO 50 J=1,JT                                               MULT 24
        DO 50 N=1,NT                                               MULT 25
        CALL FREAD (ROUT,NOUT,1)                                   MULT 26
        DO 20 I=1,JP                                               MULT 27
20      TN(I)=ROUT(I)                                              MULT 28
        CALL FREAD (ROUT,NOUT,1)                                   MULT 29
        WRITE(IOUT,22)                                             MULT 30
22      FORMAT(/)                                                  MULT 31
```

```
      TONS(J,N)=0.0                                                  MULT  32
      TNPR=C.C                                                       MULT  33
      TNVC=0.J                                                       MULT  34
      DO 30 I=1,JP                                                   MULT  35
      TONS(J,N)=TONS(J,N) + A(I)*TN(I)                               MULT  36
      TNPR= TNPR + TN(I)*ROUT(I)                                     MULT  37
30    TNVC=TNVC + TN(I)*VC(I)                                        MULT  38
50    SEL(J,N)=(TNPR-TNVC)/TONS(J,N) + VC(1)                         MULT  39
      WRITE(IOUT,60)                                                 MULT  40
60    FORMAT(70H ALL REFERENCES WILL NOW BE TO PSEUDO-TONNAGE OF THE FIRMULT  41
     *ST PRODUCT          )                                          MULT  42
      RETURN                                                         MULT  43
      END                                                            MULT  44


      SUBROUTINE EXEC                                                EXEC   1
C******************************                                      EXEC   2
C   SUBROUTINE TO MANAGE SEARCHES                                    EXEC   3
C        INSIST=-1   BASE CASE NO BUILD DECISIONS                    EXEC   4
C        INSIST=-2   OPTIMISE, NO ENFORCED 1ST DECISION              EXEC   5
C        INSIST=+VE   NO OF ALTERNATIVE ENFORCED AS 1ST DECISION     EXEC   6
C        KKEX=1    PROBABILISTIC ANALYSIS,EXPECTED NPW FROM SEARCH   EXEC   7
C        KKEX=2    DETERMINISTIC ANALYSIS,SEARCHES ONLY 1 PROBY. LEVEL EXEC 8
C   RESULTS OF SEARCH   MM,IZBB,JKK,                                 EXEC   9
C   LOCATION OF RESULTS--NPW(II,I)  II DENOTE 1ST DECISION ALTERNATIVE,- EXEC 10
C   I=1   EXPECTED NPW FROM 1 PROBABILISTIC RUN (ZERO CORRELATION)   EXEC  11
C   I=2   EXPECTED NPW FROM NKV DETERMINISTIC RUNS (TOTAL CORRELATION) EXEC 12
C   I=2+I TO  2+NKV    NPW FOR EACH DETERMINISTIC RUN                EXEC  13
C   IAV( )     CONTAINS APPROPRIATE BASECASE NPW                     EXEC  14
      DIMENSION KDMOR(20),                                           EXEC  15
     * PREISR(20),NPW(10,10),IAV(10),UTI (8)  ,VAL(8)                EXEC  16
      COMMON/MDAT1/XDAT(10,9)                                        EXEC  17
      COMMON/RWR/IOUT,IN                                             EXEC  18
      COMMON/UTIL/IU                                                 EXEC  19
      COMMON/RDAT1/REN,PRATE,VARCOS(10),LIFE,N                       EXEC  20
      COMMON/SALES/PROB(8),NKV,PDMD(9,20),PRTIS(8,20)                EXEC  21
      COMMON/OPLE/BELOW,ITRACE,UPR,INSYR,KDMD(9,20)                  EXEC  22
      COMMON/NAMES/ITI(24),INA(10,3)                                 EXEC  23
      COMMON/CONT /LW                                                EXEC  24
      COMMON/EX /INS(10),KEX                                         EXEC  25
      COMMON/SIM/CORT                                                EXEC  26
      COMMON/REF/IREF                                                EXEC  27
      COMMON/EXOP/ INSIST, MM,IZBB,JKK ,KKKK,J ,KTIME                EXEC  28
      KTIME= -1                                                      EXEC  29
      KKKK=0                                                         EXEC  30
      J=1                                                            EXEC  31
      IF(NKV.EQ.1 ) KEX=1                                            EXEC  32
      DO 2 I=1,10                                                    EXEC  33
      DO 2 II=1,10                                                   EXEC  34
2     NPW(II,I)=LW                                                   EXEC  35
      NINS=0                                                         EXEC  36
      IF(NINS.EQ.10)GO TO 10                                         EXEC  37
7     IF(INS(NINS + 1).EQ.0) GO TO 10                                EXEC  38
      NINS=NINS + 1                                                  EXEC  39
      GO TO 7                                                        EXEC  40
10    INSIST=-2                                                      EXEC  41
      L=-2                                                           EXEC  42
      KKEX=1                                                         EXEC  43
      WRITE(IOUT,6) (ITI(I), I=1,24)                                 EXEC  44
6     FORMAT(1H1,24A3)                                               EXEC  45
      IF(NKV.EQ.1)GO TO 19                                           EXEC  46
      WRITE(IOUT,20)  CORT                                           EXEC  47
      IF(KEX.LT.3)  WRITE(IOUT,22)                                   EXEC  48
      IF(KEX.GT.1)  WRITE(IOUT,23)                                   EXEC  49
19    IF(NINS.GT.0) GO TO 26                                         EXEC  50
      GO TO 999                                                      EXEC  51
26    WRITE(IOUT,24)                                                 EXEC  52
      DO 27 II=1,NINS                                                EXEC  53
```

```
          I=INS(II)                                                   EXEC   54
27        WRITE(IOUT,25)   (INA(I,MI),MI=1,3)                         EXEC   55
20        FORMAT( 54H THE VALUE OF THE DEMAND/YEAR CORRELATION COEFFICIENT(  EXEC  56
      1          ,F4.2,9H)REQUIRES ,/ 35H THE FOLLOWING SEARCHES TO BE MADE EXEC  57
      2              )                                                 EXEC   58
22        FORMAT (36X, 12H MEAN PLAN          )                       EXEC   59
23        FORMAT (36X, 33H PLAN FOR EACH PROBABILITY (STRATEGY)      )  EXEC   60
24        FORMAT( 36X, 33H WITH FOLLOWING FIRST DECISIONS           )  EXEC   61
25        FORMAT( 40X, 3A3)                                           EXEC   62
C         CALL CPLEX TO PRINT INPUT DATA  AND CALLING INITIAL SETTING UP  EXEC  63
999       IF(INSYR.GT.0) WRITE(IOUT,1215)  INSYR                      EXEC   64
          CALL CPLEX                                                  EXEC   65
C         ----------                                                  EXEC   66
          IF( KEX.EQ.3)   GO TO 450                                   EXEC   67
C  ROUTINE TO SET NEXT EVALUATION OF CPLEX                            EXEC   68
1000      L=L+1                                                       EXEC   69
          IF(L.LT.1)  GO TO 1110                                      EXEC   70
          INSIST=INS(L)                                               EXEC   71
          IREF=INSIST                                                 EXEC   72
          GO TO 1100                                                  EXEC   73
1110      INSIST=INSIST + 1                                           EXEC   74
          IF(INSIST.EQ.0.AND.NINS.GT.0)   GO TO 1000                  EXEC   75
1100      CALL CPLEX                                                  EXEC   76
C         ----------                                                  EXEC   77
          IF(INSIST.EQ.-1.AND.KKEX.EQ.1)  GO TO 200                   EXEC   78
          IF( INSIST.EQ.0 . AND.KKEX.EQ.1)   GO TO 300                EXEC   79
          IF( INSIST.GT.0.AND.KKEX.EQ.1)  GO TO 400                   EXEC   80
          IF( INSIST.EQ.-1.AND.KKEX.EQ.2)   GO TO 500                 EXEC   81
          IF( INSIST.EQ.0.AND.KKEX. EQ.2)  GO TO 600                  EXEC   82
          GO TO 700                                                   EXEC   83
C  NO DECISION PROB CASE                                              EXEC   84
200       IAV(1)=MM                                                   EXEC   85
          GO TO 1000                                                  EXEC   86
C  OPTIMUM PROB CASE                                                  EXEC   87
300       IF(IZEB.GT.0)    NPW(IZEB,1)=MM                             EXEC   88
          IF(IZEB.EQ.0.AND.MM.NE.LW)    WRITE(IOUT,310)               EXEC   89
310       FORMAT(/// 41H OPTIMUM FOR THIS CASE IS NOT TO INVEST    ///)EXEC  90
C  REMEMBER OPT DATA TO PREVENT DUPLICATE RUN                         EXEC   91
          IZ=IZEB                                                     EXEC   92
          JK=JKK                                                      EXEC   93
          IF(L.EQ.NINS)  GO TO 450                                    EXEC   94
          GO TO 1000                                                  EXEC   95
C  INSIST RUNS  PROB CASE                                             EXEC   96
400       NPW(INSIST,1)=MM                                            EXEC   97
          IF(L.LT.NINS) GO TO 1000                                    EXEC   98
450       IF(KEX.EQ.1)  GO TO 1200                                    EXEC   99
          DO 470 I=1,LIFE                                             EXEC  100
          KDMDR(I)=KDMD(1,I)                                          EXEC  101
470       PREISR(I)=PREIS(1,I)                                        EXEC  102
          PROBR=PROB(1)                                               EXEC  103
          NKVR=NKV                                                    EXEC  104
          PROB(1)=1.                                                  EXEC  105
          NKV=1                                                       EXEC  106
          KKEX=2                                                      EXEC  107
          J=1                                                         EXEC  108
          L=-2                                                        EXEC  109
          INSIST=-2                                                   EXEC  110
          GO TO 1000                                                  EXEC  111
C  TREAT EACH PROBABILITY DETERMINISTICALLY                           EXEC  112
C  NO DECISION,  DETMST. CASE                                         EXEC  113
500       IAV(J+2)=MM                                                 EXEC  114
          GO TO 1000                                                  EXEC  115
C  OPT  RUN DETMST,                                                   EXEC  116
600       IF(IZEB.GT.0)    NPW(IZEB,J+2)=MM                           EXEC  117
          IF(IZEB.EQ.0.AND.MM.NE.LW)   WRITE(IOUT,310)                EXEC  118
          IZ=IZEB                                                     EXEC  119
          JK=JKK                                                      EXEC  120
          IF(L.EQ.NINS)  GO TO 750                                    EXEC  121
```

```
          GO TO 1000                                              EXEC 122
C   INSIST RUNS  DETMST.                                          EXEC 123
700    NPW(INSIST,J+2)=000                                        EXEC 124
       IF(L.LT.NINS)  GO TO 1000                                  EXEC 125
750    IF( J.EQ.NKVR)  GO TO 1200                                 EXEC 126
       J=J+1                                                      EXEC 127
       DO 720 I=1,LIFE                                            EXEC 128
       KDMD(1,I)=KDMD(J,I)                                        EXEC 129
720    PREIS(1,I)=PREIS(J,I)                                      EXEC 130
       INSIS1=-2                                                  EXEC 131
       L=-2                                                       EXEC 132
       GO TO 1000                                                 EXEC 133
C   ALL OPLEX RUNS COMPLETED,  NOW PRODUCE THE REPORT            EXEC 134
C   RETURN CORRECT DATA IF PSEUDO DATA HAS BEEN USED             EXEC 135
1200   IF(KEX.EQ.1)  GO TO 1210                                  EXEC 136
       DO 1205 I=1,LIFE                                           EXEC 137
       KDMD(1,I)=KDMDR(I)                                         EXEC 138
1205   PREIS(1,I)=PREISR(I)                                       EXEC 139
       PROB(1)=PROBR                                              EXEC 140
       NKV =NKVR                                                  EXEC 141
1210   NKVR=NKV + 2                                               EXEC 142
       IF(KEX.EQ.1)  NKVR=1                                       EXEC 143
       WRITE(IOUT,1211)(JTI(I),I=1,24)                            EXEC 144
1211   FORMAT(1H1,24A3/30X,26H          DECISION MATRIX        /) EXEC 145
       IF(INSYR.GT.0)  WRITE (IOUT,1215)  INSYR                   EXEC 146
1215   FORMAT(  22H FIXED DECISION YEAR      ,  I5/ )             EXEC 147
C   CALC EXPECTED DETMST. NPW                                     EXEC 148
C   IF RESULT=LW  SEARCH ABANDONED, NO MEAN CAN BE OBTAINED      EXEC 149
       IF(KEX.EQ.1)  GO TO 1241                                  EXEC 150
       DO  1231  II=1,N                                          EXEC 151
       IAV(2)=0                                                  EXEC 152
       NPW(II,2)=0                                               EXEC 153
       J=0                                                       EXEC 154
       JJ=0                                                      EXEC 155
       DO 1230 I=1,NKV                                           EXEC 156
       IF(IAV(I+2).EQ.LW) J=1                                    EXEC 157
       IF(NPW(II,I+2).  EQ.LW)  JJ=1                             EXEC 158
       IAV(2)=IAV(2)+ IFIX(PROB(I) * FLOAT(IAV(I+2)))            EXEC 159
1230   NPW(II,2)=NPW(II,2)+ IFIX(PROB(I)* FLOAT(NPW(II,I+2)))    EXEC 160
       IF(J.EQ.1)  IAV(2)=LW                                     EXEC 161
1231   IF(JJ.EQ.1)  NPW(II,2) =LW                                EXEC 162
       WRITE(IOUT,1240)                                          EXEC 163
1240   FORMAT ( 16X,  43HNO YEARLY    I----------WITH TOTAL  CORRELATION--EXEC 164
      120(1H-),       /15X,   14HCORRELATION I,                  EXEC 165
      *10X,4HCERT,                      10X,  30HB Y    P R O B A EXEC 166
      2B I L I T Y    /    8H INITIAL,17X    ,85H   I MEAN   EQUIV  EXEC 167
      3 1       2        3       4     ,5H   6          )EXEC 168
1241   WRITE (IOUT ,1242)                                        EXEC 169
1242   FORMAT( 30H DECISION          NPW          ///)           EXEC 170
       XX =FLOAT(LW+2 )                                          EXEC 171
       IREF=-1                                                   EXEC 172
       DO 1250 II=1,N                                            EXEC 173
       L=0                                                       EXEC 174
       DO 1245 I=1,NKVR                                          EXEC 175
       IF(NPW(II,I).NE.LW)  L=1                                  EXEC 176
C   MODIFY ZERO POINT OF NPW AND ROUND TO NEAREST 10             EXEC 177
       IF(NPW(II,I).NE.LW.AND.IAV(I).NE.LW)                      EXEC 178
      *NPW(II,I)=NPW(II,I)-IAV(I)                                EXEC 179
       X=FLOAT(NPW(II,I))                                        EXEC 180
       IF(II.EQ.1.AND.I.GE.3)   VAL(I-2) =X                      EXEC 181
       IF(I.GE.3.AND.X.GT.VAL(I-2)) VAL(I-2)=X                   EXEC 182
       IF(I.GT.2)  UTI (I-2) =X                                  EXEC 183
       IF(X.LT.-0.01.OR.X.GT.0.01)                               EXEC 184
      *X= (X/ABS(X)) *(ABS(X)+0.499)                             EXEC 185
1245   NPW(II,I)=10*IFIX(0.1*X)                                  EXEC 186
       IF(L.EQ.0) GO TO 1250                                     EXEC 187
       CERT=FLOAT(LW)                                            EXEC 188
C   OBTAIN CERTAINTY EQUIVALENT                                  EXEC 189
       IF(IU.EQ.1.AND.KEX.GT.1)  CALL UTTL (NKV,NKV,UTI,PROB,CERT) EXEC 190
```

```
C                                                                    EXEC 191
C      PRINT OUT DECISION MATRIX                                     EXEC 192
C      ---------                                                     EXEC 193
       IF(KEX.EQ.1) WRITE(IOUT,1260) II,(INA(II,NI),NI=1,3),NPW(II,1) EXEC 194
       IF(KEX.GT.1)                                                  EXEC 195
      *WRITE(IOUT,1260) II,(INA(II,NI),NI=1,3),                      EXEC 196
      *NPW(II,1),NPW(II,2),CERT,        (NPW(II,I),I=3,NKVR)         EXEC 197
1260   FORMAT(1X,I2,1X,3A3,2I10,F12.0,8I10 /)                        EXEC 198
C  LOCATE BEST STRATEGY BY -CERT EQUIV. IFCORRELATION KEX=2          EXEC 199
C                          -EXPECTED VALUE IF NO CORRELATION  ,KEX=1 EXEC 200
       IF(IU.NE.1)   CERT=NPW(II,2)                                  EXEC 201
        IF(KEX. EQ.1)   CERT=NPW(II,1)                               EXEC 202
       IF (CERT.LE.XX) GO TO 1250                                    EXEC 203
       XX = CERT                                                     EXEC 204
       IREF = II                                                     EXEC 205
1250   CONTINUE                                                      EXEC 206
       IF(KEX.EQ.1.OR.NINS.EQ.0.OR.IREF.LE.0)    GO TO 1270          EXEC 207
       X=0.                                                          EXEC 208
       DO 1265 I=1,NKV                                               EXEC 209
1265   X=X+ PROB(I)*VAL(I)                                           EXEC 210
       X=X- FLOAT(NPW(IREF,2))                                       EXEC 211
       X=FLOAT(10*IFIX(X/10.))                                       EXEC 212
       WRITE(IOUT,1267) X                                            EXEC 213
1267   FORMAT(/// 39H BASED ON THE ABOVE FIRST DECISIONS,   / 56H THE EXEC 214
      1EXPECTED VALUE OF COMPLETE FORECAST INFORMATION IS    ,F7.0//// )EXEC 215
C  KKKK=0 FOR SEARCHES, =1 FOR STRATEGY RECOVERY                     EXEC 216
1270   KKKK=1                                                        EXEC 217
C  CALL OPLEX TO EXTRACT STRATEGY FOR IREF                           EXEC 218
       CALL OPLEX                                                    EXEC 219
C  RETURN TO MAIN PROGRAM                                            EXEC 220
       RETURN                                                        EXEC 221
       END                                                           EXEC 222


       SUBROUTINE OPLEX                                              OPLE   1
C    OPLEX SUBROUTINE MAKES D.P. SEARCH FOR PLANS                    OPLE   2
       DIMENSION MEM1(5),MR1(5),MR2(10),IRS(20,257),IRR(20,256)      OPLE   3
       DIMENSION ISV(10),ISW(10)                                     OPLE   4
       DIMENSION QINT(4,3),FLINT(4,2),PINT(4),KDINT(4)               OPLE   5
       COMMON/EXOP/ INSIST, M ,IZB ,JKK ,KKKK,JPROB,KTIME            OPLE   6
       COMMON/REMB/IST1,LI,LU                                        OPLE   7
      *         ,  IAS    (10,8),  IYS   (10,8), IAE (10,8),IYE(10,8) OPLE   8
      *,KLIFE(5),KUNITS(5)                                           OPLE   9
       COMMON/RDAT1/DEN,ERATE,VARCOS(10),LIFE,N,IDEPR,TAX,ENF,KTAX   OPLE  10
       COMMON/SALES/PROB(8),NKV,RDMD(9,20),PRFIS(8,20)               OPLE  11
       COMMON/TIME/MET,IST(10),ISTART(10,8),IEND(10,8)               OPLE  12
       COMMON/NAMES/ITI(24),INA(10,3)                                OPLE  13
       COMMON/UTIL/IU,XMIN(5)                                        OPLE  14
       COMMON/RWR/IOUT,IN                                            OPLE  15
       COMMON/MDAT1/KDAT(10,9),NR, KTDAT(100)                        OPLE  16
       COMMON/MDAT2/FAKTOR,UNTAX,MENGE,KOS(10),KVC                   OPLE  17
       COMMON/CONT/LW,ICHEK                                          OPLE  18
       COMMON/REF/IREF                                               OPLE  19
       COMMON/OPLE/BELOW,ITRACE,UPR,INSYR,KDMD(9,20)                 OPLE  20
       COMMON/INS(10),  KEX                                          OPLE  21
C      DOUBLE PRECISION IST1,IRS,IRR,NEW,IAS,IYS,IAE,IYE,JJJJ        OPLE  22
       EQUIVALENCE  (ISV(1),KDAT(1,1)),  (ISW(1),KDAT(1,8))          OPLE  23
       DATA MR1(1),MR1(2), MR1(3), MR1(4), MR1(5), MR2(1), MR2(2), MR2(3)OPLE 24
      1, MR2(4), MR2(5), MR2(6), MR2(7), MR2(8), MR2(9), MR2(10)     OPLE  25
      *         /        1,11,11,257,1    ,  10*2 /                  OPLE  26
      *,QINT(1,1), QINT(2,1), QINT(3,1), QINT(4,1), QINT(1,2),QINT(2,2), OPLE 27
      *QINT(3,2), QINT(4,2), QINT(1,3),  QINT(2,3), QINT(3,3), QINT(4,3) OPLE 28
      *    /.25781,.070312,-.054688,-.11719,.95938,                  OPLE  29
      *2*.98438,.859376,-.11719,-.054688,.070712,.25781/             OPLE  30
      *,FLINT(1,1), FLINT(2,1), FLINT(3,1), FLINT(4,1), FLINT(1,2),  OPLE  31
      *FLINT(2,2), FLINT(3,2), FLINT(4,2)                            OPLE  32
      *     /1.375,1.125,.875,.625,-.375,-.125,.125,.375/            OPLE  33
       DATA MNSR,MEM1(1), MEM1(2), MEM1(3), MEM1(4), MEM1(5)/256,5*0/ OPLE 34
```

```
C    DATA FOR ALTERNATIVES INTEGERS FOR COMPUTATIONAL EFFICIENCY        OPLE  35
C    KDAT(I,1)   OPERATING BEFORE DECN   KDAT(T,6)  CAPACITY            OPLE  36
C    KDAT(I,2)   EXTN DATA               KDAT(I,7)   FIXED COSTS        OPLE  37
C    KDAT(I,3)   EXTN DATA               KDAT(T,8) OPERATING AFTER DECN OPLE  38
C    KDAT(I,4)   CAPITAL COST            KDAT(I,9)  YEAR AVAILABLE       OPLE  39
C    KDAT(I,5)   SCRAP VALUE             KDAT(T,10)  NOT USED            OPLE  40
C    VARCOS  VARIABLE COST                                              OPLE  41
C    N   NO OF ALTERNATIVES       RATE  1+DISCOUNT RATE                 OPLE  42
C    KOMD( , )   DEMAND           SEL ( , )   SELLING PRICE             OPLE  43
      MR1(5) =-2*LW + 1                                                 OPLE  44
      RATE=FRATE+1.                                                     OPLE  45
      KTIME=KTIME+1                                                     OPLE  46
      IF(KTIME.GT.0)   GO TO 99                                         OPLE  47
C    PACK INITIAL STATE INIST1 -LIST OF ALTERNATIVES OPERATING (IST)    OPLE  48
      CALL PACK(IST1,IST,MR2,N)                                         OPLE  49
C    ALTERNATIVE NOS IN ORDER OF VARIABLE COST STORED IN KOS( )         OPLE  50
      DO 49 I=1,N                                                       OPLE  51
49    KOS(I)=I                                                          OPLE  52
      IF(N.EQ.1)  GO TO 58                                              OPLE  53
54    K=0                                                               OPLE  54
      DO 55 I=2,N                                                       OPLE  55
      M=KOS(I)                                                          OPLE  56
      L=KCS(I-1)                                                        OPLE  57
      IF(VARCOS(L).LE.VARCOS(M))GOTO 55                                 OPLE  58
C    EXCHANGE IF ORDER WRONG,THEN REPEAT                                OPLE  59
      IZ=KCS(I)                                                         OPLE  60
      KOS(I)=KOS(I-1)                                                   OPLE  61
      KOS(I-1)=IZ                                                       OPLE  62
      K=1                                                               OPLE  63
55    CONTINUE                                                          OPLE  64
      IF(K.NE.0)GOTO 54                                                 OPLE  65
58    DO 60 I=1,N                                                       OPLE  66
      DO 60 II=1,NKV                                                    OPLE  67
      IAS(I,II)=0                                                       OPLE  68
      IYS(I,II)=0                                                       OPLE  69
      IAE(I,II)=0                                                       OPLE  70
60    IYE(I,II)=0                                                       OPLE  71
      UNTAX=1.-TAX                                                      OPLE  72
      IF(TAX.EQ.0.)IDEPR=10                                            OPLE  73
C   PRINT INPUT DATA                                                    OPLE  74
      WRITE(IOUT,1023)                                                  OPLE  75
C      LIFE,CAPACITY FACTORS ETC.                                       OPLE  76
      IF(IDEPR.NE.0)WRITE(IOUT,1020)LIFE,UPR,BELOW,RRATE,TAX,IDEPR      OPLE  77
      IF(IDEPR.EQ.0)WRITE(IOUT,1020)LIFE,UPR,BELOW,RRATE,TAX            OPLE  78
      IF(MET.EQ.0)    WRITE(IOUT,1055)                                  OPLE  79
      IF(MET.GT.0)    WRITE(IOUT,1056) MET                              OPLE  80
      IF(PEN.GT.0.)WRITE(IOUT,1027)PEN                                  OPLE  81
C     DATA FOR STATE.                                                   OPLE  82
      WRITE(IOUT,1021)                                                  OPLE  83
      DO 95 J=1,N                                                       OPLE  84
      WRITE(IOUT,1022)J,(INA(J,I),I=1,3),(KDAT(J,K),K=3,7),VARCOS(J),   OPLE  85
     *KDAT(J,9)                                                         OPLE  86
95    IF(KDAT(J,4).GT.250000.OR.KDAT(J,4).LT.1000)  WRITE(IOUT,1046)    OPLE  87
C     FORECAST DATA.                                                    OPLE  88
      IF(NKV.GT.1)WRITE(IOUT,1024)(PROB(J),J=1,NKV)                     OPLE  89
      WRITE(IOUT,1025)                                                  OPLE  90
      DO 98 J=1,LIFE                                                    OPLE  91
98    WRITE(IOUT,1026) J,(PREIS(KR,J),KOMD(KR,J),KR=1,NKV)             OPLE  92
      CALL TAXF(FAKTOR,RRATE,IDEPR,TAX,ENF,KTAX)                        OPLE  93
      RETURN                                                            OPLE  94
99    IF(KKKK.EQ.1)  GO TO 900                                          OPLE  95
      IF(INSIST.EQ.-1)WRITE(IOUT,1050)                                  OPLE  96
      IF(NKV    .EQ. 1) WRITE(IOUT,1054)                                OPLE  97
      IF(NKV.EQ.1)    WRITE(IOUT,1057) JPROB                            OPLE  98
      IF(NKV    .GT. 1) WRITE(IOUT,1051)                                OPLE  99
      IF(INSIST.GT. 0) WRITE(IOUT,1052) INSIST                          OPLE 100
     *,INA(INSIST,1) ,INA(INSIST,2) ,INA(INSIST,3)                      OPLE 101
      IF(INSIST.EQ. 0) WRITE(IOUT,1053)                                 OPLE 102
```

```
C   FIND HIGHEST DEMAND,   STORE IN LU                                OPLE 103
      LI=1                                                           OPLE 104
      LU=1                                                           OPLE 105
      DO 104 J=1,NKV                                                 OPLE 106
      DO 102 I=1,LIFE                                                OPLE 107
      IF(KDMD(J,I).LE.KDMD(LI,LU))GOTO 102                           OPLE 108
      LI=J                                                           OPLE 109
      LU=I                                                           OPLE 110
102   CONTINUE                                                       OPLE 111
104   CONTINUE                                                       OPLE 112
      IF(KDMD(LI,LU).GT.100000) WRITE(IOUT,1045)                     OPLE 113
      LU=FLOAT(KDMD(LI,LU))*UPR                                      OPLE 114
C   COMPUTE LOWER CAPACITY BOUNDS,   STORE IN KDMD(9, )              OPLE 115
C   REDUCE CONSTRAINT TO SUPER CENT AFTER 70 PER CENT OF LIFE        OPLE 116
      JJ=IFIX(0.7*FLOAT(LIFE))                                       OPLE 117
      DO 110 KY=1,LIFE                                               OPLE 118
      JR=1                                                           OPLE 119
      DO 108 J=1,NKV                                                 OPLE 120
      IF(KDMD(J,KY).GE.KDMD(JR,KY))GOTO 108                          OPLE 121
      JR=J                                                           OPLE 122
108   CONTINUE                                                       OPLE 123
      IF(KDMD(JR,KY).LT.20)   WRITE(IOUT,1045)                       OPLE 124
      KDMD(9,KY)=IFIX(BELOW* FLOAT(KDMD(JR,KY)))                     OPLE 125
      IF (KY.GT.JJ) KDMD (9,KY)=IFIX(0.5*FLOAT(KDMD(9,KY)))          OPLE 126
110   CONTINUE                                                       OPLE 127
C   SKIP ANALYSIS IF BASE CASE WITH ZERO PW                          OPLE 128
      IF(IST1.NE.0 )   GO TO 100                                     OPLE 129
      IF( PEN.NE.0. )  GO TO 100                                     OPLE 130
      IF(INSIST.GT.-1)  GO TO 100                                    OPLE 131
      JKK=0                                                          OPLE 132
      IZE=0                                                          OPLE 133
      M=0                                                            OPLE 134
      WRITE(IOUT,1059)                                               OPLE 135
      RETURN                                                         OPLE 136
C   ----------                                                       OPLE 137
100   IERST=0                                                        OPLE 138
      KOUNT1=0                                                       OPLE 139
      KOUNT2=J                                                       OPLE 140
      KOUNT3=0                                                       OPLE 141
      IF(ICHEK.GT.0)   WRITE(IOUT,1016)                              OPLE 142
C   BEGIN DYNAMIC PROGRAMMING YEARLY SEARCH                          OPLE 143
      DO 215 JR=1,LIFE                                               OPLE 144
      JRR=JR-1                                                       OPLE 145
      DO 115 IWK1=1,MNSR                                             OPLE 146
C   STATE INFORMATION PACKED IN IRS( ) ,INITIALISE AT -1 SO 0 HAS MEANINGOPLE 147
115   IPS(JR,IWK1)=-1                                                OPLE 148
      KST=0                                                          OPLE 149
      IF(ICHEK.LT.2) GO TO 120                                       OPLE 150
      WRITE(IOUT,1000)JR,(KDMD(J,JR),J=1,NKV)                        OPLE 151
C   SEARCH PREVIOUS YEARS STATES,MNSR MAX NO OF REMEMBERED  STATES   OPLE 152
C   PACK IN IRS(YEAR,STATE NO) NO OF ALTERNATIVES OPERATING IN STATE OPLE 153
C   PACK IN IRR(YEAR,STATE NO) NPW FOR STATE IRS( , )                OPLE 154
120   DO 185 J=1,MNSR                                                OPLE 155
      IF(JR.GT.1)GOTO 128                                            OPLE 156
      IF(J.GT.1)GOTO 190                                             OPLE 157
      CALL UNPACK(IST1,ISV,MR2,N)                                    OPLE 158
      IWTH=0                                                         OPLE 159
      GOTO 130                                                       OPLE 160
128   IF(IPS(JRR,J).EQ.-1)GOTO 190                                   OPLE 161
      CALL UNPACK(IRS(JRR,J),ISV,MR2,M)                              OPLE 162
C   UNPACK FOR STATE J ,YEAR JRR                                     OPLE 163
C   ISV - ARRAY OF 1 OR 0 OPERATING ALTERNATIVES( TO N)              OPLE 164
C   IRR - ARRAY OF 5 VALUES (1)REPLICATED OPTIMUM PATHS  (NOT USED)  OPLE 165
C                           2)SCRAP DECISION ,ALTERNATIVE NO         OPLE 166
C                          (3) BUILD DECISION ,ALTERNATIVE NO        OPLE 167
C                          (4) STATE FOR WHICH THIS STATE HAS COME   OPLE 168
C                          (5) NPW OF THIS STATE - LOWER LIMIT       OPLE 169
      CALL UNPACK(IRR(JRR,J),MEM1,MR1,5)                             OPLE 170
      IWTH=MEM1(5)+LW                                                OPLE 171
```

```
C   MAKE EVERY DECISION WITH EVERY PREVIOUS STATE- TO CREATE ALL POSSIBLE OPLE 172
C       STATES FOR THIS YEAR                                        OPLE 172
130     M=N+1                                                       OPLE 174
        DO 180 I=1,M                                                OPLE 175
        DO 175 IK=I,M                                               OPLE 176
        NREST=0                                                     OPLE 177
        KOUNT1=KOUNT1+1                                             OPLE 178
        IS=I-1                                                      OPLE 179
        IB=IK-1                                                     OPLE 180
        IF(IS.EQ.0)GOTO 131                                         OPLE 181
        IF(KDAT(IS,1).EQ.0)GOTO 132                                 OPLE 182
131     IF(IB.EQ.0)GOTO 133                                         OPLE 183
        IF(KDAT(IB,1).EQ.0)GOTO 133                                 OPLE 184
132     IS=IB                                                       OPLE 185
        IB=I-1                                                      OPLE 186
C   DISCARD INVALID DECISIONS                                       OPLE 187
C                                                                   OPLE 188
C                                                                   OPLE 189
C   BASE CASE - ALL BUILD DECISIONS INVALID                         OPLE 190
133     IF(INSIST.GT.-1)GO TO 1129                                  OPLE 191
        IF(IB.EQ.0)GO TO 1134                                       OPLE 192
        GOTO 175                                                    OPLE 193
1129    IF(JR.EQ.1)   GO TO 1130                                    OPLE 194
C   FIRST DECISION RESTRICTIONS                                     OPLE 195
        IF(IFS(JRR,J).NE.IST1)   GO TO 1133                         OPLE 196
C   YEAR SORT- CHECK INSIST YEAR CONSTRAINTS                        OPLE 197
1130    IF(INSYR.LT.1)  GO TO 1132                                  OPLE 198
        IF(JR.NE.INSYR)  GO TO 1131                                 OPLE 199
        IF(IB.EQ.0)   GO TO 175                                     OPLE 200
        GO TO 1132                                                  OPLE 201
1131    IF(IB.NE.0)  GO TO 175                                      OPLE 202
C   DECISION SORT, CHECK ENFORCED FIRST DECISION CONSTRAINT         OPLE 203
1132    IF(INSIST.EQ.0)   GO TO 1133                                OPLE 204
        IF(IB.EQ.INSIST.OR.IB.EQ.0)  GO TO 1133                     OPLE 205
        GO TO 175                                                   OPLE 206
C   INVALID IF DECISION BEFORE ALTERNATIVE IS AVAILABLE(KDAT( ,9)   OPLE 207
1133    IF(IB.EQ.0)  GO TO 1134                                     OPLE 208
        IF(JR.LT.KDAT(IB,9))   GO TO 175                            OPLE 209
1134    KOUNT2=KOUNT2+1                                             OPLE 210
C   CHECK WHETHER DECISION VIOLATES EXTENSION STRUCTURE,VALDEC CHECKS, OPLE 211
C       ITF=0  FOR VIOLATION                                        OPLE 212
        CALL VALDEC(ITF,IS,IB,ITS)                                  OPLE 213
        IF(ITF.EQ.0)GOTO 175                                        OPLE 214
C   STATE AFTER THE DECISION TO BE STORED IN KDAT( ,8) -DONE BY NEUZ OPLE 215
        CALL NEUZ(IS,IB,ITS)                                        OPLE 216
C   CAPACITY OF NEW STATE                                           OPLE 217
        KY=IFROD(8,6)                                               OPLE 218
        IF(INSIST.EQ.-1)  GO TO 1135                                OPLE 219
        IF(KY.GT.LU.OR.KY.LT.KDMD(9,JR))GOTO 175                    OPLE 220
1135    IEWTH=0                                                     OPLE 221
C   -----------                                                     OPLE 222
C   ONLY VALID DECISIONS PASS THIS POINT                            OPLE 223
C   CASH FLOW WITH NERT,RETURNED IN NUTH                            OPLE 224
C   REPEAT FOR EACH PROBABILITY LEVEL,JEWTH IS   E(NPW)             OPLE 225
        DO 135 IZ=1,NKV                                             OPLE 226
        KOUNT3=KOUNT3+1                                             OPLE 227
        IER=0                                                       OPLE 228
        CALL NERT(IS,IB,ITS,KDMD(IZ,JR),NWTH,PREIS(IZ,JR),IER)      OPLE 229
        IEWTH=IEWTH+ IFIX(PROB(IZ)*((FLOAT( NWTH)/ RATE**(JR))+FLOAT(IWTH OPLE 230
     1)))                                                           OPLE 231
        IF(JR.LT.LIFE)GOTO 135                                      OPLE 232
C   IF FINAL YEAR,INCLUDE EXTENDED LIFE FOR  MET  YEARS             OPLE 233
        IF(RATE.EQ.0.) GO TO 137                                    OPLE 234
        IF(MET) 137,137,136                                         OPLE 235
136     R=RATE-1.                                                   OPLE 236
        IER=IFROD(8,7)                                              OPLE 237
        TERM=(FLOAT(MENGE)*PREIS(IZ,LIFE)-FLOAT(IER+KVC))*JNTAX     OPLE 238
        TERM=(TERM/(R*((1.+R)**LIFE)))*(1.-1./((1.+R)**MET))        OPLE 239
        IEWTH=IEWTH+ IFIX(PROB(IZ)*        TERM)                    OPLE 240
```

```
          GOTO 135                                                       OPLE 241
C   IF MET=0: SCRAP THE LOT AT THE END.                                 OPLE 242
137       IEWTH=IEWTH+ IF1X(PROB(IZ)* (FLOAT( IPROD(8,5))/ (RATE**(LIFE +1   OPLE 243
      1))))                                                             OPLE 244
135       CONTINUE                                                      OPLE 245
C   CHECK  NEW  WITHIN + AND - LW BOUNDS,TF +VE PRINT ERROR             OPLE 246
          NWTH=IEWTH                                                    OPLE 247
          IF(NWTH.GT.LW.AND.NWTH.LT.LW+MP1(5))   GO TO 139             OPLE 248
          IF(NWTH.LT.LW+MP1(5))    GO TO 138                            OPLE 249
          WRITE(IOUT,1044)                                             OPLE 250
138       IF(NWTH.LT.LW) NWTH=LW                                       OPLE 251
          IF(NWTH.GT.LW+MP1(5))NWTH=LW+MP1(5) -1                       OPLE 252
C   PACK STATE DETAILS IN  NEW                                         OPLE 253
139       CALL PACK (NEW,ISW,MR2,N)                                    OPLE 254
          IF(KST.EQ.1)GOTO 145                                         OPLE 255
C   IS THIS STATE ALREADY REMEMBERED THIS YEARI.E.NOOTHER SAME NEW     OPLE 256
          DO 140 JK=1,MNSR                                             OPLE 257
          IF(IRS(JR,JK).EQ.-1)GOTO 145                                 OPLE 258
          IF(IRS(JR,JK).EQ.NEW)GOTO 150                                OPLE 259
140       CONTINUE                                                     OPLE 260
C   NEW   STATE NOT ALREADY STORED----REMEMBER IT                      OPLE 261
145       KST=KST+1                                                    OPLE 262
          JK=C                                                         OPLE 263
          GOTO 160                                                     OPLE 264
C   NEW   STATE ALREADY STORED,COMPARE NPW AND REMEMBER BEST ROUTE ONLY   OPLE 265
C         -----------                                                  OPLE 266
C   ******DYNAMIC PROGRAMMING****************                          OPLE 267
C         ---------                                                    OPLE 268
150       CALL UNPACK(IRR(JR,JK),MEM1,MR1,5)                           OPLE 269
          NBEST=MEM1(1)                                                OPLE 270
          IF(NWTH-MEM1(5)-LW)175,155,162                               OPLE 271
155       IF(NBEST.LT.3)NBEST=NBEST+1                                  OPLE 272
          GOTO 165                                                     OPLE 273
160       IF(KST.GT.MNSR)GOTO 175                                      OPLE 274
162       NBEST=1                                                      OPLE 275
          IF(JK.EQ.0)JK=KST                                            OPLE 276
C   STORE STATE IN IRS                                                 OPLE 277
C   STORE CONDITIONOF STATE  NEW,REPLICATE OPTIMA,SCRAP,BUILD,PREVIOUS   OPLE 278
C         STATE,AND NPW                                                OPLE 279
          IRS(JR,JK)=NEW                                               OPLE 280
          MEM1(2)=IS                                                   OPLE 281
          MEM1(3)=IB                                                   OPLE 282
          MEM1(4)=J                                                    OPLE 283
          MEM1(5)=NWTH-LW                                              OPLE 284
165       MEM1(1)=NBEST                                                OPLE 285
C   REPLICATE OF OPTIMA NO LONGER USED.                                OPLE 286
          MEM1(1)=0                                                    OPLE 287
          CALL PACK(IRR(JR,JK),MEM1,MR1,5)                             OPLE 288
175       CONTINUE                                                     OPLE 289
180       CONTINUE                                                     OPLE 290
C   //////NEXT DECISION//////////////////////////////////////////     OPLE 291
185       CONTINUE                                                     OPLE 292
C   NEXT OF LAST YEARS STATES                                          OPLE 293
C   PRINT SEARCH RESULTS IF FINAL YEAR,OR IF DEBUG REQUESTED           OPLE 294
190       IF(ICHEK.LT.2 .AND.JR.NE.LIFE)GOTO 210                       OPLE 295
          IZ=-1                                                        OPLE 296
          DO 205 M=1,MNSR                                              OPLE 297
          IF(IRS(JR,M).EQ.-1)GOTO 210                                  OPLE 298
          CALL UNPACK(IRR(JR,M),MEM1,MR1,5)                            OPLE 299
          MEM1(5)=MEM1(5)+LW                                           OPLE 300
          IF(JR.GT.1)GOTO 195                                          OPLE 301
          CALL UNPACK(IST1,ISV,MR2,N)                                  OPLE 302
          GOTO 200                                                     OPLE 303
C   LAST YEARS STATE IS  ISV                                           OPLE 304
195       JK=MEM1(4)                                                   OPLE 305
          CALL UNPACK(IRS(JRR,JK),ISV,MR2,N)                           OPLE 306
C   THIS YEARS STATE IN ISW                                            OPLE 307
200       CALL UNPACK(IRS(JR,M),ISW,MR2,N)                             OPLE 308
```

```
        KY=IFROD)(6,3)
        IF(ICHEK.EQ.0) GO TO 205                            OPLE 309
        IF(IZ.EQ.IRS(JRR,JK))GOTO 202                       OPLE 310
        WRITE(IOUT,1012) JR,(ISV(I),I=1,N)                  OPLE 311
        IZ=IRS(JRR,JK)                                      OPLE 312
202     WRITE(IOUT,1017)MEM1(5),KY,(ISW(I),I=1,N)           OPLE 313
205     CONTINUE                                            OPLE 314
210     IF(KST.GT.0.AND.KST.LE.MNSR)GOTO 214                OPLE 315
C  TOO MANY OR NO STATES FOUND.IF NONE FOUND,ABANDON SEARCH,PRINT ERROR, OPLE 316
C       IDENTIFY CASE,RETURN TO  EXEC  TO RECORD NO RESULT  OPLE 317
        IF(KST.GT.MNSR.AND.IERST.EQ.0) WRITE(IOUT,1041) MNSR OPLE 318
        IF(KST.GT.MNSR.AND.IERST.EQ.0)IERST=1               OPLE 319
        IF(KST.GT.0)GO TO 215                               OPLE 320
        IF(KST.EQ.0)   WRITE(IOUT,1040)                     OPLE 321
        JKK=0                                               OPLE 322
        M=LW                                                OPLE 323
        IZB=INSIST                                          OPLE 324
        RETURN                                              OPLE 325
214     IF(ICHEK.LT.2) GO TO 215                            OPLE 326
        WRITE(IOUT,1019)KST                                 OPLE 327
215     CONTINUE                                            OPLE 328
C  NEXT  YEAR                                               OPLE 329
        M=LW                                                OPLE 330
        IF(KST.GT.MNSR)KST=MNSR                             OPLE 331
C  SEARCH OVER,--LOCATE TOTAL PLAN FOR BEST ENDSTATE  BY CALLING PATH OPLE 332
C       TO TRACE BACK DECISIONS--RETURNS WITH STATE NO/YEAR IN IRS(YEAR, OPLE 333
C               STATE NO), YEAR OF FIRST DECISION RETURNED IN LI OPLE 334
        DO 220 I=1,KST                                      OPLE 335
        CALL UNPACK(IRR(LIFE,I),MEM1,MR1,5)                 OPLE 336
        NWTH=MEM1(5)+LW                                     OPLE 337
        IF(NWTH.LE.M)GOTO 220                               OPLE 338
        M=NWTH                                              OPLE 339
        IS=I                                                OPLE 340
220     CONTINUE                                            OPLE 341
        LI=IS                                               OPLE 342
        CALL PATH(LI,IRS,IRR,20,MEM1,MR1,LIFE,MNSR+1)       OPLE 343
C                                                           OPLE 344
C   PRINT SEARCH RESULTS FOR TOTAL OPTIMUM PLAN             OPLE 345
C                                                           OPLE 346
        CALL UNPACK(IST1,ISV,MR2,N)                         OPLE 347
        JR=0                                                OPLE 348
        WRITE(IOUT,1028)(ISV(I),I=1,N)                      OPLE 349
        JKK=0                                               OPLE 350
        IZB=0                                               OPLE 351
        DO 221  I=1,N                                       OPLE 352
        ISTART(I,1) = ISV(I)                                OPLE 353
221     IEND  (I,1) =0                                      OPLE 354
        DO 225 JR=1,LIFE                                    OPLE 355
        IWK1=IRS(JR,MNSR+1)                                 OPLE 356
        CALL UNPACK(IRS(JR,IWK1),ISV,MR2,N)                 OPLE 357
        CALL UNPACK(IRR(JR,IWK1),MEM1,MP1,5)                OPLE 358
        DO 223  I=1,N                                       OPLE 359
        IF(ISV(I  ).EQ.1.AND.ISTART(I,1).EQ.0)  ISTART(I,1) = JR OPLE 360
223     IF(ISV(I  ).EQ.0.AND.ISTART(I,1).GT.0.AND.IEND(I,1).EQ.0) OPLE 361
       *IEND(I,1) = JR                                      OPLE 362
        MEM1(5)=MEM1(5)+LW                                  OPLE 363
        KAPY=IPROD(1,6)                                     OPLE 364
        WRITE(IOUT,1029) JR,MEM1(3),MEM1(2),       KAPY,MEM1(5), OPLE 365
       *KDMD(1,JR),(ISV(I),I=1,N)                           OPLE 366
        IF(JKK .EQ.0.AND.  MEM1(3).NE.0)      IZB=MEM1(3)   OPLE 367
        IF(JKK .EQ.0.AND.  MEM1(3).NE.0)      JKK=JR        OPLE 368
225     CONTINUE                                            OPLE 369
        IREF=IZB                                            OPLE 370
C  PLAN LOCATED,PRINTED AND INITIAL DECISION STORED IN IZB,JKK OPLE 371
C PACK THE STRATEGY EXCLUDING INITIAL AND END STATES        OPLE 372
C  FOR LATER STRATEGY SELECTION IN--                        OPLE 373
C       IAE ---BUILD ALTERNATIVE,  IYB-- BUILD YEAR         OPLE 374
C       IAS ----SCRAP ALTERNATIVE, IYS ---SCRAP YEARS       OPLE 375
                                                            OPLE 376
```

```
C     IREF=PLAN PACKING FLAG                                        OPLE 377
C          NO OF DECISIONS PACKABLE IN WORD LENGTH    =5            OPLE 378
      IF(IREF.LE.0.OR.IREF.GT.10)GOTO 226                           OPLE 379
      DO 329 I=1,5                                                  OPLE 380
      KUNITS(I)= N+1                                                OPLE 381
      KLIFE(I)=LIFE + 1                                             OPLE 382
      ISV(I)=0                                                      OPLE 383
329   ISW(I)=0                                                      OPLE 384
      JJ =0                                                         OPLE 385
      DO 320   I =1,N                                               OPLE 386
      ISTART(I,1)=ISTART(I,1)-IST(I)                                OPLE 387
      IF (ISTART(I,1). EQ. 0 ) GO TO 320                            OPLE 388
      JJ= JJ +1                                                     OPLE 389
      ISV(JJ) = I                                                   OPLE 390
      ISW(JJ) = ISTART(I,1)                                         OPLE 391
  320 CONTINUE                                                      OPLE 392
      IF (JJ.LE. 5 ) GO TO  350                                     OPLE 393
351   WRITE (IOUT,360)                                              OPLE 394
      IREF =-1                                                      OPLE 395
      GO TC 226                                                     OPLE 396
360   FORMAT(/48H PLAN TOO COMPLEX TO PACK ,TRANSFER NOT POSSIBLE   OPLE 397
     */1X,60(1H*)/1X,60(1H*),/  66H CHECK THAT A ZERO PLAN HAS NOT BEEN OPLE 398
     *TRANSFERRED FOR EVALUATION         //////)                   OPLE 399
350   CALL   PACK(JJJJ,ISW,KLIFE,5)                                 OPLE 400
      IYS(IREF,JPROB)=JJJJ                                          OPLE 401
      CALL PACK(JJJJ,ISV,KUNITS,5)                                  OPLE 402
      IAS(IREF,JPROB)  =JJJJ                                        OPLE 403
      DO 352  I=1,5                                                 OPLE 404
      ISV(I)=0                                                      OPLE 405
352   ISW(I)=0                                                      OPLE 406
      JJ= 0                                                         OPLE 407
      DO 340 I=1,N                                                  OPLE 408
      IF (IEND (I,1).EQ . 0 ) GO TO 340                             OPLE 409
      JJ=JJ+1                                                       OPLE 410
      ISV (JJ) =I                                                   OPLE 411
      ISW(JJ) =IEND(I,1)                                            OPLE 412
  340 CONTINUE                                                      OPLE 413
      IF (JJ.LE. 5  ) GO TO  370                                    OPLE 414
      GO TO 351                                                     OPLE 415
370   CALL   PACK(JJJJ,ISW,KLIFE,5)                                 OPLE 416
      IYE(IREF,JPROB) =JJJJ                                         OPLE 417
      CALL   PACK (JJJJ,ISV,KUNITS,5)                               OPLE 418
      IAE(IREF,JPROB) =JJJJ                                         OPLE 419
C OPTIMUM PLAN PACKED                                               OPLE 420
226   M=MEM1(5)                                                     OPLE 421
      L=1                                                           OPLE 422
      IS=JKK                                                        OPLE 423
      JK=IS                                                         OPLE 424
      LI=IS                                                         OPLE 425
      IF(JKK.EQ.LIFE.OR.IZB.EQ.0)  GO TO 270                        OPLE 426
      IK=1                                                          OPLE 427
      NBEST=LW                                                      OPLE 428
      KLOW=LW + MR1(5)                                              OPLE 429
C  LOCATE OPTIMUM QUARTER FOR FIRST DECISION                        OPLE 430
C  BUILD DECISION MADE SUCCESSIVELY IN QUARTER 1,2,3,4 AND THE NPW  OPLE 431
C       EVALUATED.HIGHEST NPW SUBJECT TO CAPACITY CONSTRAINT GIVES  OPLE 432
C     OPTIMUM QUARTER  (IK)                                         OPLE 433
C  N.B.   SEARCH NPW IS NOT REVISED  TO ALLOW FOR EXACT QUARTER OF START OPLE 434
      DO 260 I=1,4                                                  OPLE 435
      IEWTH=0                                                       OPLE 436
      DO 250 KV=1,NKV                                               OPLE 437
C  INTERPOLATED PRICE + DEMAND FIGURES FOR QUARTERS,IN PINT( )AND KDINT(OPLE 438
C   QUADRATIC INTER. UNLESS FIRST YEAR, QINT( )INTERPOLATION CONSTANTS OPLE 439
      IF(JK-1) 270,242,235                                          OPLE 440
235   DO 240 K=1,4                                                  OPLE 441
      PINT(K)=0.                                                    OPLE 442
      KDINT(K)=0                                                    OPLE 443
      DO 240 IR=1,3                                                 OPLE 444
```

```
          IS=JK+IB-2                                                    OPLE 445
          PINT(K)=PINT(K)+QINT(K,IB)*PREIS(KV,IS)                       OPLE 446
          KDINT(K)=KDINT(K)+ IFIX(QINT(K,IB) *FLOAT(KDMD(KV,IS)))       OPLE 447
240       CONTINUE                                                      OPLE 448
          GOTO 246                                                      OPLE 449
C   LINEAR INTERPOLATION IN FIRST YEAR FLINT( ) INTERPOLATION CONSTANTS OPLE 450
242       DO 244 K=1,4                                                  OPLE 451
          PINT(K)=FLINT(K,1)*PREIS(KV,1)+FLINT(K,2)*PREIS(KV,2)         OPLE 452
          KDINT(K)= IFIX(FLINT(K,1)* FLOAT(KDMD(KV,1)) +FLINT(K,2)*FLOAT(KDMOPLE 453
          1D(KV,2)))                                                    OPLE 454
244       CONTINUE                                                      OPLE 455
246       IZ=IRS(JK,MNSR+1)                                            OPLE 456
          IF(ICHEK.EQ.2.AND.I.EQ.1)WRITE(IOUT,1028)PINT,KDINT          OPLE 457
          CALL UNPACK(IRS(JK,IZ),ISW,MR2,N)                            OPLE 458
          IF(JK.EQ.1)GOTO 248                                          OPLE 459
          IZ=IRS(JK-1,MNSR+1)                                          OPLE 460
          CALL UNPACK(IRS(JK-1,IZ),ISV,MR2,N)                          OPLE 461
          GOTO 249                                                      OPLE 462
248       CALL UNPACK(IST1,ISV,MR2,N)                                  OPLE 463
249       CALL VALDEC(IZ,MEM1(2),MEM1(3),ITS)                          OPLE 464
          CALL KERT(MEM1(2),MEM1(3),ITS,KDINT,NWTH,PINT,I)             OPLE 465
          IEWTH=IEWTH + IFIX(PROD(KV) * FLOAT(NNTH))                   OPLE 466
          IF(KLOW.GT.KDINT(I))    KLOW=KDINT(I)                        OPLE 467
250       CONTINUE                                                      OPLE 468
          KAPY=JPROD(1,6)                                              OPLE 469
          X= FLOAT(KLOW) *BELOW                                        OPLE 470
          KLOW=IFIX(X)                                                 OPLE 471
          IF(ICHEK.EQ.2 )WRITE(IOUT,1006)I,IEWTH                       OPLE 472
C   QUARTER MUST MEET LOWER CAPACITY CONSTRAINT                        OPLE 473
          IF(KAPY.LT.KLOW)    GO TO 260                                OPLE 474
          IF(IEWTH.LT.NBEST)GOTO 260                                   OPLE 475
          NBEST=IEWTH                                                   OPLE 476
          IK=I                                                         OPLE 477
260       CONTINUE                                                      OPLE 478
C   END OF QUARTERS EVALUATION ,OPTIMUM QUARTER GIVEN IN IK            OPLE 479
C   SUBROUTINE USE COUNTERS                                            OPLE 480
          WRITE(IOUT,1035 )   IK                                       OPLE 481
270       IF(ICHEK.GT.0)WRITE(IOUT,1018)KOUNT1,KOUNT2,KOUNT3           OPLE 482
          RETURN                                                        OPLE 483
C   RETURN TO EXEC HAVING COMPLETED THE SEARCH                         OPLE 484
C   AFTER ALL D.P. SEARCHES,OPTIMUM STRATEGY IS RECOVERED FROM PACKED  OPLE 485
C   DATA AND PLACED IN ISTART( , ) AND TEND( , ) FOR EVALUATION        OPLE 486
C   1 PLAN PER PROBABILITY LEVEL                                       OPLE 487
C   TRANSFER CANNOT BE DONE IF IREF =-1                                OPLE 488
900       IF(IREF.EQ.-1) GO TO 463                                     OPLE 489
          DO 400   I =1,N                                              OPLE 490
          DO 400 JJ =1,NKV                                             OPLE 491
          ISTART (I,JJ ) = 0                                           OPLE 492
400       IEND   (I,JJ ) = 0                                           OPLE 493
C RECOVER BEST STRATEGY                                                OPLE 494
          DO 440  JJ = 1,NKV                                           OPLE 495
          JI=JJ                                                        OPLE 496
          IF(KEX. EQ.1 )   JI=1                                        OPLE 497
          JJJJ=IYS(IREF,JI)                                            OPLE 498
          CALL UNPACK(JJJJ,ISW,KLIFE,5)                                OPLE 499
          JJJJ=IAS(IREF,JI)                                            OPLE 500
          CALL UNPACK (JJJJ,ISV,KUNITS,5)                              OPLE 501
          DO 420  JJJ =1,5                                             OPLE 502
          J = ISV(JJJ)                                                 OPLE 503
          IF(J.EQ.0)GOTO 420                                           OPLE 504
          ISTART(J,JJ)=ISW(JJJ)                                       OPLE 505
420       CONTINUE                                                      OPLE 506
          JJJJ=IYE(IREF,JJ)                                            OPLE 507
          CALL UNPACK(JJJJ,ISW,KLIFE,5)                                OPLE 508
          JJJJ=IAE(IREF,JJ)                                            OPLE 509
          CALL UNPACK (JJJJ,ISV,KUNITS,5)                              OPLE 510
          DO 440  JJJ =1,5                                             OPLE 511
          J = ISV(JJJ)                                                 OPLE 512
```

```
      IF(J.EQ.0)GOTO 440                                          OPLE 513
      IEND(J,JJ)=ISW(JJJ)                                         OPLE 514
440   CONTINUE                                                    OPLE 515
      IF(KEX.GT.1)GO TO 461                                       OPLE 516
C        FILL IN A STRATEGY WHEN IT IS NOT GIVEN                  OPLE 517
      DO 460 JJ =2,NKV                                            OPLE 518
      DO 460 I =1,N                                               OPLE 519
      ISTART(I,JJ)=ISTART(I,1)                                    OPLE 520
460   IEND(I,JJ)=IEND(I,1)                                        OPLE 521
C  LAST RETURN TO  EXEC    CALCULATION MOVES TO EVALUATION PHASE  OPLE 522
461   IREF=KEX-1                                                  OPLE 523
      RETURN                                                      OPLE 524
463   WRITE(IOUT,1058)                                            OPLE 525
      RETURN                                                      OPLE 526
1000  FORMAT(I4,41X,8(I6,1X))                                     OPLE 527
1002  FORMAT( I4,2X, 2(5I1,1X))                                   OPLE 528
1006  FORMAT(10I8)                                                OPLE 529
1008  FORMAT(    18X,14HOPTIMAL POLICY//7H YEAR,3X,8HDECISION,3X, OPLE 530
     *       4HCAPY,5X,5HWORTH,5X 6HDEMAND,10X,5HSTATE/            OPLE 531
     *8X,8HELD SCRP,22X,7H1 LEVEL,///                             OPLE 532
     *,43X,19HINITIAL   ,8(5I1,1X))                               OPLE 533
1009  FORMAT(2X,I2,4X,I2,2X,I2,      3(2X,I8),9X,8(5I1,1X))       OPLE 534
1016  FORMAT  (24H LIST OF  ALL   STATES   /56H YEAR O.D STATE  NPWOPLE 535
     * CAPACITY NEW STATE    DEMANDS        /)                    OPLE 536
1017  FORMAT( 17X, 2I8,2X,  2(5I1,1X))                            OPLE 537
1018  FORMAT   (41H NO OF SUBROUTINE CALLS:(CREATED STATES   ,    OPLE 538
     *         I6,6H,)NEUZ ,I6,6H, WERT ,I6/  )                   OPLE 539
1019  FORMAT(1H ,I3,14H STATES STORED)                            OPLE 540
1020  FORMAT(/6H LIFE ,I2,20H,UPPER/LOWER BOUNDS ,F5.2,1X,F5.2/   OPLE 541
     *15H DISCOUNT RATE ,2PF6.2,2X,9HTAX RATE ,2PF5.2,1X, 3H(PERCENT) OPLE 542
     *,2X,I2,15H-YEAR WRITE-OFF)                                  OPLE 543
1021  FORMAT(/ 62H   UNIT    EXTN CAPITAL SCRAP CAPACITY FIXED VARIABOPLE 544
     *LE YEAR   / 15X,  17HOF   COST   VALUE ,6X,2(4X,4HCOST) ,2X , OPLE 545
     * 12  AVAILABLE    )                                         OPLE 546
1022  FORMAT(1X,I2,1X,3A3,2X,I2,2X,I6,1X,I6,2X,I6,2X,I5,2X,F6.2,2X,I5 ) OPLE 547
1023  FORMAT(/22H DATA FOR THIS PROBLEM/1H ,21(1H-))              OPLE 548
1024  FORMAT(22H DEMAND PROBABILITIES ,/6F16.2)                   OPLE 549
1025  FORMAT(21H TIME    PRICE/DEMAND    /    )                   OPLE 550
1026  FORMAT(1X,I2,2X,3(F8.2,1H/,I6))                             OPLE 551
1027  FORMAT(26H PENALTY FOR LOW CAPACITY ,F9.2)                  OPLE 552
1028  FORMAT(15H INTRP. PRICES ,4(F6.2,1X)/17X,5HDMDS ,4(I6,1X))  OPLE 553
1035  FORMAT (/ 45H FIRST DECISION SHOULD OPERATE IN QUARTER   ,I1/) OPLE 554
1040  FORMAT(//56H NO ALLOWED STATES FOUND,-CONSTRAINTS TOO RESTRICTIVE OPLE 555
     *,     /////////)                                            OPLE 556
1041  FORMAT(6(2H *),39H NO OF STATES GENERATED GREATER THAN   ,I6,/ OPLE 557
     *42H THE ALLOWED NO. BEST STATE MAY BE LOST    /)            OPLE 558
1044  FORMAT(6(2H *),45H SOME NPW OUT OF RANGE,VALUES SET ON LIMITS, ,OPLE 559
     *37H BEST STATE MAY BE LOST          ///)                    OPLE 560
1045  FORMAT(6(2H *),45H DEMAND OUTSIDE RECOMMENDED RANGE (20-100000) )OPLE 561
1046  FORMAT(    72H INVESTMENT OUTSIDE RECOMMENDED RANGE (250000-10000OPLE 562
     *)      )                                                    OPLE 563
1050  FORMAT(1H1,   14X, 32HBASE CASE - NO BUILD DECISIONS     )  OPLE 564
1051  FORMAT( /  13X, 24HNO YEARLY CORRELATIONS    /)             OPLE 565
1052  FORMAT (10X, 3CHFIRST DECISION RESTRICTED TO    ,I3,4X,3A3//) OPLE 566
1053  FORMAT(17X, 45HOPTIMUM WITHOUT FIRST DECISION RESTRICTIONS  //)OPLE 567
1054  FORMAT( /  13X, 30HCOMPLETE YEARLY CORRELATIONS   /)        OPLE 568
1055  FORMAT( 1X,22HSCRAPPING ALL AT END   /)                     OPLE 569
1056  FORMAT(1X, 13HNO SCRAPPING ,  ,I5, 22H YEARS EXTENDED LIFE  /) OPLE 570
1057  FORMAT(17X,18HPROBABILITY LEVEL       ,I3  )                OPLE 571
1058  FORMAT( 52H STRATEGY CANNOT BE CHOSEN FOR TRANSFER TO EVALUATE )OPLE 572
1059  FORMAT(  ///28H BASE-CASE CAN BE OMITTED      ///)          OPLE 573
      END                                                         OPLE 574
```

```
      SUBROUTINE WERT(IS,IB,ITS,KD,NUT,PREIS,IEQ)            WERT    1
C********************************                            WERT    2
C   SUBROUTINE TO EVALUATE NEW FOR 1 YEAR, DISCOUNTED TO BEGIN OF YEAR 1  WERT  3
C     WHEN IEQ.GT.0, THEN IEQ GIVES QUARTER IN WHICH DECISION FALLS  WERT  3
      COMMON/MDAT1/KDAT(10,9),N,KTDAT(100)                   WERT    4
      COMMON/MDAT2/FAKTOR,UMTAX,MENGE,KOS(10),KV             WERT    5
      COMMON/KDAT3/PEN,FRATE,VARCOS(10)                      WERT    6
      DIMENSION KD(4),PREIS(4)                               WERT    7
      RATE=FRATE+1.                                          WERT    8
      YNCH=0.                                                WERT    9
      CAPL=0.                                                WERT   10
      SCRAP=0.                                               WERT   11
      FCOS=0.                                                WERT   12
      IF(IEQ.GT.0)GOTO 2                                     WERT   13
      FCOS=IPROD(8,7)                                        WERT   14
      GOTO 7                                                 WERT   15
C  FIXED COSTS FOR QUARTERS CALCULATION WITH DECISION AT BEGIN OF  WERT  16
C     QUARTER  IEQ IN FCOS                                   WERT   17
2     YB=0.25 * FLOAT(IEQ-1)                                 WERT   18
      RYB=RATE ** YB                                         WERT   19
      KOL=1                                                  WERT   20
      DO 5 I=1,4                                             WERT   21
      IF(I.GE.IEQ)KOL=8                                      WERT   22
      FCOS=FCOS+FLOAT(IPROD(7,KOL))*(RATE**(1.-.25*FLOAT(I)))  WERT  23
5     CONTINUE                                               WERT   24
      FCOS=FCOS/4.                                           WERT   25
C  CAPITAL COSTS                                             WERT   26
7     IF(IB.EQ.0)GOTO 9                                      WERT   27
      CAPL= FAKTOR*FLOAT( KDAT(IB,4))                        WERT   28
      IF(IEQ.GT.0)CAPL=CAPL/RYB                              WERT   29
C  SCRAPPING                                                 WERT   30
9     IF(IS.EQ.0)GOTO 20                                     WERT   31
      IF(ITS.NE.0)GOTO 12                                    WERT   32
C  ONE ALTERNATIVE SCRAPPED                                  WERT   33
      SCRAP=KDAT(IS,5)                                       WERT   34
      GOTO 20                                                WERT   35
C  WHOLE BRANCH SCRAPPED                                     WERT   36
12    DO 15 I=1,N                                            WERT   37
      IF(KDAT(I,1).EQ.0)GOTO 15                              WERT   38
      IP=KDAT(I,2)                                           WERT   39
      IF(IP.EQ.0)GOTO 15                                     WERT   40
      IP=IAES(KTDAT(IP))+IP+1                                WERT   41
      IF(KTDAT(IP).EQ.IS)SCRAP=SCRAP+FLOAT(KDAT(I,5))        WERT   42
15    CONTINUE                                               WERT   43
20    IF(IEQ.GT.0)SCRAP=SCRAP/RYB                            WERT   44
C  VARIABLE COSTS AND SALES INCOME                           WERT   45
      KOL=1                                                  WERT   46
      DO 50 IP=1,4                                           WERT   47
      IF(IP.GE.IEQ)KOL=8                                     WERT   48
      V=0.                                                   WERT   49
      MENGE=0                                                WERT   50
      YB=0.                                                  WERT   51
C  BUILD UP OPERATING PLANTS IN ORDER OF VARIABLE COSTS,UNTIL CAPACITY  WERT  52
C     IS OVERSTEPPED (ORDER GIVEN BY  KOS )                  WERT   53
      DO 30 I=1,N                                            WERT   54
      KU=KOS(I)                                              WERT   55
      IF(KDAT(KU,KOL).EQ.0)GOTO 30                           WERT   56
      KAD=KDAT(KU,6)                                         WERT   57
      IF(KAD.GT.KD(IP)-MENGE)KAD=KD(IP)-MENGE                WERT   58
      V=V+VARCOS(KU)*FLOAT(KAD)                              WERT   59
      MENGE=MENGE+KAD                                        WERT   60
      IF(MENGE.GE.KD(IP))GOTO 40                             WERT   61
30    CONTINUE                                               WERT   62
      YB=-PEN*FLOAT(KD(IP)-MENGE)                            WERT   63
40    YB=YB+(PREIS(IP)*FLOAT(MENGE))-V                       WERT   64
C  SALES INCOME AND VAR. COSTS NOW IN YB                     WERT   65
      IF(IEQ.GT.0)GOTO 42                                    WERT   66
      YNCH=YB                                                WERT   67
                                                            WERT   68
```

```
      GOTO 55                                                     WERT  69
42    RYB=(Y3*.25)*(RATE**(.25*(4.-FLOAT(IP))))                   WERT  70
      YNCM=YNCM+RYB                                               WERT  71
50    CONTINUE                                                    WERT  72
C  SUM OF EACH  END-OF-YEAR CASH FLOW AFTER TAX IN NWT            WERT  73
C  CAPITAL 2 YEARS UP-COUNTEDTO REPRESENT EXPENSE 1 YEAR BEFORE START-UPWERT  74
55    YNCM=YNCM-FCOS+SCRAP                                        WERT  75
      YNCM=YNCM+UNTAX                                             WERT  76
      NWT=YNCM-CAPL +0.5                                          WERT  77
      KV=V+0.5                                                    WERT  78
      RETURN                                                      WERT  79
C  THESE MATHEMATICAL OPERATIONS WITH INTEGERS CAUSE ROUND-OFF ERRORS  WERT  80
C  OF ABOUT+-5. HENCE NPW QUOTED TO NEAREST 10.                  WERT  81
      END                                                         WERT  82


      SUBROUTINE VALDEC(ITF,IS,IB,ITS)                            VALD   1
C  FINDS IF DECISION VIOLATES EXTENSION STRUCTURE,ITF=0 MEANS VIOLATION VALD   2
C  IS    -SCRAP DECISION, IB   --BUILD DECISION,  VALUE GIVES ALTERNATIVVALD   3
C  INVALID DECISIONS ARE:                                        VALD   4
C          -BUILD EXTENSION BEFORE PARENT PLANT                  VALD   5
C          -SCRAP EXTENSION WITHOUT SCRAPPING PARENT PLANT       VALD   6
C  IF PARENT PLANT IS SCRAPPED,ITS FLAGSNEUZ AND WEPT TO SCRAP ALL VALD   7
C    ASSOCIATED EXTENSIONS                                       VALD   8
C  ANALYSIS TAKEN FROM NETWORK THEORY                            VALD   9
      COMMON/MDAT1/KDAT(10,9),N,KTDAT(100)                       VALD  10
      ITS=0                                                       VALD  11
      IF(IS+IB.EQ.0)GOTO 80                                       VALD  12
C  AT LEAST 1 DECISION NOT ZERO                                  VALD  13
      IPS=0                                                       VALD  14
      IPB=0                                                       VALD  15
      IF(IS.NE.0)IPS=KDAT(IS,2)                                   VALD  16
      IF(IB.NE.0)IPB=KDAT(IB,2)                                   VALD  17
      IF(IB.EQ.0.OR.IS.EQ.0)GOTO 40                               VALD  18
      IF(KDAT(IB,1).EQ.KDAT(IS,1))GOTO 90                         VALD  19
40    IF(IPB.EQ.0.OR.IPS.EQ.0)GOTO 50                             VALD  20
C  ARE THEY ON THE SAME BRANCH                                   VALD  21
      LS=IABS(KTDAT(IPS))+IPS+1                                   VALD  22
      LB=IABS(KTDAT(IPB))+IPB+1                                   VALD  23
      IF(KTDAT(LS).EQ.KTDAT(LB))GOTO 90                           VALD  24
50    IF(IPS.EQ.0)GOTO 60                                         VALD  25
      IF(KDAT(IS,3).NE.0)GOTO 90                                  VALD  26
      ITS=1                                                       VALD  27
60    IF(IPB.EQ.0)GOTO 80                                         VALD  28
      IF(KDAT(IB,3).EQ.0)GOTO 80                                  VALD  29
      NZ=KDAT(IB,3)                                               VALD  30
      IF(KDAT(NZ,1).EQ.0)GOTO 90                                  VALD  31
      NZ=KDAT(NZ,2)                                               VALD  32
      IF(KTDAT(NZ))70,90,80                                       VALD  33
70    LS=-KTDAT(NZ)+NZ                                            VALD  34
      LB=1+NZ                                                     VALD  35
      DO 75 IPS=LB,LS                                             VALD  36
      IPB=KTDAT(IPS)                                              VALD  37
      IF(KDAT(IPB,1).EQ.1)GOTO 90                                 VALD  38
75    CONTINUE                                                    VALD  39
80    ITF=1                                                       VALD  40
C  VALID RETURNS                                                 VALD  41
      RETURN                                                      VALD  42
90    ITF=0                                                       VALD  43
C  INVALID RETURNS                                               VALD  44
      RETURN                                                      VALD  45
      END                                                         VALD  46
```

```
      FUNCTION IPROD(I,J)                                          IPRO    1
C  CALCULATE SUM OF PRODUCT OF 2 VECTORS IN AN ARRAY              IPRO    2
C        WHEN I=8 ANDJ=6   SUM IS CAPACITY OF STATE               IPRO    3
C        WHEN I=9 AND J=5  SUM IS SCRAP VALUE OF STATE            IPRO    4
      COMMON/KDAT1/KDAT(10,9),N                                   IPRO    5
      L=0                                                         IPRO    6
      DO 10 K=1,N                                                 IPRO    7
10    L=L+KDAT(K,I)*KDAT(K,J)                                     IPRO    8
      IPROD=L                                                     IPRO    9
      RETURN                                                      IPRO   10
      END                                                         IPRO   11


      SUBROUTINE PATH(K,IRS,IRR,MY,MEM1,MR1,LIFE,MNSRP1)          PATH    1
C  FINDS PATH (=POLICY) LEADING TO GIVEN FINAL-YEAR ENTRY         PATH    2
C  IN RESULTS MATRICES, AS SET OF POINTERS IN COL OF IRS.         PATH    3
C  ALSO FINDS FIRST DECISION ON THAT PATH.                        PATH    4
C  ON ENTRY K POINTS TO PATH END.                                 PATH    5
C  ON RETURN K IS YEAR OF 1ST DECISION OR 0 IF NONE.             PATH    6
C  UNWINDS PLAN, SINCE MEM(4) FOR YEARJR IS K FOR YEAR (JR-1)     PATH    7
      DIMENSION IRS(MY,251),IRR(MY,250),MEM1(5),MR1(1)           PATH    8
C     DOUBLE PRECISION  IRS,IRR                                   PATH    9
      JR=LIFE+1                                                   PATH   10
10    JR=JR-1                                                     PATH   11
      IF(JR.LT.1)GOTO 20                                          PATH   12
      IRS(JR,MNSRP1)=K                                            PATH   13
      CALL UNPACK(IRR(JR,K),MEM1,MR1,5)                           PATH   14
      K=MEM1(4)                                                   PATH   15
      GOTO 10                                                     PATH   16
C  NOW FIND FIRST DECISION,  PUT IN K                             PATH   17
20    DO 30 JR=1,LIFE                                             PATH   18
      K=IRS(JR,MNSRP1)                                            PATH   19
      CALL UNPACK(IRR(JR,K),MEM1,MR1,5)                           PATH   20
      IF(MEM1(2).NE.0.OR.MEM1(3).NE.0)GOTO 40                     PATH   21
30    CONTINUE                                                    PATH   22
C     NO DECISION FOUND                                           PATH   23
      K=0                                                         PATH   24
      RETURN                                                      PATH   25
C     DECISION IN YEAR JR                                         PATH   26
40    K=JR                                                        PATH   27
      RETURN                                                      PATH   28
      END                                                         PATH   29


      SUBROUTINE PACK(NPAK,NMBRS,KRAD,N)                          PACK    1
      DIMENSION KRAD(N),NMBRS(N)                                  PACK    2
C                                                                 PACK    3
C  ROUTINE TO PACK N NUMBERS INTO 1 WORD                          PACK    4
C  THE NUMBERS ARE IN NMBRS(I),I=1,N AND ARE PACKED INTO NPAK.    PACK    5
C  EACH NMBR(I) HAS A VALUE BETWEEN 0 AND KRAD(I)-1 (INCLUSIVE).  PACK    6
C  KRAD(1) CAN HAVE ANY VALUE AS IT IS MULTIPLIED BY 0.           PACK    7
C  NO VALIDITY CHECKS ARE MADE. ALL NUMBERS ARE INTEGERS.         PACK    8
C                                                                 PACK    9
      IT=0                                                        PACK   10
      DO 10 I=1,N                                                 PACK   11
10    IT=IT*KRAD(I)+NMBRS(I)                                      PACK   12
      NPAK=IT                                                     PACK   13
      RETURN                                                      PACK   14
      END                                                         PACK   15


      SUBROUTINE UNPACK(NPAK,NMBRS,KRAD,N)                        UPAC    1
      DIMENSION NMBRS(N),KRAD(N)                                  UPAC    2
C                                                                 UPAC    3
C  ROUTINE TO UNPACK, INTO NMBRS(I),I=1,N, NUMBERS PACKED         UPAC    4
C  INTO NPAK BY SUBROUTINE PACK, WITH RADICES KRAD(I)             UPAC    5
C  ALL NUMBERS A E INTEGERS.                                      UPAC    6
C  UNPACKED NO.. IS REMAINDER ON DIVISION OF PREVIOUS QUOTIENT    UPAC    7
C  BY CURRENT RADIX.                                              UPAC    8
```

```
      KWOT1=NPAK                                              UPAC   9
      I=N                                                     UPAC  10
10    IF(I.EQ.1)  GO TO 20                                    UPAC  11
      KWOT=KWOT1                                              UPAC  12
      KWOT1=KWOT/KRAD(I)                                      UPAC  13
      NMBRS(I)=KWOT-(KWOT1*KRAD(I))                           UPAC  14
      I=I-1                                                   UPAC  15
      GOTO 10                                                 UPAC  16
20    NMBRS(1)=KWOT1                                          UPAC  17
      RETURN                                                  UPAC  18
      END                                                     UPAC  19


      SUBROUTINE DPAK (PAK,NMBRS,KRAD,N)                      DPAC   1
C     SUBROUTINE PACK(PAK,NMBRS,KRAD,N)                       DPAC   2
C  PACK SUBROUTINE WITH DOUBLE PRECISION                      DPAC   3
      DOUBLE PRECISION PAK                                    DPAC   4
      DIMENSION KRAD(N),NMBRS(N)                              DPAC   5
      PAK=0.D0                                                DPAC   6
      DO 10 I=1,N                                             DPAC   7
10    PAK=PAK*DBLE(FLOAT(KRAD (I))) + DBLE(FLOAT(NMBRS(I)))   DPAC   8
      RETURN                                                  DPAC   9
      END                                                     DPAC  10


      SUBROUTINE DUNPAK (PAK,NMBRS,KRAD,N)                    DUPA   1
C     SUBROUTINE UNPACK (PAK,NMBRS,KRAD,N)                    DUPA   2
C  UNPACK SUBROUTINE WITH DOUBLE PRECISION                    DUPA   3
      DOUBLE PRECISION PAK ,KWOT,KWOT1 ,DKRAD                 DUPA   4
      DIMENSION KRAD(N),NMBRS(N)                              DUPA   5
      KWOT1=PAK                                               DUPA   6
      I=N                                                     DUPA   7
10    IF(I.EQ.1)  GO TO 20                                    DUPA   8
      KWOT=KWOT1                                              DUPA   9
      DKRAD=DBLE(FLOAT(KRAD(I)))                              DUPA  10
      KWOT1=KWOT/DKRAD                                        DUPA  11
      KWOT1=KWOT1 -DMOD((KWOT1 + 0.1),1.D0)                   DUPA  12
      NMBRS(I)= IFIX(SNGL(KWOT-(KWOT1*DKRAD)))                DUPA  13
      I=I-1                                                   DUPA  14
      GO TO 10                                                DUPA  15
20    NMBRS(1)= IFIX(SNGL(KWOT1))                            DUPA  16
      RETURN                                                  DUPA  17
      END                                                     DUPA  18


      SUBROUTINE NEUZ(IS,IB,ITS)                              NEUZ   1
C     COMPUTES NEW STATE IN COL. 8 OF KDAT.                   NEUZ   2
C  GIVEN THE BUILD AND SCRAP DECISIONS(IB+IS)                 NEUZ   3
      COMMON/MDAT1/KDAT(10,9),N,KTDAT(100)                    NEUZ   4
      DO 5 I=1,N                                              NEUZ   5
5     KDAT(I,8)=KDAT(I,1)                                     NEUZ   6
      IF(IB.GT.0)KDAT(IB,8)=1                                 NEUZ   7
      IF(ITS.EQ.0)GOTO 20                                     NEUZ   8
C  SCRAP A TREE, IF PARENT PLANT IS NOT OPERATING             NEUZ   9
      DO 10 I=1,N                                             NEUZ  10
      IF(KDAT(I,1).EQ.0)GOTO 10                               NEUZ  11
      IP=KDAT(I,2)                                            NEUZ  12
      IF(IP.E(.0)GOTO 10                                      NEUZ  13
      IP=IAES(KTDAT(IP))+IP+1                                 NEUZ  14
      IF(KTDAT(IP).EQ.IS)KDAT(I,8)=0                          NEUZ  15
10    CONTINUE                                                NEUZ  16
      RETURN                                                  NEUZ  17
20    IF(IS.GT.0)KDAT(IS,8)=0                                 NEUZ  18
      RETURN                                                  NEUZ  19
      END                                                     NEUZ  20
```

```
       SUBROUTINE INVEST                                      INVE    1
C*********************************                            INVE    2
C   INVESTMENT EVALUATION    L.H.ROSE                         INVE    3
C   SUBROUTINE TO PREPARE DATA AND MANAGE ECONOMIC EVALUATION AND   INVE    4
C   SENSITIVITY                                               INVE    5
       COMMON/INV /SENS(5)                                    INVE    6
       COMMON/TRAPS/PCAPY(10,4),PCAPL(10,4),PFOP(10,4),PVOP(10,4),   INVE    7
      *PSCRAP(10,4),PFINE(4)              ,PENE(4),NC         INVE    8
       COMMON/BLO1/PN,K,KS,P(8),S(5),CAPL(20),PKB            INVE    9
       COMMON/RDAT1/FINE,DR,UVOP(10),NT,NUT,ND,TX ,ENF,KTAX,ALT(10,3)   INVE   10
       COMMON/RDAT2/UFOP(10),UCAPY(10),UCAPL(10),USCRAP(10),ADDN(20)   INVE   11
       COMMON/SALES/PEOB(8),JT,TCNS(9,20),SEL(8,20),WOFK(8,20)   INVE   12
       COMMON/TIME/MIEND,IST(10),ISTART(10,8),IEND(10,8)     INVE   13
       COMMON/NAMES/ISW(24),INA(10,3)                        INVE   14
       COMMON/UTIL/IU,XMIN(5)                                INVE   15
       COMMON/RWR/IOUT,IN                                    INVE   16
       COMMON/SIM/CORT,CORS,KMC,ERR                          INVE   17
       COMMON/CONT/LN,ICHEK                                  INVE   18
       COMMON/STRAT/IFIR(8),ISEC(8),MPLAN                    INVE   19
       COMMON/REF/IREF                                       INVE   20
       WRITE(IOUT,6)   (ISW(I),I=1,24)                       INVE   21
6      FORMAT(1H1,24A3)                                      INVE   22
       NN=NT-1                                               INVE   23
C   DETERMINE WHETHER MONTE-CARLO OR PROBABILISTIC EVALUATION REQUIRED.   INVE   24
C     FLAG:  KMC=10 OR 0 RESPECTIVELY.                       INVE   25
       KMC=10                                                INVE   26
       IF( CORT.GT.0.99.AND.CORS.GT.0.99. AND. MC.EQ.0      )   INVE   27
      *KMC=0                                                 INVE   28
       IF(JT.EQ.1. AND. MC.EQ.0           )    KMC=0         INVE   29
       IF(IREF.NE.-1)  GO TO 10                              INVE   30
       WRITE(IOUT,5)                                         INVE   31
       RETURN                                                INVE   32
5      FORMAT( 48H NO OPERATE DATA, EVALUATION CANNOT PROCEED    )   INVE   33
10     IF(K .GT.0)GOTO 60                                    INVE   34
       WRITE(IOUT,61) DR                                     INVE   35
       GO TO 515                                             INVE   36
60     WRITE(IOUT,50)                                        INVE   37
50     FORMAT(26H DCF CALCULATION REQUESTED )                INVE   38
61     FORMAT (45H NPW CALCULATION REQUESTED:  DISCOUNT RATE  = ,2PF6.2)   INVE   39
515    JJJ=1                                                 INVE   40
       WRITE(IOUT,57) NT                                     INVE   41
57     FORMAT ( 16H PROJECT LENGTH   ,I8,7H YEARS   )        INVE   42
       IF(MIEND.EQ.0)    WRITE(IOUT,40)                      INVE   43
40     FORMAT( 1X,22HSCRAPPING ALL AT END     )             INVE   44
       IF(MIEND.GT.0)  WRITE(IOUT,41)  MIEND                 INVE   45
41     FORMAT( 1X, 16HNO SCRAPPING,  ,I7, 22H YEARS EXTENDED LIFE   )INVE   46
       WRITE(IOUT,58) ND                                     INVE   47
58     FORMAT ( 18H DEPRECIATED OVER  ,I6, 7H YEARS )        INVE   48
       WRITE(IOUT,525) TX                                    INVE   49
525    FORMAT ( 12H TAX RATE = ,9X,2PF6.2)                   INVE   50
       IF(IU.EQ.1)WRITE(IOUT,577)XMIN(1)                     INVE   51
577    FORMAT(13H MAXIMUM LOSS,1X, F13.2)                    INVE   52
       IF(KMC.EQ.0)  GO TO 400                               INVE   53
       WRITE(IOUT,410)                                       INVE   54
410    FORMAT( 47H A MONTE-CARLO SIMULATION WILL BE CARRIED OUT    /   INVE   55
      1 68H THE 4 DATA SETS FOR THE TRAPEZOIDAL DISTRIBUTIONS ARE AS FOLLIWE   INVE   56
       20WS:        )                                        INVE   57
       WRITE(IOUT,326) PENE                                  INVE   58
       WRITE(IOUT,330) PFINE                                 INVE   59
       WRITE(IOUT,541)                                       INVE   60
       DO 430 N=1,NUT                                        INVE   61
       DO 430 J=1,4                                          INVE   62
430    WRITE(IOUT,33)N,(INA(N,JJ),JJ=1,3),PCAPY(N,J),PCAPL(N,J),   INVE   63
      1PFOP(N,J),PVOP(N,J),PSCRAP(N,J)                       INVE   64
       GO TO 535                                             INVE   65
400    IF(ENF.GT.0.)WRITE(IOUT,526) ENF                      INVE   66
526    FORMAT(26H FRACTIONAL INFLATION RATE        ,5F6.2)   INVE   67
       WRITE(IOUT,330) FINE                                  INVE   68
```

```
         WRITE(IOUT,531)                                                      INVE  69
         WRITE(IOUT,532)((ADDN(N) ),N=1,MN)                                   INVE  70
531      FORMAT( 36H ADDITIONAL YEARLY COSTS BY YEAR:              )          INVE  71
532      FORMAT( /1X,1.F6.0/1X,1 F6.0)                                        INVE  72
         WRITE(IOUT,541)                                                      INVE  73
541      FORMAT(//7X,                                                         INVE  74
     *           64H UNIT   CAPACITY   CAPITAL   FIXED COST   VARIABLE COSINVE  75
     1T   SCRAP   //)                                                         INVE  76
         WRITE(IOUT,83) ((N,(INA(N,JJ),JJ=1,3),UCAPY(N),UCAPL(N),UFOP(N),     INVE  77
     1UVOP(N),USCRAP(N) ) , N=1,NUT)                                          INVE  78
83       FORMAT(I3,1X,3A3,3F10.1,5X,F10.3,3X,F10.1)                           INVE  79
530      FORMAT(26H SHORTFALL PENALTY PER TON    , 4F 8.2)                    INVE  80
535      WRITE(IOUT,543)                                                      INVE  81
543      FORMAT(//21H  OPERATING STRATEGY   /,11X,                           INVE  82
     *43HEXISTING   START/END PAIRS, BY PROBABILITY   /)                      INVE  83
         DO 545 J=1,NUT                                                       INVE  84
         DO 599 JJ=1,JT                                                       INVE  85
         IF(IEND(J,JJ).EQ.0.AND.IST(J).EQ.1)IEND(J,JJ)=NT+MIEND +1            INVE  86
         IF(IEND(J,JJ).EQ.0.AND.ISTART(J,JJ).NE.0)IEND(J,JJ)=NT+MIEND+1       INVE  87
599      IF(IEND(J,JJ).GE.(NT+1))                                            INVE  88
     *IEND(J,JJ) =NT+MIEND +1                                                 INVE  89
545      WRITE(IOUT,544)J,(INA(J,N),N=1,3),IST(J),(( ISTART(J,JJ),IEND(J,JJINVE  90
     *)),JJ=1,JT)                                                            INVE  91
544      FORMAT(I3,1X, 3A3,I3,5X,  8(I3,I3,2X))                               INVE  92
C COLLECTION OF 2ND DECISION IN ISEC( )                                      INVE  93
         DO 75 JJ=1,JT                                                        INVE  94
         ISEC(JJ)=NT+1                                                        INVE  95
         I=NT+1                                                               INVE  96
         DO 70 J=1,NUT                                                        INVE  97
         IF(ISTART(J,JJ).LT.I. AND.ISTART(J,JJ). NE.0)                        INVE  98
     *I=ISTART(J,JJ)                                                          INVE  99
70       IF(IEND(J,JJ).LT.I.AND.IEND(J,JJ).NE.0)                             INVE 100
     *I=IEND(J,JJ)                                                           INVE 101
C   I IS NOW   1ST DECISION                                                  INVE 102
         IFIR(JJ)=I                                                           INVE 103
         IF(JJ.EQ.1.OR.KMC.EQ. 0)    GO TO 74                                 INVE 104
         IF(IFIR(JJ).EQ.IFIR(JJ-1))GO TO 74                                   INVE 105
         WRITE(IOUT,77)                                                       INVE 106
77       FORMAT(//65H FIRST DECISIONS OF ALL PROBABILITY LEVEL PLANS NOT IDINVE 107
     2ENTICAL,      /28H SIMULATION CANNOT PROCEED      )                     INVE 108
         RETURN                                                              INVE 109
74       DO 75  J=1,NUT                                                       INVE 110
         IF(ISTART(J,JJ).NE.I.AND.ISTART(J,JJ).LT.ISEC(JJ).AND.              INVE 111
     *ISTART(J,JJ). NE.0)   ISEC(JJ)=ISTART(J,JJ)                            INVE 112
75       IF(IEND(J,JJ).NE.I.AND.IEND(J,JJ).LT.ISEC(JJ).AND.IEND(J,JJ).NE.0)INVE 113
     *ISEC(JJ)=IEND(J,JJ)                                                    INVE 114
C   KS IS SENSITIVITY COUNTER                                                INVE 115
         KS=0                                                                 INVE 116
         IF(JT.GT.1)                                                         INVE 117
     *WRITE(IOUT,576)CORT,CORS                                               INVE 118
576      FORMAT(//                                                           INVE 119
     1 29H DEMAND/YEAR  CORRELATION IS  ,F5.3/                               INVE 120
     2 29H PRICE/DEMAND CORRELATION IS  ,F5.3 )                              INVE 121
         JJJ=1                                                               INVE 122
         IF(JT-3)12,13,13                                                    INVE 123
12       J=JT                                                                INVE 124
         GO TO 66                                                            INVE 125
13       J=3                                                                 INVE 126
66    WRITE(IOUT,68)                                                         INVE 127
68       FORMAT(//                                                           INVE 128
     172H YR DEMAND PRICE WORK CAP/ DEMAND PRICE WORK CAP/DEMAND PRICE WINVE 129
     *ORK CAPL              )                                                INVE 130
         WRITE(IOUT,67) (JJ,JJ=JJJ,J)                                        INVE 131
67       FORMAT( 5X,2HNO, 3(I7,16X))                                         INVE 132
         WRITE(IOUT,65)(PROB(JJ),JJ=JJJ,J)                                   INVE 133
65       FORMAT(5X,4HPROB, 3(F9.4,14X)/)                                     INVE 134
         DO 59 N=1,NT                                                        INVE 135
      59 WRITE(IOUT,64) N, ((TONS(JJ,N),SEL(JJ,N),WORK(JJ,N)),JJ=JJJ,J)      INVE 136
```

```
64     FORMAT(I5,3(F6.0,F9.3,F6.0,2X))                          INVE 137
       IF(J-JT)62,63,63                                         INVE 138
62     JJJ=JJJ+3                                                INVE 139
       J=J+3                                                    INVE 140
       IF(J.GT.JT)  J=JT                                        INVE 141
       GO TO 66                                                 INVE 142
63     IF(KMC.GT.0)  GO TO 600                                  INVE 143
       CALL FLOW                                                INVE 144
C      ---------                                                INVE 145
       DIS=100.*DR                                              INVE 146
210    IF(KS)550,550,551                                        INVE 147
C  PRINTOUT NORMAL RESULT                                       INVE 148
550    IF(KMC.GT.0)WRITE(IOUT,555)                              INVE 149
       IF(KMC.EQ.0)WRITE(IOUT,560)PW,DIS,PKB,(P(J),J=1,JT)      INVE 150
555    FORMAT(20X,6HNORMAL    /////)                            INVE 151
560    FORMAT(  24H NORMAL                    ,F12.1,F8.2,F5.1, 8F9.0)  INVE 152
C   RETURN FOR SENSITIVITY ANALYSIS                             INVE 153
562    KS=1                                                     INVE 154
       IF(SENS(1).LT.0.0000001) GO TO 571                       INVE 155
       S(1)=1.+SENS(1)/100.                                     INVE 156
       GO TO 63                                                 INVE 157
551    SL=(S(KS)-1.0)*100.                                      INVE 158
       IF(KMC.EQ.0)GOTO 800                                     INVE 159
       GOTO(710,720,730,740,750)KS                              INVE 160
710    WRITE(IOUT,711)SL                                        INVE 161
711    FORMAT(F6.0,20H 0/0 PRICE INCREASE ,F10.1,F8.2,F5.1, 8F9.0)   INVE 162
       GOTO 565                                                 INVE 163
720    WRITE(IOUT,721)SL                                        INVE 164
721    FORMAT(F6.0,20H 0/0 VOLUME INCREASE,F10.1,F8.2,F5.1, 8F9.0)   INVE 165
       GOTO 565                                                 INVE 166
730    WRITE(IOUT,731)SL                                        INVE 167
731    FORMAT(F6.0,20H 0/0 CAPITAL INCR   ,F10.1,F8.2,F5.1, 8F9.0)   INVE 168
       GOTO 565                                                 INVE 169
740    WRITE(IOUT,741)SL                                        INVE 170
741    FORMAT(F6.0,20H 0/0 FIXED COSTS    ,F10.1,F8.2,F5.1, 8F9.0)   INVE 171
       GOTO 565                                                 INVE 172
750    WRITE(IOUT,751)SL                                        INVE 173
751    FORMAT(F6.0,20H 0/0 VARIABLE COSTS ,F10.1,F8.2,F5.1, 8F9.0)   INVE 174
       GOTO 565                                                 INVE 175
800    GOTO(810,820,830,840,850)KS                              INVE 176
810    WRITE(IOUT,711)SL,PW,DIS,PKB,(P(J),J=1,JT)               INVE 177
       GOTO 565                                                 INVE 178
820    WRITE(IOUT,721)SL,PW,DIS,PKB,(P(J),J=1,JT)               INVE 179
       GOTO 565                                                 INVE 180
830    WRITE(IOUT,731)SL,PW,DIS,PKB,(P(J),J=1,JT)               INVE 181
       GOTO 565                                                 INVE 182
840    WRITE(IOUT,741)SL,PW,DIS,PKB,(P(J),J=1,JT)               INVE 183
       GOTO 565                                                 INVE 184
850    WRITE(IOUT,751)SL,PW,DIS,PKB,(P(J),J=1,JT)               INVE 185
C   ITERATE ON NEXT SENSITIVITY CASE                            INVE 186
565    IF(S(KS).    LT.1.0)  GO TO 571                          INVE 187
       S(KS)=1.-(SENS(KS)/100.)                                 INVE 188
       GO TO 63                                                 INVE 189
571    S(KS)=1.0                                                INVE 190
       KS=KS+1                                                  INVE 191
       IF( KS.GT.5)  RETURN                                     INVE 192
       IF(SENS(KS).LT.0.0000001)     GO TO 571                  INVE 193
       S(KS)=1.0+(SENS(KS)/100.)                                INVE 194
       GO TO 63                                                 INVE 195
C  TO MONTE-CARLO EVALUATION WHEN KMC.GT.0                      INVE 196
600    CALL MONTE                                               INVE 197
C      ---------                                                INVE 198
       WRITE(IOUT,610)                                          INVE 199
610    FORMAT(// 60H THIS STUDY HAS BEEN FOR THE FOLLOWING SENSITIVITY   INVE 200
      1LEVEL               )                                    INVE 201
       GO TO 210                                                INVE 202
       END                                                      INVE 203
```

```
      SUBROUTINE FLOW                                        FLOW    1
C************************                                    FLOW    2
C  GIVEN ALTERNATIVE AND COMMERCIAL DATA,WILL EVALUATE NPW OR DCF    FLOW    3
C  WITH SENSITIVITY ANALYSIS                                 FLOW    4
C  ---CAN BE USED WITH INVEST OR MONTE,                      FLOW    5
C  EVALUATES PROBABILISTICALLY,AND (EXCEPT FROM MONTE)  IT EVALUATES FLOW    6
C   -PAYBACK,AND YEARLY CASH FLOW                            FLOW    7
      DIMENSION CAPY(20),FOP(20),VOP(20),PA(2),D(2),PUND(8,20)   FLOW    8
      DIMENSION JORDER(10), PTONS(20),  PSEL(20) ,BCCAPL(8),FWORK(20)  FLOW    9
      COMMON/RDAT1/FINT,OR,UVOP(10),NT,NUT,ND,TX ,ENF,KTAX,ALT(10,3)   FLOW   10
      COMMON/RDAT2/UFOP(10),UCAPY(10),UCAPL(10),USCRAP(10),ADDN(20)    FLOW   11
      COMMON/SALES/PROB(8),JT,TONS(9,20),SEL(8,20),WORK(8,20)   FLOW   12
      COMMON/TIME/MTEND,IST(10),ISTART(10,8),IEND(10,8)      FLOW   13
      COMMON/BLO1/PW,K,KS,P(8),S(5),CAPL(20),PKC            FLOW   14
      COMMON/SIM/CORT,CORS,KMC,ERR,IYEAR,JMC               FLOW   15
      COMMON/RHS/IOUT,IN                                    FLOW   16
      COMMON/CONT/LW,ICHEK                                  FLOW   17
      COMMON/STRAT/IFIR(8),ISEC(8),MPLAN                   FLOW   18
      COMMON/UTIL/IU,XMIN(5)                                FLOW   19
      COMMON/REF/IREF                                       FLOW   20
      COMMON/TEST/LEVD(8),LEVS(8)                           FLOW   21
      JUMP=0                                                FLOW   22
      IF(JMC.GT.1)  GO TO 5                                 FLOW   23
      DO 1 I=1,JT                                           FLOW   24
      LEVD(I)=0                                             FLOW   25
1     LEVS(I)=0                                             FLOW   26
5     CONTINUE                                              FLOW   27
C  INITIALISE STRATEGIES                                    FLOW   28
                MPLAN=(JT+1)/2                              FLOW   29
      ISTR= MPLAN                                           FLOW   30
      IREAL= MPLAN                                          FLOW   31
      KSEC= 0                                               FLOW   32
C  EXISTING PLANT CAPACITY AND VAR COST                     FLOW   33
      BFOP=0.                                               FLOW   34
      BCAPY=0.                                              FLOW   35
      BVOP=0.                                               FLOW   36
      DO 60 I=1,NUT                                         FLOW   37
      BFOP=BFOP+UFOP(I)*FLOAT(IST(I))                       FLOW   38
      BVOP=BVOP+  UVOP(I)*UCAPY(I) *FLOAT(IST(I))           FLOW   39
60    BCAPY=BCAPY + UCAPY(I) *FLOAT(IST(I))                 FLOW   40
      IF(BCAPY.NE.0.0)                                      FLOW   41
     *BVOP=BVOP/BCAPY                                       FLOW   42
C  REMEMBER ORDER OF UNITS IN DECENDING ORDER OF VARIABLE COST  FLOW   43
      JORDER(1)=1                                           FLOW   44
      IF(NUT.EQ.1)  GO TO 43                                FLOW   45
      DO 40 J=1,NUT                                         FLOW   46
40    JORDER(J)=J                                           FLOW   47
      KO=0                                                  FLOW   48
      JJJ=NUT-1                                             FLOW   49
41    DO 42 J=1,JJJ                                         FLOW   50
      JOR=JORDER(J+1)                                       FLOW   51
      JORD=JORDER(J)                                        FLOW   52
      IF(UVCP( JOR         ).GT.UVOP(JORD      ))  GO TO 44 FLOW   53
      GO TO 42                                              FLOW   54
44    JJ=JORDER(J)                                          FLOW   55
      JORDER(J)=JORDER(J+1)                                 FLOW   56
      JORDER(J+1)=JJ                                        FLOW   57
      KO=1                                                  FLOW   58
42    CONTINUE                                              FLOW   59
      IF(KO.EQ.0)  GO TO 43                                 FLOW   60
      KO=0                                                  FLOW   61
      GO TO 41                                              FLOW   62
209   CONTINUE                                              FLOW   63
43    KT=1                                                  FLOW   64
C  SUBROUTINE TAXF  GIVES FACTOR F WHICH INCLUDES ALL TAX ALLOWANCE,  FLOW   65
C    INFLATION,AND TIMING OF CAPITAL EXPENCES                FLOW   66
210   CALL TAXF(F,OR,ND,TX,ENF,KTAX)                        FLOW   67
C************************                                    FLOW   68
```

```
      IF(ICHEK.EQ.2)  WRITE(IOUT,211)  KT,DR,F                    FLOW  69
211   FORMAT( / 17H DCF ITERATION    ,I3, 10H DISCOUNT    ,F9.4,   FLOW  70
     1 12H TAX FACTOR  ,F9.4)                                     FLOW  71
C              EACH PROBABILITY                                   FLOW  72
      DO 100 J=1,JT                                               FLOW  73
      BCCAFL(J)=0.0                                               FLOW  74
      IF(KMC.LE.0)  ISTR=J                                        FLOW  75
      PW=0.                                                       FLOW  76
      WORK (J,NT) =0.0                                            FLOW  77
      DF=1.                                                       FLOW  78
      IF(KT.GT.1 )  GO TO 118                                     FLOW  79
      IF(KMC.GT.0.AND.ICHEK.LT.2)   GO TO 118                     FLOW  80
      IF(ICHEK.EQ.0)GOTO 118                                      FLOW  81
      WRITE(IOUT,117)J                                            FLOW  82
117   FORMAT(///3X,17HPROBABILITY LEVEL,I3,//,                    FLOW  83
     1         75H YEAR CAPACITY    CAPITAL    FIXED COST   PRODCTN NETFLOW  84
     2INCOME VAR.COST DEPRCN.                   //)               FLOW  85
C              EACH YEAR                                          FLOW  86
118   DO 110 N=1,NT                                               FLOW  87
      IF(N.LT.NT)  WORK(J,NT)=WORK(J,NT)-WORK(J,N)                FLOW  88
C  IF MONTE CARLO, SIMULATE YEARLY FORECAST DATA BY RANDOM  GENERATOR  FLOW  89
C   PLUS CORRELATION                                              FLOW  90
C ***   YEARLY DEMAND CORRELATION CORT,                          FLOW  91
C ***    PRICE/SALES CORRELATION  CORS                           FLOW  92
      IF(J.GT.1.AND.IREF.EQ.0)   GO TO 145                        FLOW  93
      IF(KT.GT.1.AND.IREF.EQ.0)  GO TO 145                        FLOW  94
      IF(KT.GT.1.AND.JT.EQ.1)  GO TO 145                          FLOW  95
      IF(KMC.LE.0)  GO TO 65                                      FLOW  96
C  MAKE SECOND DECISION IF LEVEL AGREES WITH TIMING FOR SECOND DECISION FLOW  97
      IF (KSEC .GT.0 )GO TO  65                                   FLOW  98
      ISTR =IREAL                                                 FLOW  99
      DO 140  JJ=1,JT                                             FLOW 100
      IF(N. LT.ISEC(JJ) )  GO TO 140                              FLOW 101
      IF (JJ. NE.IREAL )  GO TO 140                               FLOW 102
      KSEC=1                                                      FLOW 103
      GO TO 65                                                    FLOW 104
140   CONTINUE                                                    FLOW 105
C   SUMMARISING  YEARLY CASH FLOWS                                FLOW 106
65    M=N                                                         FLOW 107
      CAPY(M)=0.0                                                 FLOW 108
      CAPL(M)=0.                                                  FLOW 109
      VOP(M)=0.                                                   FLOW 110
      FOP(M)=0.                                                   FLOW 111
      DO 76 I=1,NUT                                               FLOW 112
      IF(M.LT.ISTART(I,ISTR). OR.M.GE. IEND(I,ISTR))    GO TO 76  FLOW 113
      CAPY(M)=CAPY(M)+UCAPY(I)                                    FLOW 114
C  VARIABLE COST AS A MEAN FOR ALL OPERATING UNITS                FLOW 115
      IF(CAPY(M).NE.0.0)                                          FLOW 116
     *VOP(M)=(VOP(M)*(CAPY(M)-UCAPY(I))+UVOP(I)*UCAPY(I))/(CAPY(M)) FLOW 117
      FOP(M)=FOP(M)+UFOP(I)                                       FLOW 118
C   IS IT SECOND DECISION                                         FLOW 119
      IF(KSEC.EQ.1)  GO TO 68                                     FLOW 120
      IF( M.EQ.ISTART(I,ISTR))                                    FLOW 121
     1CAPL(M)=CAPL(M)+UCAPL(I)*FLOAT(1-IST(I))                    FLOW 122
C---SUBSTRACT SCRAP VALUES FROM FIXED COSTS                       FLOW 123
      IF(M.EQ.IEND(I,ISTR)) FOP(M)=FOP(M )-USCRAP(I)              FLOW 124
      IF(M.EQ.NT.AND.IEND(I,ISTR).EQ.(NT+1)) FOP(M)=FOP(M)-USCRAP(I) FLOW 125
      GO TO 76                                                    FLOW 126
C  COLLECT ANY BYPASSED DECISIONS BY STRATEGY CHANGE              FLOW 127
68    KSEC=2                                                      FLOW 128
      IF(M .GE.ISTART(I,ISTR).AND.ISTART(I,ISTR).GT.IFIR(ISTR))   FLOW 129
     *CAPL(M) =CAPL(M) +UCAFL(I)*FLOAT(1-IST(I))                  FLOW 130
70    IF(M .GE.  IEND(I,ISTR).AND.  IEND(I,ISTR).GT.IFIR(ISTR))   FLOW 131
     1                     FOP (M) = FOP(M)-USCRAP(I)             FLOW 132
76    CONTINUE                                                    FLOW 133
C  MAKE YEARLY CORRECTIONS TO FIXED COSTS                         FLOW 134
      FOP(M)=FOP(M) + ADDN(M)                                     FLOW 135
145   IF(KMC.LT.1.OR.JT.EQ.1)GOTO 129                             FLOW 136
```

```
      IF(KT.GT.1)    GO TO 129                                        FLOW 137
      PTONS(N)= TONS(1,N)                                             FLOW 138
      PWORK(N) =WORK(1,N)                                             FLOW 139
      PSEL(N) = SEL(1,N)                                              FLOW 140
      IF(N.EQ.1)   JL=IYEAR                                           FLOW 141
      CALL       HIST (JL,JJ,PROB,CORT,CORS,JMC,N,JT)                 FLOW 142
      LEVD(JL)=LEVD(JL) + 1                                           FLOW 143
      LEVS(JJ)=LEVS(JJ)+1                                             FLOW 144
      TONS(1,N)=TONS(JL,N)                                            FLOW 145
      WORK(1,N)= WORK(JL,N)                                           FLOW 146
      SEL(1,N)= SEL(JJ,N)                                             FLOW 147
      IREAL=JL                                                        FLOW 148
131   IF(ICHEK .EQ.2)  WRITE(IOUT,128)  SEL(1,N), TONS(1,N),JL,IREAL,ISTFLOW 149
     *R                                                               FLOW 150
128   FORMAT( 6H PRICE, F6.2, 8H  DEMAND,  F7.0,  11H  DMDLEVEL  ,    FLOW 151
     *I2, 12H  CUMULLEVEL  ,I2,  11H   STRATEGY ,I2)                  FLOW 152
C  MONTE -CARLO SALES DATA NOW GENERATED                             FLOW 153
129   DF=CF/(DR+1.)                                                   FLOW 154
      IF(CAPY(N) - TONS(J,N)*S(2))121,121,130                         FLOW 155
 121  X=CAPY(N)                                                       FLOW 156
      V=VCF(N)*X                                                      FLOW 157
      SHFL=TONS(J,N)-CAPY(N)                                          FLOW 158
      GO TO 150                                                       FLOW 159
 130  X=TONS(J,N) *S(2)                                               FLOW 160
      SHFL=0.                                                         FLOW 161
C   SELECT PLANTS TO OPERATE                                          FLOW 162
      CAV=CAPY(N)                                                     FLOW 163
      VAV=VCF(N)                                                      FLOW 164
      DO 120 JH=1,NUT                                                 FLOW 165
      JJ=JORDER(JH)                                                   FLOW 166
      IF(N.LT. ISTART(JJ,ISTR) .OR. N .GT.IEND(JJ,ISTR))  GO TO 120   FLOW 167
      VAV=(VAV*CAV-UVOP(JJ)*UCAPY(JJ))                                FLOW 168
      CAV=CAV-UCAPY(JJ)                                               FLOW 169
      VAV=VAV/(CAV +0.00001)                                          FLOW 170
      IF(CAV-X          )       122,122,120                           FLOW 171
120   CONTINUE                                                        FLOW 172
122   V=VAV*CAV+(X-CAV)*UVOP(JJ)                                      FLOW 173
C         ANNUAL PROFIT BEFORE TAX                                    FLOW 174
150   CASH=X*SEL(J,N)*S(1)-V*S(5)-FOP(N)*S(4)   -SHFL*FINE            FLOW 175
      X1=X                                                            FLOW 176
      CASH1=CASH                                                      FLOW 177
C   REMOVE BASE CASE SALES AND VAR COST                              FLOW 178
      V5=V*S(5)                                                       FLOW 179
      IF(BCAPY.LT.0.0001)  GO TO 160                                  FLOW 180
      IF(BCAPY  -  TONS(J,N)*S(2))151,151,152                         FLOW 181
151   X=BCAPY                                                         FLOW 182
      V=BVCF*X                                                        FLOW 183
      SHFL=TONS(J,N) - BCAPY                                          FLOW 184
      GO TO 155                                                       FLOW 185
152   X=TONS(J,N) *S(2)                                               FLOW 186
      SHFL=0.                                                         FLOW 187
C  SELECT PLANTS TO OPERATE IN BASE CASE                             FLOW 188
      CAV=BCAPY                                                       FLOW 189
      VAV=BVCP                                                        FLOW 190
      DO 153  JH=1,NUT                                                FLOW 191
      JJ=JORDER(JH)                                                   FLOW 192
      IF( IST(JJ).E1.0)   GO TO 153                                   FLOW 193
      VAV=(VAV*CAV-UVOP(JJ)*UCAPY(JJ))                                FLOW 194
      CAV=CAV-UCAPY(JJ)                                               FLOW 195
      VAV=VAV/(CAV+.00001)                                            FLOW 196
      IF(CAV-X  ) 154,154,153                                         FLOW 197
153   CONTINUE                                                        FLOW 198
154   V=VAV*CAV+(X-CAV) *UVOP(JJ)                                     FLOW 199
 155  CASH=CASH-X*SEL(J,N)*S(1) +V*S(5) +SHFL*FINE+BFOP*S(4)          FLOW 200
C  DISCOUNTED CACLN,ALLOW FOR DEPRECIATION TAX ALLOWANCE BY FACTOR F  FLOW 201
C   CAPITAL SPENT 2 YEARS BEFORE FIRST INCOME                         FLOW 202
C   ANNUAL INCOMES AND EXPENDITURES OCCUR AT END OF YEAR             FLOW 203
C---CASH ADJUSTMENT    --SUBSTRACTED AFTER TAX TO REPRESENT WORKING CAPIFLOW 204
160   PW=PW+(CASH*(1.-TX)-((APL(N)*F*S(3)      )-  WORK(J,N))*DF      FLOW 205
```

```
C      DEPRECIATION ALLOWANCE   FOR PAYBACK CALCULATION           FLOW 206
C   CASH FLOW AND PAYBACK CALCULATION(NOT WITH MONTE-CARLO)       FLOW 207
       IF( KMC .GT.0 .AND.ICHEK .LT.2)    GO TO 1111              FLOW 208
       BCCAPL(J)= BCCAPL(J)+ CAPL(N) *S(7)*DF*(DR+1.)**2          FLOW 209
       IF(KT.GT.1         ) GO TO 1111                           FLOW 210
       CALL TAXD(DEP,N,ND,NT,CAPL,S,MIEND,ENF,KTAX)              FLOW 211
C      ---------                                                FLOW 212
       IF(ICHEK.EQ.0)GOTO 105                                    FLOW 213
       WRITE(IOUT,82)  N,CAPY(N),CAPL(N),FOP(N),X1,CASH1,VG,DEP  FLOW 214
    82 FORMAT (I5,   7F10.2)                                     FLOW 215
C      PROFIT AFTER TAX   FOR PAYBACK                           FLOW 216
  105  CASHP=CASH-(CASH-DEP)*TX -WOFK(J,N)                       FLOW 217
C  CALCULATION HAS BEEN CARRIED OUT IN CONSTANT MONEY. FOR CASH FLOW ANDFLOW 218
C    PAYBACK,                                                   FLOW 219
C   NOW CONVERT TO CURRENT MONEY                                FLOW 220
C  CUMULATIVE CASH FLOW = LAST YEARS CUM. FLOW + THIS YEARS FLOW +NEXT FLOW 221
C    YEARS CAPITAL                                              FLOW 222
       PUND(J,N)=    (CASHP*(1.+ ENF)**N)                        FLOW 223
       IF(N.EQ.1) PUND(J,1) =PUND(J,1)- CAPL(1)*S(3)             FLOW 224
       IF(N.NE.1) PUND(J,N-1)= PUND(J,N-1) -CAPL(N)*S(3)*(1.+ENF)**(N-1) FLOW 225
       IF(N.GT.2)  PUND(J,N-1) =PUND(J,N-1) + PUND(J,N-2)        FLOW 226
       IF(N.EQ.NT)                                               FLOW 227
      *PUND(J,NT) =PUND(J,NT)+ PUND(J,NT-1)                      FLOW 228
 1111 CONTINUE                                                   FLOW 229
  110 CONTINUE                                                   FLOW 230
C END CORRECTION FOR FURTHER YEARS                              FLOW 231
       IF(ABS(DR)-1.E-6) 111,111,112                             FLOW 232
  111  PW=PW+(CASH*(1.-TX)*FLOAT(MIEND)   )                      FLOW 233
       GO TO 113                                                 FLOW 234
  112  PW= PW+ CASH*( 1.-TX)  * (1.-1./ ((1.+DR)  **FLOAT( MIEND   ))) FLOW 235
      1/ (DR*((1.+DR )** FLOAT(NT)))                             FLOW 236
  113  P (J)=PW                                                  FLOW 237
       IF(KMC.GT.0)GOTO 191                                      FLOW 238
  100 CONTINUE                                                   FLOW 239
       PW=0.                                                     FLOW 240
       DO 190 J=1,JT                                             FLOW 241
  190   PW=P (J)*PROB(J)+PW                                      FLOW 242
  191  IF(K)280,280,290                                          FLOW 243
C   ITERATION AND CONVERGENCE FOR DCF  (IF K=1)                 FLOW 244
C       SS=1.0/(2.0+R)                                          FLOW 245
C       WHERE  R  IS  DCF-RATE-OF-RETURN                        FLOW 246
C       SS  LIES BETWEEN  0  (R= INF)                           FLOW 247
C                    AND  1  (R=-1.)                            FLOW 248
C       S1   MOST RECENT ESTIMATE OF   SS                       FLOW 249
C       S2   PREVIOUS   SS                                      FLOW 250
C       F1   NPW FOR S1                                         FLOW 251
C       F2   NPW FOR S2                                         FLOW 252
C       SU   UPPER LIMIT FOR  SS  (NPW+VE)                      FLOW 253
C       SL   LOWER LIMIT FOR  SS  (NPW-VE)                      FLOW 254
C       JUMP  MARKS WHERE TO GO TO AFTER NPW CALCULATED         FLOW 255
C**                                                             FLOW 256
C                                                               FLOW 257
  290  IF(JUMP)700,700,719                                       FLOW 258
  700  S1=1.0/(2.0+DR)                                           FLOW 259
       P1=PW                                                     FLOW 260
       KT=2                                                      FLOW 261
       IF(PW)701,280,704                                         FLOW 262
  701  SL=S1                                                     FLOW 263
       SS=1.0/(1.9 + DR)                                         FLOW 264
       JUMP=1                                                    FLOW 265
       GOTO 713                                                  FLOW 266
  702  IF(PW)703,707,707                                         FLOW 267
  703  SL=S1                                                     FLOW 268
       SS=SS*0.8+0.2                                             FLOW 269
       GOTO 718                                                  FLOW 270
  704  SU=S1                                                     FLOW 271
       SS=1.0/(2.1 + DR)                                         FLOW 272
       JUMP=2                                                    FLOW 273
       GOTO 713                                                  FLOW 274
```

```
705     IF(PW)708,718,706                                            FLOW 275
706     SU=S1                                                        FLOW 276
        SS=2.0*SS                                                    FLOW 277
        GOTO 718                                                     FLOW 278
707     SU=S1                                                        FLOW 279
        GOTO 709                                                     FLOW 280
708     SL=S1                                                        FLOW 281
709     DEN=F2-P1                                                    FLOW 282
        IF(DEN)710,712,710                                           FLOW 283
710     SS=(S1*P2-S2*P1)/DEN                                         FLOW 284
        IF(SS-SU)711,711,712                                         FLOW 285
711     IF(SL-SS)713,713,712                                         FLOW 286
712     SS=0.5*(SL+SU)                                               FLOW 287
713     IF(KMC.GT.0)                                                 FLOW 288
       *IF(ABS(SS-S1)-0.0001)280,714,714                             FLOW 289
        IF(ABS(SS-S1)-.000001)280,714,714                            FLOW 290
714     JUMP=3                                                       FLOW 291
        GOTO 718                                                     FLOW 292
715     IF(P1)716,280,717                                            FLOW 293
716     SL=SS                                                        FLOW 294
        GOTO 709                                                     FLOW 295
717     SU=S1                                                        FLOW 296
        GOTO 709                                                     FLOW 297
718     DR=1.0/SS-2.0                                                FLOW 298
        GOTO 210                                                     FLOW 299
719     S2=S1                                                        FLOW 300
        P2=P1                                                        FLOW 301
        S1=SS                                                        FLOW 302
        P1=PW                                                        FLOW 303
        KT=KT+1                                                      FLOW 304
        IF(KT-20)720,720,380                                         FLOW 305
720     GOTO(702,705,715),JUMP                                       FLOW 306
  380   WRITE(IOUT,390)                                              FLOW 307
  390   FORMAT(34H NO CONVERGENCE IN 20 ITERATIONS        )          FLOW 308
C   DCF COMPLETED,EVALUATE YEARLY CASH -FLOW (PUND),AND PAYBACK (PKB) FLOW 309
280     IF(KMC.LT.1.OR.JT.EQ.1)  GO TO 281                           FLOW 310
        DO 279 N=1,NT                                                FLOW 311
        TONS(1,N)= PTONS(N)                                          FLOW 312
        WORK(1,N)= PWORK (N)                                         FLOW 313
279     SEL(1,N)= PSEL (N)                                           FLOW 314
281     CONTINUE                                                     FLOW 315
        IF(KMC.GT.0)  GO TO 545                                      FLOW 316
        IF(KS.NE.0)GOTO 409                                          FLOW 317
        WRITE(IOUT,401)                                              FLOW 318
401     FORMAT(//54H  CUMULATIVE UNDISCOUNTED CASH FLOW  (CURRENT MONEY) FLOW 319
       *          ,                        //45H        YEAR         FLOW 320
       * PRCE1       PROB2       FROB3          //)                  FLOW 321
        DO 403  I=1,NT                                               FLOW 322
403     WRITE(IOUT,402)  I ,(FUND(J,I),J=1,JT)                       FLOW 323
402     FORMAT(I10,10F10.2)                                          FLOW 324
C   LOCATION OF PAY BACK YEAR                                        FLOW 325
409     PKB=0.                                                       FLOW 326
        Y=-CAFL(1)*S(3)                                              FLOW 327
        DO 400 N=1,NT                                                FLOW 328
        X=0.                                                         FLOW 329
        DO 410 J=1,JT                                                FLOW 330
  410   X=X+FUND(J,N)*PROB(J)                                        FLOW 331
        IF(X)430,430,420                                             FLOW 332
420     PKB= FLOAT(N) - X/(X-Y +0.00001)                            FLOW 333
        GO TO 540                                                    FLOW 334
  430   Y=X                                                          FLOW 335
  400   CONTINUE                                                     FLOW 336
        WRITE(IOUT,404)                                              FLOW 337
404     FORMAT(//36H PAYBACK GREATER THAN PROJECT LIFE      )        FLOW 338
540     DIS=100.*DR                                                  FLOW 339
        IF(KS.GT.0)  GO TO 545                                       FLOW 340
        IF(KMC.EQ.0)                                                 FLOW 341
       *WRITE(IOUT,620)  PW,DIS                                      FLOW 342
        IF(PKB.GT.0.001)WRITE(IOUT,630)PKB                           FLOW 343
```

```
C    COLLECT MEAN DISCOUNTED CAPITAL IN X                          FLOW 344
     X=0.0                                                          FLOW 345
     DO 610 J=1,JT                                                  FLOW 346
610  X=X + BCCAPL(J)*PROB(J)                                        FLOW 347
     IF(K.EQ.0)                                                     FLOW 348
     *CALL EAVBC (X,NT,DR,PW)                                       FLOW 349
     WRITE(IOUT,615)  X                                             FLOW 350
615  FORMAT(      /39H MEAN DISCOUNTED CAPITAL INVESTMENT IS     ,F9.1/ )FLOW 351
620  FORMAT(/33H             MEAN NEW IS      ,F10.1/ 33H           FLOW 352
     1       FOR DISCOUNT RATE       ,F6.2  )                       FLOW 353
630  FORMAT(33H           MEAN PAYBACK YEAR IS       ,F6.2)         FLOW 354
     IF(IU.EQ.1.AND.KMC.EQ.0.AND.JT.GT.1)                           FLOW 355
     *CALL UTIL   ( JT,JT,P,PROB,CERT)                              FLOW 356
C    ------------                                                   FLOW 357
     IF(IU.EQ.1.AND.KMC.EQ.0.AND.JT.GT.1)                           FLOW 358
     *WRITE(IOUT,563) CERT                                          FLOW 359
563  FORMAT(  33H   CERTAINTY EQUIVALENT EQUALS      ,F10.2    )     FLOW 360
     IF(KMC.GT.-1)  WRITE(IOUT,119)                                 FLOW 361
119  FORMAT(///50X,  20H DISTRIBUTION OF NPW  /,/9H     CASE,15X,52H FLOW 362
     1 EXP NPW   DIS    FBK      1      2      3   //)               FLOW 363
545  RETURN                                                         FLOW 364
     END                                                            FLOW 365


     SUBROUTINE EAVBC (X,NT,DR,PW)                                  EAVB  1
C EVALUATION OF EQUIVALENT ANNUAL VALUE AND BENIFITS/COST RATIO,    EAVB  2
C   GIVEN NEW (PW),MEAN CAPITAL (X),PROJECT LENGTH (NT)             EAVB  3
C   DR=DISCOUNT RATE,    PRINTS RESULT, NOTHING RETURNED.           EAVB  4
     COMMON/RWR/IOUT                                                EAVB  5
     DF=1.                                                          EAVB  6
     EAV=0.                                                         EAVB  7
     DO 1 I=1,NT                                                    EAVB  8
     DF=DF/(1.+DR)                                                  EAVB  9
     EAV=EAV+DF                                                     EAVB  10
1    CONTINUE                                                       EAVB  11
     EAV=PW/EAV                                                     EAVB  12
     BCR=0.0                                                        EAVB  13
     IF(X.GT.0.0000001)   BCR=PW/X                                  EAVB  14
     WRITE(IOUT,2)EAV,BCR                                           EAVB  15
2    FORMAT( 33H   EQUIVALENT ANNUAL VALUE IS    ,F10.2,            EAVB  16
     *    26H   BENEFITS COST RATIO IS   ,F 8.2  )                  EAVB  17
     RETURN                                                         EAVB  18
     END                                                            EAVB  19


     SUBROUTINE MONTE                                               MONT  1
C*************************                                          MONT  2
C  GENERATES A SET OF DATA FOR MONTE-CARLO ANALYSIS,TO BE EVALUATED BY MONT  3
C  FLOW                                                             MONT  4
C  NLEV GIVES THE NUMBER OF PROBABILITY LEVELS FOR THE RESULTS      MONT  5
     DIMENSION X(5),RES(500),PROBLV(10)                            MONT  6
     COMMON/TRAPS/PCAPY(10,4),PCAPL(10,4),PFOP(10,4),PVOP(10,4),   MONT  7
     *PSCRAP(10,4),PFINE(4)          ,PENF(4),MC                    MONT  8
     COMMON/BLO1/PW,K                                               MONT  9
     COMMON/RDAT1/FINE,DR,UVOP(10),NT,NUT,MD,TX ,ENF                MONT  10
     COMMON/RDAT2/UFOP(10),UCAPY(10),UCAPL(10),USCRAP(10)          MONT  11
     COMMON/UTIL/IU                                                 MONT  12
     COMMON/SIM/CORT,CORS,KMC,ERR,IYEAR,JMC                         MONT  13
     COMMON/RWR/IOUT                                                MONT  14
     COMMON/CONT/LN,ICHEK                                           MONT  15
     COMMON/NAMES/ISN(24),INA(10,3)                                MONT  16
     COMMON/SALES/ PROB(8)                                         MONT  17
     COMMON/TEST/LEVD(8),LEVS(8)                                   MONT  18
     DATA                                                           MONT  19
     *PROBLV(1), PROBLV(2), PROBLV(3), PROBLV(4), PROBLV(5),       MONT  20
     *PROBLV(6), PROBLV(7), PROBLV(8), PROBLV(9), PROBLV(10)       MONT  21
     *          /.025,.05,.1,.2,.50,.80,.90,.95,.975,0./          MONT  22
     NLEV=9                                                        MONT  23
```

```
      DO 1  I=1,530                                          MONT  24
1     RES(I)=FLOAT( LW)                                      MONT  25
      RAN=1137.                                              MONT  26
C  RANSET(X) IS A LIBRARY SUBROUTINE WHICH INITIALISES THE 0 TO 1 RANDOM NO
C  GENERATOR (RANF).BY INCLUDING THIS INITIALISER REPRODUCABLE RANDOMSERIES
C  ARE PRODUCED.
      CALL RANSET(RAN)                                       MONT  27
      IYEAR=1                                                MONT  28
C  INITIAL RUNS DONE WITH A CENTRE LEVEL                     MONT  29
      IF (JT.GT.1)  IYEAR =2                                 MONT  30
      JMC=0                                                  MONT  31
10    JMC=JMC + 1                                            MONT  32
C  OBTAIN DATA FOR ALTERNATIVES FROM TRAPZ SUBROUTINE        MONT  33
      DO 30 N=1,NUT                                          MONT  34
      DO 20 I=1,4                                            MONT  35
20    X(I)=FCAPY(N,I)                                        MONT  36
      CALL TRAPZ(X)                                          MONT  37
      UCAPY(N)=X(5)                                          MONT  38
      DO 21 I=1,4                                            MONT  39
21    X(I)=FCAPL(N,I)                                        MONT  40
      CALL TRAPZ(X)                                          MONT  41
      UCAPL(N)=X(5)                                          MONT  42
      DO 22 I=1,4                                            MONT  43
22    X(I)=FFOP(N,I)                                         MONT  44
      CALL TRAPZ(X)                                          MONT  45
      UFOP(N)=X(5)                                           MONT  46
      DO 23 I=1,4                                            MONT  47
23    X(I)=FVOP(N,I)                                         MONT  48
      CALL TRAPZ(X)                                          MONT  49
      UVOP(N)=X(5)                                           MONT  50
      DO 24 I=1,4                                            MONT  51
24    X(I)=FSCRAP(N,I)                                       MONT  52
      CALL TRAPZ(X)                                          MONT  53
      USCRAP(N)=X(5)                                         MONT  54
30    CONTINUE                                               MONT  55
C  DATA FOR INFLATION                                        MONT  56
      DO 31 I=1,4                                            MONT  57
31    X(I)=FENF(I)                                           MONT  58
      CALL TRAPZ(X)                                          MONT  59
      ENF=X(5)                                               MONT  60
C  OBTAIN DATA FOR PENALTY                                   MONT  61
      DO 32 I=1,4                                            MONT  62
32    X(I)=FFINE(I)                                          MONT  63
      CALL TRAPZ(X)                                          MONT  64
      FINE=X(5)                                              MONT  65
C  EVALUATE WITH THIS SET OF DATA  .                         MONT  66
      CALL FLOW                                              MONT  67
C  STORE RESULTS OF FLOW( K=1-DCF, K=0-NPW                   MONT  68
      IF(K.EQ.1)   RES(JMC)= 100.*DR                         MONT  69
      IF( K.EQ.0)  RES(JMC) = PW                             MONT  70
      IF(JMC.EQ.500)  GO TO 106                              MONT  71
C  LIMIT MC CYCLES FOR TESTING                               MONT  72
      IF(ICHEK. EQ. 2.AND. JMC. EQ.10)  GO TO 106            MONT  73
      IF(JMC.LT.KMC) GO TO 10                                MONT  74
      IF(KMC.GT.10)  GO TO  104                              MONT  75
      R=0.                                                   MONT  76
      A=0.                                                   MONT  77
C  ROUGH ESTIMATE OF VARIANCE (IN A) TO EVALUATE NO OF ITERATIONS REQD. MONT  78
      DO 102 I=1,JMC                                         MONT  79
      R=R+RES(I)                                             MONT  80
102   A=A+RES(I)*RES(I)                                      MONT  81
      R=(A- R*R/FLOAT(JMC))/FLOAT(JMC)    +0.00000001        MONT  82
      RO=R                                                   MONT  83
      IF(ERR.LT.0.00001)  GO TO 103                          MONT  84
      R=(1.96*SQRT(R) /ERR)                                  MONT  85
      R=R*R                                                  MONT  86
      KMC=IFIX(R)                                            MONT  87
      IF(KMC.LT.50)  KMC=50                                  MONT  88
      IF(KMC.GT.500)   KMC=500                               MONT  89
```

```
        GO TC 101                                               MONT    90
C  IF ERROR NOT SPECIFIED SET 100 ITERATIONS                    MONT    91
103    KMC=100                                                  MONT    92
C   STRATIFICATION OF PROBABILITY SELECTION TO IMPROVE MC EFFICIENCY  MONT  93
101    IF (JT.EQ.1) GO TO 10                                    MONT    94
       RPROB=PROB(2)                                            MONT    95
       IF (RPROB.LT.0.022) GO TO 90                             MONT    96
       KMC=IFIX(FLOAT(KMC)*RPROB)                               MONT    97
       IF(JMC.GE.KMC)  KMC=JMC+1                                MONT    98
       GO TO 10                                                 MONT    99
90     WRITE(IOUT,91)                                           MONT   100
       RETURN                                                   MONT   101
91     FORMAT( 64H SIMULATION CANNOT PROCEED,SECOND PROBABILITY LEVEL LG.MONT 102
      *022         )                                            MONT   103
104    IF(IYEAR.EQ.1 )  GO TO 106                               MONT   104
       IF (IYEAR.EQ.JT) IYEAR =1                                MONT   105
       IF(IYEAR.LT.JT.AND.IYEAR.GT.1)                           MONT   106
      *IYEAR=IYEAR+1                                            MONT   107
       KMC=IFIX(FLOAT(KMC)/RPROB)                               MONT   108
       RPROB=RPROB+PROB(IYEAR)                                  MONT   109
       KMC=IFIX(FLOAT(KMC)*RPROB)                               MONT   110
       GO TC 10                                                 MONT   111
C   ITERATIONS COMPLETED,-ANALYSE AND DISPLAY RESULTS.          MONT   112
C   ORDER RESULTS IN INCREASING VALUE    RES(1)-LOW             MONT   113
106    JJJ=JMC-1                                                MONT   114
       KMC=JMC                                                  MONT   115
105    KO=0                                                     MONT   116
       DO 110 J=1,JJJ                                           MONT   117
       IF(RES(J).GT.RES(J+1))    GO TO 107                      MONT   118
       GO TO 110                                                MONT   119
107    R=RES(J+1)                                               MONT   120
       RES(J+1)=RES(J)                                          MONT   121
       RES(J)=R                                                 MONT   122
       KO=1                                                     MONT   123
110    CONTINUE                                                 MONT   124
       IF(KC.EQ.1)  GO TO 105                                   MONT   125
       IF(ICHEK.EQ.0)GOTO 85                                    MONT   126
C   RANF(0.0) IS A LIBRARY FUNCTION TO PROVIDE THE NEXT RANDOM NO. IN THE SERIES
       Z=RANF(0.0)                                              MONT   127
       WRITE(IOUT,112)  KMC,Z                                   MONT   128
112    FORMAT(//I9,26H EVALUATIONS CARRIED OUT     /,           MONT   129
      *  26H NEXT RANDOM NUMBER IS        ,F7.5//)              MONT   130
C   DATA NOW AVAILABLE FOR PLOTTING AS   RES(J),J/KMC           MONT   131
       WRITE(IOUT,119)                                          MONT   132
119    FORMAT(////// 30H SUMMARY OF ALL SIMULATIONS        /)   MONT   133
       WRITE(IOUT,81) (RES(I), I=1,KMC)                         MONT   134
81     FORMAT(1X,7F10.2)                                        MONT   135
       LD=0                                                     MONT   136
       LS=0                                                     MONT   137
       DO 111 I=1,JT                                            MONT   138
       LD=LD+LEVD(I)                                            MONT   139
111    LS=LS+LEVS(I)                                            MONT   140
       DO 113 I=1,JT                                            MONT   141
       LEVD(I)=(100*LEVD(I))/LD                                 MONT   142
113    LEVS(I)=(100*LEVS(I))/LS                                 MONT   143
       WRITE(IOUT,114) ( LEVD(I),I=1,JT          )              MONT   144
114    FORMAT(//54H PERCENT EACH DEMAND PROBABILITY LEVEL SIMULATED:  MONT  145
      */8I8/)                                                   MONT   146
       WRITE(IOUT,115)(LEVS(I),I=1,JT)                          MONT   147
115    FORMAT(32H AND PRICE PROBABILITY LEVEL:          /8I8//) MONT   148
85     WRITE(IOUT,6)   (ISW(I),I=1,24)                          MONT   149
6      FORMAT(1H1,24A3)                                         MONT   150
       CALL DIAGR(RES,KMC,K)                                    MONT   151
C    ------------------                                         MONT   152
C   COLLECT AND TABULATE RESULTS FOR VARIOUS CONFIDENCE LEVELS. MONT   153
       WRITE(IOUT,125)                                          MONT   154
125    FORMAT(/73H TABULAR SUMMARY: PERCENT PROBY.THAT RESULT WILL BE LESMONT 155
      1S THAN 'VALUE'                   )                       MONT   156
```

```
      DO 120 I=1,NLEV                                              MONT 157
      IR  =IFIX(FLOAT(KMC)*PROBLV(I)-0.499) + 1                    MONT 158
      RESPRO=RES(IR)                                               MONT 159
      PRO=10).*FLOAT(IR    )/FLOAT(KMC)                            MONT 160
120   WRITE(IOUT,130)PRO,RESPRO                                    MONT 161
130   FORMAT(22X,F9.2,9X,F10.2)                                    MONT 162
C   AND NOW THE EXPECTED VALUE AND STANDARD DEVIATION.             MONT 163
      R=0.                                                         MONT 164
      RR=0.                                                        MONT 165
      A=0.                                                         MONT 166
      DO 135 I=1,KMC                                               MONT 167
      RR=RR+RES(I)*RES(I)                                          MONT 168
135   R=R+RES(I)                                                   MONT 169
      RR=(RR-R*R/FLOAT(KMC))/FLOAT(KMC-1)                          MONT 170
      R=R/FLOAT(KMC)                                               MONT 171
      IF(RC.GT.0.)                                                 MONT 172
     *A=SQRT(3.8416*RO/FLOAT(KMC))                                 MONT 173
      WRITE(IOUT,140) R,A                                          MONT 174
140   FORMAT(/ 15X ,18HEXPECTED VALUE IS     ,F10.2,5H   +-,F10.1) MONT 175
      RR=SQRT(RR)                                                  MONT 176
      WRITE(IOUT,143)RR                                            MONT 177
143   FORMAT( 11X, 22HSTANDARD DEVIATION IS       ,F10.2)          MONT 178
      IF(IU.EQ.0)  GO TO 150                                       MONT 179
      DO 142 I=1,KMC                                               MONT 180
C  EVALUATION OF THE CERTY EQUIVALENT OF THE DISTRIBUTION          MONT 181
142   CALL UTIL    (KMC,1,RES,R,CERT)                              MONT 182
      WRITE(IOUT,145)  CERT                                        MONT 183
145   FORMAT( 33H  CERTAINTY EQUIVALENT EQUALS      ,F10.2   //)   MONT 184
150   CONTINUE                                                     MONT 185
      KMC=-1                                                       MONT 186
      WRITE(IOUT,6)  (ISW(I),I=1,24)                               MONT 187
      WRITE(IOUT,160)                                              MONT 188
160   FORMAT(  64H FINAL RUN THROUGH EVALUATION TO PRODUCE CASH-FLOW,EAVMONT 189
     1 AND E/C        /46H     USING FIXED PLANS AND MEAN DATA     )MONT 190
C  RETURN TO FLOW FOR MEAN PAYBACK,P/C,ETC.-KMC=-1 APPROPRIATE FLAG.MONT 191
      CALL MEAN(PENF(1),PENF(2),PENF(3),PENF(4),ENF)               MONT 192
      CALL MEAN(PFINE(1),PFINE(2),PFINE(3),PFINE(4),FINE)          MONT 193
      DO 591 J=1,NUT                                               MONT 194
      CALL  MEAN (PCAPY(J,1),PCAPY(J,2),PCAPY(J,3),PCAPY(J,4), UCAPY(J))MONT 195
      CALL  MEAN(PCAPL(J,1),PCAPL(J,2),PCAPL(J,3),PCAPL(J,4),UCAPL(J)) MONT 196
      CALL  MEAN(PFOP(J,1),FFOP(J,2),PFOP(J,3),PFOP(J,4),JFOP(J))  MONT 197
      CALL  MEAN(PVOP(J,1),FVOP(J,2),PVOP(J,3),PVOP(J,4),UVOP(J))  MONT 198
591   CALL  MEAN (PSCRAP(J,1),PSCRAP(J,2), PSCRAP(J,3), PSCRAP(J,4), USCRMONT 199
     *RAP(J))                                                      MONT 200
      CALL FLOW                                                    MONT 201
      KMC=10                                                       MONT 202
C  END OF MONTE-CARLO ANALYSIS.                                    MONT 203
      RETURN                                                       MONT 204
      END                                                          MONT 205


      SUBROUTINE DIAGR(RES,KMC,K)                                  DIAG   1
C***********************************                               DIAG   2
C  THIS SUBROUTINE PRINTS THE CUMULATIVE DISTRIBUTION PLOT FOR THE MC DIAG 3
      DIMENSION RES(KMC), IRE(62), SCALE(6)                        DIAG   4
      COMMON/RWR/IOUT,IR                                           DIAG   5
      DATA IH,JH,KH/ 1H ,1H+,1HS  /                                DIAG   6
      ABST=RES(KMC)-RES(1)                                         DIAG   7
      WRITE(IOUT,150)                                              DIAG   8
150   FORMAT(/46H CUMULATIVE DISTRIBUTION    GRAPHICAL SUMMARY    )DIAG   9
      SCA=ABST/5.                                                  DIAG  10
      SCALE(1)=RES(1)                                              DIAG  11
      DO 210 I=2,6                                                 DIAG  12
210   SCALE(I)=SCALE(I-1)+SCA                                      DIAG  13
      IF(K.GT.0)GO TO 265                                          DIAG  14
      DO 160 I =1,6                                                DIAG  15
160   SCALE(I)=FLOAT( 10*IFIX(SCALE(I)/ 10.))                      DIAG  16
265   WRITE (IOUT,250) SCALE                                       DIAG  17
250   FORMAT(/ 1X,F8.1, 5F12.1)                                   DIAG  18
```

```
260      FORMAT(6X,20II,1)(6H-----1),2H-I)              DIAG  19
         WRITE(IOUT,260)                                DIAG  20
         IC=2                                           DIAG  21
         ICC=2                                          DIAG  22
         WRITE(IOUT,272)                                DIAG  23
272      FORMAT(2X,5H110 I,60X,3H: I)                   DIAG  24
         STEP=ABST/60.                                  DIAG  25
         NTIM=95                                        DIAG  26
         WRITE(IOUT,310)                                DIAG  27
         DO 300 I=1,19                                  DIAG  28
         DO 280 J=1,62                                  DIAG  29
280      IRE(J)=IH                                      DIAG  30
         ICC=ICC+1                                      DIAG  31
         GOTO(282,282,282,277),ICC                      DIAG  32
277      DC 220 J=13,60,12                              DIAG  33
220      IRE(J)=JH                                      DIAG  34
         ICC=0                                          DIAG  35
282      IB=(NTIM*KMO/100)                              DIAG  35
         IF(IB.LE.0)IB=1                                DIAG  36
         IPOSI=1+IFIX((RES(IB)-RES(1))/STEP+0.5)        DIAG  37
         IRE(IPOSI)=KH                                  DIAG  38
         GOTO(285,290),IC                               DIAG  39
290      WRITE(IOUT,180)IRE                             DIAG  40
         IC=1                                           DIAG  41
         GOTO 295                                       DIAG  42
285      WRITE(IOUT,155)NTIM,IRE                        DIAG  43
155      FORMAT(1X,I4,2H I,62A1,1HI)                    DIAG  44
180      FORMAT(6X,1HI,62A1,1HI)                        DIAG  45
         IC=2                                           DIAG  46
295      WRITE(IOUT,310)                                DIAG  47
         NTIM=NTIM-5                                     DIAG  48
300      CONTINUE                                       DIAG  49
310      FORMAT(6X,1HI,62X,1HI)                         DIAG  50
         WRITE(IOUT,375)                                DIAG  51
375      FORMAT(4X,4H0 I:,61X,1HI)                      DIAG  52
         WRITE(IOUT,260)                                DIAG  53
         WRITE(IOUT,250) SCALE                          DIAG  54
         RETURN                                         DIAG  55
         END                                            DIAG  56
                                                        DIAG  57


      SUBROUTINE TRAPZ(X)                               TRAP  1
C*********************                                  TRAP  2
C  RANDOMLY CHOOSE A VALUE (IN X(5)) FROM A DISTRIBUTION DEFINED BY THE  TRAP  3
C  4 POINTS OF A TRAPIZIODAL DISTRIBUTION( X(1)- X(4))  TRAP  4
      DIMENSION X(5), CP(4)                             TRAP  5
      IF((X(4)-X(1)).LT.0.0001)  GO TO 50               TRAP  6
      Y=0.5*(X(3)+X(4)-X(1)-X(2))                       TRAP  7
      CP(2)=0.5*(X(2)-X(1))/Y                           TRAP  8
C  RANF(0.0) IS A LIBRARY FUNCTION TO PROVIDE THE NEXT RANDOM NO. IN THE SERIES
      Z=RANF( 0.)                                       TRAP  9
C  3 SECTIONS TO CUMULATIVE   DISTRIBUTION              TRAP  10
      IF(Z.LT.CP(2))   GO TO 10                         TRAP  11
      CP(3)=CP(2)+  (X(3)-X(2))/Y                       TRAP  12
      IF(Z.LT.CP(3))   GO TO 20                         TRAP  13
C  EQUATION FOR THE LAST SECTION                        TRAP  14
      X(5)=X(4)-(2.*(1.-Z)*Y*(X(4)-X(3)))**.5           TRAP  15
      GO TO 40                                          TRAP  16
C  EQUATION FOR FIRST SECTION                           TRAP  17
10    X(5)=X(1) + (2.*Z*Y*(X(2)-X(1)))**0.5             TRAP  18
      GO TO 40                                          TRAP  19
C  EQUATION FOR MIDDLE SECTION                          TRAP  20
20    X(5)=X(2)+Y*(Z-CP(2))                             TRAP  21
40    CONTINUE                                          TRAP  22
      RETURN                                            TRAP  23
50    X(5)=X(1)                                         TRAP  24
      GO TO 40                                          TRAP  25
      END                                               TRAP  26
```

```
      SUBROUTINE HIST(JL,JJ,PROB,CORT,CORS,JNC,N,JT)            HIST    1
C    SUBROUTINE TO DISTORT ORIGIONAL PROBABILITIES BY THE CORRELATION   HIST    2
C    COEFFICIENT AND THEN RANDOMLY CHOOSE THE NEXT LEVEL.       HIST    3
C    NEXT LEVELS RETURNED IN JL AND JJ.                         HIST    4
      DIMENSION PROB(8)                                         HIST    5
      COMMON/HIS/CORDE(8,8),CORSA(8,8)                          HIST    6
      COMMON/RHR/IOUT                                           HIST    7
      COMMON/CONT/LN,ICHEK                                      HIST    8
      IF(JNC.GT.1.OR.N.GT.1)GOTO 17                             HIST    9
      DO 5 I=1,JT                                               HIST   10
      DO 5 J=1,JT                                               HIST   11
      CORDE(I,J)=0.0                                            HIST   12
      CORSA(I,J)=0.0                                            HIST   13
  5   CONTINUE                                                  HIST   14
C                                                               HIST   15
C  FIRST GENERATE *RANDOM*  STEPPING SEQUENCE                   HIST   16
C                                                               HIST   17
C   DEMAND ONLY ALLOWED TO JUMP ONE LEVEL                       HIST   18
      IF(JT-2)42,41,40                                          HIST   19
 40   DO 35 I=2,JT                                              HIST   20
      IF(PROB(I-1)*3.0-PROB(I))31,31,32                         HIST   21
 31   CORDE(I-1,I)=0.5                                          HIST   22
      CORDE(I,I-1)=PROB(I-1)*0.5/PROB(I)                        HIST   23
      GOTO 35                                                   HIST   24
 32   IF(PROB(I-1)-3.0*PROB(I))34,33,33                         HIST   25
 33   CORDE(I,I-1)=0.5                                          HIST   26
      CORDE(I-1,I)=PROB(I)*0.5/PROB(I-1)                        HIST   27
      GOTO 35                                                   HIST   28
 34   CORDE(I,I-1)=PROB(I-1)/((PROB(I-1)+PROB(I))*1.5)          HIST   29
      CORDE(I-1,I)=PROB(I)/((PROB(I-1)+PROB(I))*1.5)            HIST   30
 35   CONTINUE                                                  HIST   31
      LOOP=JT-1                                                 HIST   32
      DO 36 I=2,LOOP                                            HIST   33
      CORDE(I,I)=1.0-CORDE(I,I-1)-CORDE(I,I+1)                  HIST   34
 36   CONTINUE                                                  HIST   35
      CORDE(1,1)=1.0-CORDE(1,2)                                 HIST   36
      CORDE(JT,JT)=1.0-CORDE(JT,JT-1)                           HIST   37
      GOTO 43                                                   HIST   38
 41   CORDE(1,1)=PROB(1)                                        HIST   39
      CORDE(2,1)=PROB(1)                                        HIST   40
      CORDE(1,2)=PROB(2)                                        HIST   41
      CORDE(2,2)=PROB(2)                                        HIST   42
      GOTO 43                                                   HIST   43
 42   CORDE(1,1)=1.0                                            HIST   44
C   PRICE STEPPING SEQUENCE, PRICE ALLOWED TO JUMP TO ANY LEVEL HIST   45
 43   DO 44 I=1,JT                                              HIST   46
      DO 44 J=1,JT                                              HIST   47
      CORSA(I,J)=PROB(J)                                        HIST   48
 44   CONTINUE                                                  HIST   49
C  NOW  GET COMBINED MATRIX FOR CORRELATION LEVELS  CORT AND CORS HIST  50
      ORT=1.0-CORT                                              HIST   51
      ORS=1.0-CORS                                              HIST   52
      DO 46 I=1,JT                                              HIST   53
      DO 46 J=1,JT                                              HIST   54
      CORSA(I,J)=CORSA(I,J)*ORS                                 HIST   55
      CORDE(I,J)=CORDE(I,J)*ORT                                 HIST   56
 46   CONTINUE                                                  HIST   57
      DO 47 I=1,JT                                              HIST   58
      CORDE(I,I)=CORDE(I,I)+CORT                                HIST   59
      CORSA(I,I)=CORSA(I,I)+CORS                                HIST   60
 47   CONTINUE                                                  HIST   61
      IF(ICHEK.EQ.0)  GO TO 17                                  HIST   62
      WRITE(IOUT,102)                                           HIST   63
 102  FORMAT(/47H THE DERIVED PROBABILITY MATRIX FOR DEMAND IS  /)HIST  64
      DO 8 J=1, JT                                              HIST   65
 8    WRITE(IOUT,100) (CORDE(J,K),K=1,JT)                       HIST   66
 100  FORMAT(1X,8F10.6)                                         HIST   67
```

```
         WRITE(IOUT,101)                                              HIST  68
101      FORMAT(                //15H PRICE MATRIX              /)     HIST  69
103      FORMAT(/ 70H (P(ROW,COLUMN) GIVES THE PROBABILITY OF MOVING FROM HIST  70
        1ROW TO COLUMN)           //)                                HIST  71
         DO 9 J=1,JT                                                 HIST  72
9        WRITE(IOUT,100) (CORSA(J,K),K=1,JT                          HIST  73
         WRITE(IOUT,103)                                             HIST  74
17       IFLAG=0                                                     HIST  75
C  RANF(0.0) IS A LIBRARY FUNCTION TO PROVIDE THE NEXT RANDOM NO. IN THE SERIES
18       Z=RANF(0.0)                                                 HIST  76
         J=1                                                         HIST  77
         X=0.0                                                       HIST  78
10       IF(IFLAG.EQ.0) X=X+CORDE(JL,J)                              HIST  79
         IF(IFLAG.EQ.1) X=X+CORSA(JL,J)                              HIST  80
         IF(Z.LT.X) GO TO 20                                         HIST  81
         J=J+1                                                       HIST  82
         GOTO 10                                                     HIST  83
20       IF(IFLAG.EQ.1)    GO TO 30                                  HIST  84
         JL=J                                                        HIST  85
         IFLAG=1                                                     HIST  86
         GOTO 18                                                     HIST  87
30       JJ=J                                                        HIST  88
         RETURN                                                      HIST  89
         END                                                         HIST  90


         SUBROUTINE TAXD(DEP,N,ND,NT,CAPL,S,MIEND,ENF,KTAX)          TAXD   1
C*********************                                               TAXD   2
C                                                                    TAXD   3
C           DEP  =RETURNED DEPRECIATION FOR YEAR N                   TAXD   4
C           N    = YEARS OF PROJECT                                  TAXD   5
C           ND   = YEARS OF STRAIGHT LINE DEPRECIATION               TAXD   6
C           NT   = LIFE OF PROJECT                                   TAXD   7
C           CAPL(20)*S(3)  : CAPITAL BY YEAR                         TAXD   8
C           MIEND= YEARS EXTENDED LIFE                               TAXD   9
C           ENF  = INFLATION RATE                                    TAXD  10
C           KTAX = TAX CALCULATION FLAG  = 0 FOR STRAIGHT LINE DEPR. TAXD  11
C                                                                    TAXD  12
C---YEARLY DEPRECIATION ALLOWANCE FOR CASH FLOW PROFILE AND PAYBACK  TAXD  13
C---STRAIGHT LINE DEPRECIATION FOR  ND  YEARS                        TAXD  14
C---ANY REMAINING AT END LUMPED IN   LAST YEAR ALLOWANCE            TAXD  15
C    ENF=INFLATION RATE,DEPRECIATION GIVEN IN CONSTANT VALUE MONEY   TAXD  16
C    TERMS,BASE YEAR 0                                               TAXD  17
         DIMENSION CAPL(20),S(5)                                     TAXD  18
         DEP=0.                                                      TAXD  19
         JJJ=N-ND                                                    TAXD  20
         IF(JJJ)101,102,102                                          TAXD  21
101      JJJ=0                                                       TAXD  22
102      JJJ=JJJ+1                                                   TAXD  23
         DO 90 J=JJJ,N                                               TAXD  24
C  IF LAST YEAR + SCRAP,ALLOW REMAINING DEPRECIATION                 TAXD  25
         IREMDE  =1                                                  TAXD  26
         IF(N.EQ.NT.AND.MIEND.EQ.0)  IREMDE = ND-(NT-JJ)            TAXD  27
         DEP= DEP + (CAPL(JJ)*S(3)/FLOAT(ND))*FLOAT(IREMDE)          TAXD  28
90       CONTINUE                                                    TAXD  29
         DEP =DEP /((1.+ENF)**N)                                     TAXD  30
         RETURN                                                      TAXD  31
         END                                                         TAXD  32


         SUBROUTINE TAXF(F,DR,ND,TX,ENF,KTAX)                        TAXF   1
C*********************                                               TAXF   2
C---SUBROUTINE TO CALCULATE THE TAX FACTOR FOR DISCOUNTING           TAXF   3
C---TX = TAX RATE (FRACTIONAL)                                       TAXF   4
C---ENF= INFLATION RATE (FRACTIONAL)                                 TAXF   5
C---KTAX= CALC. FLAG, FOR ALTERNATIVES TO 10 YR STRAIGHT LINE        TAXF   6
C---F = FACTOR RETURNED                                              TAXF   7
C---DR = DISCOUNT FACTOR                                             TAXF   8
```

```
C---ND = YEARS OF DEPRECIATION                                        TAXF   9
C---CREDIT TAX ALLOWANCE TO CAPITAL                                   TAXF  10
C---FIRST TAX ALLOWANCE RECIEVED 2 YEARS AFTER CAPITAL SPENT          TAXF  11
C---HENCE  ( 1+JJ)                                                    TAXF  12
C  ENF=INFLATION RATE,EVALUATION IN CONSTANT VALUE MONEY-BASE YEAR 0  TAXF  13
       F=0.                                                           TAXF  14
       FF=1.0/((1.+DR)*(1.+ENF))                                      TAXF  15
       DO 46  JJ=1,ND                                                 TAXF  16
       F=FF**(1.+FLOAT(JJ))+F                                         TAXF  17
46     CONTINUE                                                       TAXF  18
       F=1.0-TX*F/FLOAT(ND)                                           TAXF  19
C  CONVERT TO CONSTANT MONEY BASE YEAR 0                              TAXF  20
C  CAPITAL UNITS CURRENT MONEY IN YEAR -1                             TAXF  21
       F=F*(1.+ENF)                                                   TAXF  22
C  UP-COUNT ALL CAPITAL 2 YEARS,BECAUSE IT IS SPENT 2 YEARS BEFORE    TAXF  23
C    OTHER YEARLY FLOWS                                               TAXF  24
       F=F*(1. + DR)*(1. + DR)                                        TAXF  25
       RETURN                                                         TAXF  26
       END                                                            TAXF  27


       SUBROUTINE UTIL    (N,NP,V,PROP,CERT)                          UTIL   1
C      *********************                                          UTIL   2
C  TAKES NPWS OR DCFS AND PROBABILITIES,  GIVES CERTAINTY EQUIVALENTS UTIL   3
C  TAKES N VALUES (V)  WITH EITHER PROBABILITY PROB( ) OR 1/N         UTIL   4
C  (FLAGED BY NP=0 OR 1),PLUS MIN PERMITTED PROFIT XMIN(1), RETURNS CERTUTIL 5
       DIMENSION V(N),PROB(NP)                                        UTIL   6
       COMMON/UTIL/IU,XMIN(5)                                         UTIL   7
       COMMON/ CONT/ LW,ICHEK                                         UTIL   8
       IF(N.LE.0)GO TO 20                                             UTIL   9
       A=1./FLOAT(N)                                                  UTIL  10
       FLLW =FLOAT(LW +1)                                             UTIL  11
       CERT=1.0                                                       UTIL  12
       DO 5 I=1,N                                                     UTIL  13
       IF(V(I).LT.FLLW   ) GO TO 20                                   UTIL  14
       IF(V(I).LT.XMIN(1))  GO TO 30                                  UTIL  15
       IF(NP.NE.1)  A=PROB(I)                                         UTIL  16
5        CERT=CERT*((V(I)- XMIN(1))**A)                               UTIL  17
       CERT=CERT+XMIN(1)                                              UTIL  18
C  SUCCESSFUL RETURN WHEN CERT.GT.LW + 1                              UTIL  19
       RETURN                                                         UTIL  20
20     CERT=FLOAT( LW )                                               UTIL  21
C        CERT=LW        -INCOMPLETE DATA,-NO EVALUATION.              UTIL  22
       RETURN                                                         UTIL  23
30     CERT=FLLW                                                      UTIL  24
C        CERT=LW + 1   -SOME RESULTS BELOW MAX LOSS.                  UTIL  25
       RETURN                                                         UTIL  26
       END                                                            UTIL  27


       SUBROUTINE MEAN(A,B,C,D,Z)                                     MEAN   1
C    PROVIDES MEAN VALUE FOR A TRAPEZOIDAL DISTRIBUTION.              MEAN   2
C  AND CHECKS THAT THE ORDER IS CORRECT                               MEAN   3
1      IF(A.LE.B)  GO TO 10                                           MEAN   4
       Z=A                                                            MEAN   5
       A=B                                                            MEAN   6
       B=Z                                                            MEAN   7
10     IF(B.LE.C)  GO TO 20                                           MEAN   8
       Z=B                                                            MEAN   9
       B=C                                                            MEAN  10
       C=Z                                                            MEAN  11
       GO TO 1                                                        MEAN  12
20     IF(C.LE.D)  GO TO 30                                           MEAN  13
       Z=C                                                            MEAN  14
       C=D                                                            MEAN  15
       D=Z                                                            MEAN  16
       GO TO 1                                                        MEAN  17
30     Z=A*A+A*B+B*B                                                  MEAN  18
       Z=D*D+D*C+C*C-Z                                                MEAN  19
       DEN=D+C-A-B                                                    MEAN  20
       IF(DEN. LT.0.0000001.AND.DEN.GT.-0.0000001)  GO TO 40         MEAN  21
```

```
      Z=Z/(3.C*DEN)                                        MEAN  22
      RETURN                                               MEAN  23
   40 Z=0.5*(B+C)                                          MEAN  24
      RETURN                                               MEAN  25
      END                                                  MEAN  26


      SUBROUTINE POINT(A,B,C,D,Z)                          POIN   1
C     PROVIDES TRAPEZOIDAL DATA FOR A SINGLE POINT.        POIN   2
      A=Z                                                  POIN   3
      B=Z                                                  POIN   4
      C=Z                                                  POIN   5
      D=Z                                                  POIN   6
      RETURN                                               POIN   7
      END                                                  POIN   8


      SUBROUTINE FREAD(ROUT,IOUT,ICO)                      FREA   1
C********************************                          FREA   2
C     FREE READ SUBROUTINE                                 FREA   3
C*****ICO  CONTROL VALUE FOR READING                       FREA   4
C***   = 0  NO TEXT IN LINE                                FREA   5
C***   = 1  WITH TEXT IN LINE, TEXT IS RETURNED IN IOUT(1) FREA   6
C    NUMBERS SEPARATED BY BLANK OR COMMA,NUMBERS MAY OR MAY NOT HAVE A  FREA  7
C     DECIMAL POINT,  ALL DATA TAKEN AS REAL,RETURNED IN ROUT  FREA  8
C     TEXT IN 3A3 FORMAT IN FIRST 9 COLUMNS         RETURNED IN IOUT FREA 9
      DIMENSION INP(80),IOUT(3),ROUT(20),L(15)             FREA  10
      COMMON/RWR/IW,IR                                     FREA  11
      DATA L(1), L(2), L(3), L(4), L(5), L(6), L(7), L(8), L(9), L(10), FREA 12
     *L(11), L(12), L(13), L(14), L(15)                    FREA  13
     *      /1H0,1H1,1H2,1H3,1H4,1H5,1H6,1H7,1H8,1H9,1H.,1H ,1H-,1H,,1H+FREA 14
     1/                                                    FREA  15
      IC=ICO+1                                             FREA  16
      DO 5 NU=1,3                                          FREA  17
    5 IOUT(NU)=L(12)                                       FREA  18
      DO 6 NU=1,20                                         FREA  19
    6 ROUT(NU)=0.                                          FREA  20
      DO 7 NU=1,80                                         FREA  21
    7 INP(NU)=L(12)                                        FREA  22
C     N COUNTS THE POSITION IN THE VECTOR  INP             FREA  23
      N=0                                                  FREA  24
      NEG=0                                                FREA  25
      ITRANS=0                                             FREA  26
      EX=1.                                                FREA  27
C     K COUNTS THE POSITION IN THE VECTOR  ROUT  FOR OUTPUT FREA  28
      K=1                                                  FREA  29
C     IRK   =0  INTEGER NUMBER DISCOVERED                  FREA  30
C     IRK   =1  REAL NUMBER DISCOVERED                     FREA  31
      IRK=0                                                FREA  32
      GOTO(13,11),IC                                       FREA  33
   11 READ(IR,12)(IOUT(I),I=1,3),(INP(M),M=10,80)          FREA  34
      WRITE(IW,89) (IOUT(I),I=1,3) , (INP(M), M=10,80)     FREA  35
   12 FORMAT(3A3,71A1)                                     FREA  36
      N=10                                                 FREA  37
      GOTO 16                                              FREA  38
   13 READ(IR,14)INP                                       FREA  39
      WRITE(IW,88) INP                                     FREA  40
   14 FORMAT(80A1)                                         FREA  41
   15 N=N+1                                                FREA  42
   16 IF(N.EQ.81.AND.INP(N-1).EQ.L(12)) GO TO 90           FREA  43
      IF(N.EQ.81.AND.INP(80).NE.L(12)) GO TO 65            FREA  44
      IF(N.GE.82)GOTO 90                                   FREA  45
      DO 20 I=1,14                                         FREA  46
      IF(INP(N).EQ.L(I))GOTO 30                            FREA  47
      IF(I.EQ.14)GOTO 90                                   FREA  48
   20 CONTINUE                                             FREA  49
C     40 NUMBERS BETWEEN  0.  AND 9.                       FREA  50
   30 IF(I-11)40,38,39                                     FREA  51
   38 IRK=1                                                FREA  52
C     POINT DISCOVERED                                     FREA  53
      TRANS=FLOAT(ITRANS)                                  FREA  54
```

```
      GOTC 15                                                    FREA   55
C     39  :  3 POSSIBILITIES  :  COMMA,BLANK,MINUS               FREA   56
39    IF(I.EQ.13)NEG=1                                           FREA   57
      IF(I.EQ.12.OR.I.EQ.14)GOTO 60                              FREA   58
      GOTO 15                                                    FREA   59
C     40  L IS BETWEEN 1 AND 10   ONLY NUMBERS                   FREA   60
40    IF(IRK-1)45,55,55                                          FREA   61
45    ITRANS=ITRANS*10+(I-1)                                     FREA   62
C      INTEGER NUMBER OR FIRST PART OF A REAL NUMBER             FREA   63
      GOTO 15                                                    FREA   64
55    EX=EX/10.                                                  FREA   65
C      REAL NUMBER SECOND PART OF THE NUMBER. (AFTER THE POINT)  FREA   66
      TRANS=TRANS+FLOAT(I-1)*EX                                  FREA   67
      GOTO 15                                                    FREA   68
60    IF(INP(N-1).EQ.L(12).OR.INP(N-1).EQ.L(14).OR.N.EQ.1)  GO TO 15  FREA   69
C                                                                FREA   70
65    IF(IRK.EQ.1)GOTO 70                                        FREA   71
      IF(NEG.EQ.1)ITRANS=-ITRANS                                 FREA   72
      ROUT(K)=FLOAT(ITRANS)                                      FREA   73
      K=K+1                                                      FREA   74
      GOTO 75                                                    FREA   75
70    IF(NEG.EQ.1)TRANS=-TRANS                                   FREA   76
      ROUT(K)=TRANS                                              FREA   77
      K=K+1                                                      FREA   78
      IRK=C                                                      FREA   79
      EX=1.                                                      FREA   80
75    ITRANS=0                                                   FREA   81
      NEG=C                                                      FREA   82
      GOTC 15                                                    FREA   83
80    DO 87 NU=1,80                                              FREA   84
87    INP(NU)=L(12)                                              FREA   85
      INP(N)=L(15)                                               FREA   86
      WRITE(IW,88)INP                                            FREA   87
88    FCRMAT(1X,80A1)                                            FREA   88
89    FORMAT( 1X,3A3,71A1)                                       FREA   89
      WRITE(IW,85)                                               FREA   90
85    FCRMAT( 1X, 26HUNACCEPTABLE DATA INPUT      ,15(1H*))      FREA   91
90    RETURN                                                     FREA   92
      END                                                        FREA   93


      BLOCK DATA                                                 BLK.    1
      COMMCN/EX /INS(10),KEX                                     BLK.    2
      COMMCN/TRAPS/PCAPY(10,4),PCAPL(10,4),PEOP(10,4),PVOP(10,4),  BLK.    3
     *PSCRAP(10,4),PFINE(4)  ,          PENF(4),MC              BLK.    4
      COMMCN/BLO1/PN,K,KS,P(3),S(5),RIEN(21)                    BLK.    5
      COMMCN/RDAT1/FINE,DR,UVOP(10),NT,NUT,ND,TX ,ENF,KTAX       BLK.    6
      COMMCN/RDAT2/UFOP(10),UCAPY(10),UCAPL(10),USCRAP(10),ADDN(20)  BLK.    7
      COMMCN/SALES/PROB(8),JT,DUMM(340)                         BLK.    8
      COMMCN/TIME/NIEND,IST(10),ISTART(10,8),IEND(10,8)         BLK.    9
      COMMCN/UTIL/IU,XMEN(5)                                    BLK.   10
      COMMCN/RWR/IOUT,IN                                        BLK.   11
      COMMCN/SIM/CORT,CORS,KMC,ERR,IYEAR,JMC                    BLK.   12
      COMMCN/CONT/LW,ICHEK                                      BLK.   13
      COMMCN/MDAT1/KDAT(10,9),NR,KTDAT(100)                     BLK.   14
      COMMON/MDAT2/FAKTOR,UNTAX                                 BLK.   15
      COMMCN/INV /SENS(5)                                       BLK.   16
      COMMCN/REF/IREF                                           BLK.   17
      COMMCN/OPLE/BELOW,ITRACE,UPR                              BLK.   18
      DATA EELOW,UPR / .5  , 1.75/                              BLK.   19
      DATA SENS(1), SENS(2), SENS(3), SENS(4), SENS(5)/5*0.0/  ,  BLK.   20
     *ENF,FENF(1), PENF(2), PENF(3), PENF(4) /5*0.0/            BLK.   21
      DATA IOUT,IN/2,1/                                         BLK.   22
      DATA DR,TX,ND,JT,PROB(1),FINE     ,TU/ .15,  .50,  10,  1,  1.0,  BLK.   23
     * 0.0,           0/                                        BLK.   24
      DATA K,KMC,S(1),  S(2), S(3), S(4), S(5),                 BLK.   25
     1             ERR,KEX/0,0         , 5*1.0, 0.0,3  /        BLK.   26
      DATA FFINE(1), PFINE(2), PFINE(3), PFINE(4) /4*0.0/       BLK.   27
      DATA LW,ICHEK / -50000, 0 /                               BLK.   28
      DATA NIEND,UNTAX,NT / 5,  0.50,  15 /                     BLK.   29
      DATA NR              ,CORS,CORT,MC,IREF/1  ,1.0,1.0,0  , -1/  BLK.   30
      END                                                       BLK.   31
```

```
HEAD                                                                    1
STANDARD TEST
TAX                                                                     4
25 10
UTILITY                                                                 5
-200000
LIFE                                                                    6
5 10
PROBABILITY                                                             7
.33 .34 .33
DEMAND                                                                  8
1400. 1680. 2100. 2700. 3181. 3470. 3630. 3750. 3830. 3910.
1200. 1400. 1630. 1810. 2081. 2340. 2580. 2760. 2780. 2984.
1040. 1160. 1300. 1470. 1620. 1790. 1890. 1960. 1980. 2000.
PRICE                                                                   9
40.0      40.0      40.0      40.0      40.0
40.0      40.0      40.0      40.0      40.0
40.0      40.0      40.0      40.0      40.0
CORRELATIONS                                                           10
0.5 0.5
ALTERNATIVES                                                           14
SM.PLT     2000. 143000. 27500. 3.5 0. 0 1
2ND SM     2000. 143000. 27500. 3.5 C. 0 1
PILOT PLT 1200. 110000. 20000. 3.5 0. 0 1
END
BOUNDARIES                                                            16
0.5      3.
EXTENSIONS                                                            17
2ND SM    SM.PLT
END
TRAPZ
10000
CHECK                                                                 20
1
PLAN                                                                   EX
1
SM.PLT
END
EVALUATE
COR
0.98    1.0
DCF
EVALUATE                                                               EX
WORKING
5.00 5000 5000 5000 5000 5000 5000 5000
4000 4000 4000 4000 4000 4000 4000 4000 4000 4000
3000 3000 3000 3000 3000 3000 3000 3000
LIFE
5
PROB
0.5 0.5
MULTI
2
1  2
3.5    7.0
P1,Y1,D     700 350
P1,Y1,P      40   80
P1,Y2,D     840 420
P1,Y2,P      40   80
P1,Y3,D    1050  525
P1,Y3,P      40   80
P1,Y4,D    1350   675
P1,Y4,P      40   80
P1,Y5,D     1590  800
P1,Y5,P      40   80
P2,Y1,D     1200
```

```
P2,Y1,P      40
P2,Y2,D    14C0
P2,Y2,P      40
P2,Y3,D    160J
P2,Y3,P      40
P2,Y4,D    181J
P2,Y4,P      4C
P2,Y5,D    208D
P2,Y5,P      40
NOTRAPZ
EVALUATE
STCF
```

APPENDIX 2
TABLES

CONTENTS   <u>Page</u>

Table 1 - DISCOUNT FACTOR TABLES

$$\text{Factor} = \frac{1}{(1+r)^n}$$

r is Discount Rate
n is years discounted

Discount Rate 1 - 15%

Years n	1%	2%	3%	4%	5%	6%	7%	8%	9%	10%	11%	12%	13%	14%	15%
0	1.0000	1.0000	1.0000	1.0000	1.0000	1.0000	1.0000	1.0000	1.0000	1.0000	1.0000	1.0000	1.0000	1.0000	1.0000
1	.9901	.9804	.9709	.9615	.9524	.9434	.9346	.9259	.9174	.9091	.9009	.8929	.8850	.8772	.8696
2	.9803	.9612	.9426	.9246	.9070	.8900	.8734	.8573	.8417	.8264	.8116	.7972	.7831	.7695	.7561
3	.9706	.9423	.9151	.8890	.8638	.8396	.8163	.7938	.7722	.7513	.7312	.7118	.6931	.6750	.6575
4	.9610	.9238	.8885	.8548	.8227	.7921	.7629	.7350	.7084	.6830	.6587	.6355	.6133	.5921	.5718
5	.9515	.9057	.8626	.8219	.7835	.7473	.7130	.6806	.6499	.6209	.5935	.5674	.5428	.5194	.4972
6	.9420	.8880	.8375	.7903	.7462	.7050	.6663	.6302	.5963	.5645	.5346	.5066	.4803	.4556	.4323
7	.9327	.8706	.8131	.7599	.7107	.6651	.6227	.5835	.5470	.5132	.4817	.4523	.4251	.3996	.3759
8	.9235	.8535	.7894	.7307	.6768	.6274	.5820	.5403	.5019	.4665	.4339	.4039	.3762	.3506	.3269
9	.9143	.8368	.7664	.7026	.6446	.5919	.5439	.5002	.4604	.4241	.3909	.3606	.3329	.3075	.2843
10	.9053	.8203	.7441	.6756	.6139	.5584	.5083	.4632	.4224	.3855	.3522	.3220	.2946	.2697	.2472
11	.8963	.8043	.7224	.6496	.5847	.5268	.4751	.4289	.3875	.3505	.3173	.2875	.2607	.2366	.2149
12	.8874	.7885	.7014	.6246	.5568	.4970	.4440	.3971	.3555	.3186	.2858	.2567	.2307	.2076	.1869
13	.8787	.7730	.6810	.6006	.5303	.4688	.4150	.3677	.3262	.2897	.2575	.2292	.2042	.1821	.1625
14	.8700	.7579	.6611	.5775	.5051	.4423	.3878	.3405	.2992	.2633	.2320	.2046	.1807	.1597	.1413
15	.8613	.7430	.6419	.5553	.4810	.4173	.3624	.3152	.2745	.2394	.2090	.1827	.1599	.1401	.1229
16	.8528	.7284	.6232	.5339	.4581	.3936	.3387	.2919	.2519	.2176	.1883	.1631	.1415	.1229	.1069
17	.8444	.7142	.6050	.5134	.4363	.3714	.3166	.2703	.2311	.1978	.1696	.1456	.1252	.1078	.0929
18	.8360	.7002	.5874	.4936	.4155	.3503	.2959	.2502	.2120	.1799	.1528	.1300	.1108	.0946	.0808
19	.8277	.6864	.5703	.4746	.3957	.3305	.2765	.2317	.1945	.1635	.1377	.1161	.0981	.0829	.0703
20	.8195	.6730	.5537	.4564	.3769	.3118	.2584	.2145	.1784	.1486	.1240	.1037	.0868	.0728	.0611
25	.7798	.6095	.4776	.3751	.2953	.2330	.1842	.1460	.1160	.0923	.0736	.0588	.0471	.0378	.0304
30	.7419	.5521	.4120	.3083	.2314	.1741	.1314	.0994	.0754	.0573	.0437	.0334	.0256	.0196	.0151
35	.7059	.5000	.3554	.2534	.1813	.1301	.0937	.0676	.0490	.0356	.0259	.0189	.0139	.0102	.0075
40	.6717	.4529	.3066	.2083	.1420	.0972	.0668	.0460	.0318	.0221	.0154	.0107	.0075	.0053	.0037
45	.6391	.4102	.2644	.1712	.1113	.0727	.0476	.0313	.0207	.0137	.0091	.0061	.0041	.0027	.0019
50	.6080	.3715	.2281	.1407	.0872	.0543	.0339	.0213	.0134	.0085	.0054	.0035	.0022	.0014	.0009

Table 1 - DISCOUNT FACTOR TABLES   (cont'd.)

Discount Rate 16 - 50%

Years n	16%	17%	18%	19%	20%	21%	22%	23%	24%	25%	30%	35%	40%	45%	50%
0	1.0000	1.0000	1.0000	1.0000	1.0000	1.0000	1.0000	1.0000	1.0000	1.0000	1.0000	1.0000	1.0000	1.0000	1.0000
1	.8621	.8547	.8475	.8403	.8333	.8264	.8197	.8130	.8065	.8000	.7692	.7407	.7143	.6897	.6667
2	.7432	.7305	.7182	.7062	.6944	.6830	.6719	.6610	.6504	.6400	.5917	.5487	.5102	.4756	.4444
3	.6407	.6244	.6086	.5934	.5787	.5645	.5507	.5374	.5245	.5120	.4552	.4064	.3644	.3280	.2963
4	.5523	.5337	.5158	.4987	.4823	.4665	.4514	.4369	.4230	.4096	.3501	.3011	.2603	.2262	.1975
5	.4761	.4561	.4371	.4190	.4019	.3855	.3700	.3552	.3411	.3277	.2693	.2230	.1859	.1560	.1317
6	.4104	.3898	.3704	.3521	.3349	.3186	.3033	.2888	.2751	.2621	.2072	.1652	.1328	.1076	.0878
7	.3538	.3332	.3139	.2959	.2791	.2633	.2486	.2348	.2218	.2097	.1594	.1224	.0949	.0742	.0585
8	.3050	.2848	.2660	.2487	.2326	.2176	.2038	.1909	.1789	.1678	.1226	.0906	.0678	.0512	.0390
9	.2630	.2434	.2255	.2090	.1938	.1799	.1670	.1552	.1443	.1342	.0943	.0671	.0484	.0353	.0260
10	.2267	.2080	.1911	.1756	.1615	.1486	.1369	.1262	.1164	.1074	.0725	.0497	.0346	.0243	.0173
11	.1954	.1778	.1619	.1476	.1346	.1228	.1122	.1026	.0938	.0859	.0558	.0368	.0247	.0168	.0116
12	.1685	.1520	.1372	.1240	.1122	.1015	.0920	.0834	.0757	.0687	.0429	.0273	.0176	.0116	.0077
13	.1452	.1299	.1163	.1042	.0935	.0839	.0754	.0678	.0610	.0550	.0330	.0202	.0126	.0080	.0051
14	.1252	.1110	.0985	.0876	.0779	.0693	.0618	.0551	.0492	.0440	.0254	.0150	.0090	.0055	.0034
15	.1079	.0949	.0835	.0736	.0649	.0573	.0507	.0448	.0397	.0352	.0195	.0111	.0061	.0038	.0023
16	.0930	.0811	.0708	.0618	.0541	.0474	.0415	.0364	.0320	.0281	.0150	.0082	.0041	.0023	.0015
17	.0802	.0693	.0600	.0520	.0451	.0391	.0340	.0296	.0258	.0225	.0101	.0054	.0029	.0016	.0009
18	.0691	.0592	.0508	.0437	.0376	.0323	.0279	.0241	.0208	.0180	.0077	.0039	.0021	.0011	.0006
19	.0596	.0506	.0431	.0367	.0313	.0267	.0229	.0196	.0168	.0144	.0059	.0029	.0015	.0008	.0004
20	.0514	.0433	.0365	.0308	.0261	.0221	.0187	.0159	.0135	.0115	.0045	.0021	.0012	.0005	.0003
25	.0245	.0197	.0160	.0129	.0105	.0085	.0069	.0057	.0046	.0038	.0012	.0005	.0002	.0001	.0000
30	.0116	.0090	.0070	.0054	.0042	.0033	.0026	.0020	.0016	.0012	.0003	.0001	.0000	.0000	.0000
35	.0055	.0041	.0030	.0023	.0017	.0013	.0009	.0007	.0005	.0004	.0001	.0000	.0000	.0000	.0000
40	.0026	.0019	.0013	.0010	.0007	.0005	.0004	.0002	.0002	.0001	.0000	.0000	.0000	.0000	.0000
45	.0013	.0009	.0006	.0004	.0003	.0002	.0001	.0001	.0001	.0000	.0000	.0000	.0000	.0000	.0000
50	.0006	.0004	.0003	.0002	.0001	.0001	.0000	.0000	.0000	.0000	.0000	.0000	.0000	.0000	.0000

Table 2 - CUMULATIVE DISCOUNT FACTOR TABLES

$$\text{Factor} = \sum_{i=0}^{n} \frac{1}{(1+r)^i}$$

r = Discount Rate
n = Years discounted

Discount Rate 1 - 15%

Year n	1%	2%	3%	4%	5%	6%	7%	8%	9%	10%	11%	12%	13%	14%	15%
0	1.0000	1.0000	1.0000	1.0000	1.0000	1.0000	1.0000	1.0000	1.0000	1.0000	1.0000	1.0000	1.0000	1.0000	1.0000
1	1.9901	1.9804	1.9709	1.9615	1.9524	1.9434	1.9346	1.9259	1.9174	1.9091	1.9009	1.8929	1.8850	1.8772	1.8696
2	2.9704	2.9416	2.9135	2.8861	2.8594	2.8334	2.8080	2.7833	2.7591	2.7355	2.7125	2.6901	2.6681	2.6467	2.6257
3	3.9410	3.8839	3.8286	3.7751	3.7232	3.6730	3.6243	3.5771	3.5313	3.4869	3.4437	3.4018	3.3612	3.3216	3.2832
4	4.9020	4.8077	4.7171	4.6299	4.5459	4.4651	4.3872	4.3121	4.2397	4.1699	4.1024	4.0373	3.9745	3.9137	3.8550
5	5.8534	5.7135	5.5797	5.4518	5.3295	5.2124	5.1002	4.9927	4.8897	4.7908	4.6959	4.6048	4.5172	4.4331	4.3522
6	6.7955	6.6014	6.4172	6.2421	6.0757	5.9173	5.7665	5.6229	5.4859	5.3553	5.2305	5.1114	4.9976	4.8887	4.7845
7	7.7282	7.4720	7.2303	7.0021	6.7864	6.5824	6.3893	6.2064	6.0330	5.8684	5.7122	5.5638	5.4226	5.2883	5.1604
8	8.6517	8.3255	8.0197	7.7327	7.4632	7.2098	6.9713	6.7466	6.5348	6.3349	6.1461	5.9676	5.7988	5.6389	5.4873
9	9.5660	9.1622	8.7861	8.4353	8.1078	7.8017	7.5152	7.2469	6.9952	6.7590	6.5370	6.3283	6.1316	5.9464	5.7716
10	10.4713	9.9826	9.5302	9.1109	8.7217	8.3601	8.0236	7.7101	7.4177	7.1446	6.8892	6.6502	6.4262	6.2161	6.0188
11	11.3676	10.7868	10.2526	9.7605	9.3064	8.8869	8.4987	8.1390	7.8052	7.4951	7.2065	6.9377	6.6869	6.4527	6.2337
12	12.2551	11.5753	10.9540	10.3851	9.8632	9.3838	8.9427	8.5361	8.1607	7.8137	7.4924	7.1944	6.9176	6.6603	6.4206
13	13.1337	12.3484	11.6350	10.9856	10.3935	9.8527	9.3577	8.9038	8.4869	8.1034	7.7499	7.4235	7.1218	6.8424	6.5832
14	14.0037	13.1062	12.2961	11.5631	10.8986	10.2950	9.7455	9.2442	8.7862	8.3667	7.9819	7.6282	7.3025	7.0021	6.7245
15	14.8650	13.8493	12.9379	12.1184	11.3796	10.7122	10.1079	9.5595	9.0607	8.6061	8.1909	7.8109	7.4624	7.1422	6.8474
16	15.7179	14.5777	13.5611	12.6523	11.8377	11.1059	10.4467	9.8514	9.3126	8.8237	8.3792	7.9740	7.6039	7.2651	6.9542
17	16.5622	15.2919	14.1661	13.1657	12.2740	11.4773	10.7632	10.1216	9.5436	9.0216	8.5488	8.1196	7.7291	7.3729	7.0472
18	17.3983	15.9920	14.7535	13.6593	12.6896	11.8276	11.0591	10.3719	9.7556	9.2014	8.7016	8.2497	7.8399	7.4674	7.1280
19	18.2260	16.6785	15.3238	14.1339	13.0853	12.1581	11.3356	10.6036	9.9501	9.3649	8.8393	8.3658	7.9380	7.5504	7.1982
20	19.0455	17.3514	15.8775	14.5903	13.4622	12.4699	11.5940	10.8181	10.1285	9.5136	8.9633	8.4694	8.0248	7.6231	7.2593
25	23.0232	20.5235	18.4131	16.6221	15.0939	13.7834	12.6536	11.6748	10.8226	10.0770	9.4217	8.8431	8.3300	7.8729	7.4641
30	26.8077	23.3965	20.6004	18.2920	16.3725	14.7648	13.4090	12.2578	11.2737	10.4269	9.6938	9.0552	8.4957	8.0027	7.5660
35	30.4085	25.9986	22.4872	19.6646	17.3742	15.4982	13.9477	12.6546	11.5668	10.6442	9.8552	9.1755	8.5856	8.0700	7.6166
40	33.8347	28.3555	24.1148	20.7928	18.1591	16.0463	14.3317	12.9246	11.7574	10.7791	9.9511	9.2438	8.6344	8.1050	7.6418
45	37.0945	30.4902	25.5187	21.7200	18.7741	16.4558	14.5955	13.1084	11.8812	10.8628	10.0079	9.2825	8.6609	8.1232	7.6543
50	40.2001	32.4236	26.7298	22.4822	19.2559	16.7619	14.8007	13.2335	11.9617	10.9148	10.0417	9.3045	8.6752	8.1327	7.6605

Table 2 - CUMULATIVE DISCOUNT FACTOR TABLES (cont'd.)

Discount Rate 16 - 50%

Years n	16%	17%	18%	19%	20%	21%	22%	23%	24%	25%	30%	35%	40%	45%	50%
0	1.0000	1.0000	1.0000	1.0000	1.0000	1.0000	1.0000	1.0000	1.0000	1.0000	1.0000	1.0000	1.0000	1.0000	1.0000
1	1.8621	1.8547	1.8475	1.8403	1.8333	1.8264	1.8197	1.8130	1.8065	1.8000	1.7692	1.7407	1.7143	1.6897	1.6667
2	2.6052	2.5852	2.5656	2.5465	2.5278	2.5095	2.4915	2.4740	2.4568	2.4400	2.3609	2.2894	2.2245	2.1653	2.1111
3	3.2459	3.2096	3.1743	3.1399	3.1065	3.0739	3.0422	3.0114	2.9813	2.9520	2.8161	2.6959	2.5889	2.4933	2.4074
4	3.7982	3.7432	3.6900	3.6386	3.5887	3.5404	3.4936	3.4483	3.4043	3.3616	3.1662	2.9969	2.8492	2.7195	2.6049
5	4.2743	4.1993	4.1271	4.0577	3.9906	3.9260	3.8636	3.8035	3.7454	3.6893	3.4356	3.2199	3.0352	2.8755	2.7366
6	4.6847	4.5892	4.4976	4.4098	4.3255	4.2446	4.1669	4.0923	4.0205	3.9514	3.6427	3.3851	3.1680	2.9831	2.8244
7	5.0386	4.9224	4.8115	4.7057	4.6046	4.5080	4.4155	4.3271	4.2423	4.1611	3.8021	3.5075	3.2628	3.0573	2.8829
8	5.3436	5.2071	5.0775	4.9544	4.8372	4.7256	4.6193	4.5180	4.4212	4.3289	3.9247	3.5981	3.3306	3.1085	2.9220
9	5.6065	5.4505	5.3030	5.1634	5.0310	4.9055	4.7863	4.6732	4.5655	4.4631	4.0190	3.6653	3.3790	3.1438	2.9480
10	5.8332	5.6586	5.4940	5.3390	5.1925	5.0541	4.9232	4.7993	4.6819	4.5705	4.0915	3.7150	3.4136	3.1681	2.9653
11	6.0286	5.8364	5.6560	5.4865	5.3271	5.1770	5.0354	4.9019	4.7757	4.6564	4.1473	3.7519	3.4383	3.1849	2.9769
12	6.1971	5.9884	5.7932	5.6105	5.4392	5.2785	5.1274	4.9853	4.8514	4.7251	4.1903	3.7792	3.4559	3.1965	2.9846
13	6.3423	6.1183	5.9095	5.7147	5.5327	5.3624	5.2028	5.0531	4.9124	4.7801	4.2233	3.7994	3.4685	3.2045	2.9897
14	6.4675	6.2293	6.0080	5.8023	5.6106	5.4318	5.2646	5.1082	4.9616	4.8241	4.2487	3.8144	3.4775	3.2100	2.9931
15	6.5755	6.3242	6.0915	5.8759	5.6755	5.4891	5.3153	5.1531	5.0013	4.8593	4.2682	3.8254	3.4839	3.2138	2.9954
16	6.6685	6.4053	6.1623	5.9377	5.7296	5.5365	5.3568	5.1895	5.0333	4.8874	4.2832	3.8337	3.4885	3.2164	2.9970
17	6.7487	6.4746	6.2223	5.9897	5.7746	5.5757	5.3908	5.2191	5.0591	4.9099	4.2948	3.8398	3.4918	3.2182	2.9980
18	6.8178	6.5339	6.2731	6.0333	5.8122	5.6080	5.4187	5.2432	5.0799	4.9279	4.3037	3.8443	3.4941	3.2194	2.9986
19	6.8775	6.5845	6.3162	6.0702	5.8435	5.6348	5.4416	5.2628	5.0967	4.9424	4.3105	3.8476	3.4958	3.2203	2.9991
20	6.9288	6.6278	6.3527	6.1009	5.8696	5.6569	5.4604	5.2787	5.1103	4.9539	4.3158	3.8501	3.4970	3.2209	2.9994
25	7.0971	6.7662	6.4669	6.1951	5.9476	5.7216	5.5140	5.3233	5.1474	4.9849	4.3286	3.8556	3.4994	3.2220	2.9999
30	7.1772	6.8294	6.5168	6.2347	5.9789	5.7465	5.5339	5.3392	5.1601	4.9950	4.3321	3.8568	3.4999	3.2222	3.0000
35	7.2153	6.8582	6.5386	6.2512	5.9915	5.7562	5.5413	5.3448	5.1644	4.9984	4.3330	3.8571	3.5000	3.2222	3.0000
40	7.2335	6.8713	6.5482	6.2582	5.9966	5.7599	5.5440	5.3468	5.1659	4.9995	4.3332	3.8571	3.5000	3.2222	3.0000
45	7.2421	6.8773	6.5523	6.2611	5.9986	5.7613	5.5450	5.3475	5.1664	4.9998	4.3333	3.8571	3.5000	3.2222	3.0000
50	7.2463	6.8801	6.5542	6.2623	5.9995	5.7618	5.5454	5.3478	5.1666	4.9999	4.3333	3.8571	3.5000	3.2222	3.0000

TABLE 3 - NORMAL DISTRIBUTION TABLES

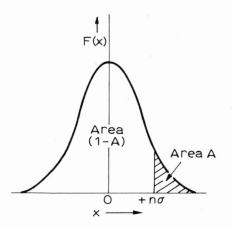

Standard Normal Distribution:

Total area   =   1

Mean   =   0

Standard Deviation   =   1

A   =   Area outside range   +nσ to ∞

(1-A) =   Area in range   -∞ to nσ

n	Outside Limit A	Inside Limit (1-A)
∞	0	1.0000
5.0	0.0000003	0.9999997
4.0	0.00003	0.99997
3.1	0.0010	0.9990
3.0	0.0013	0.9987
2.5	0.0062	0.9938
2.33	0.0100	0.9900
2.0	0.0227	0.9773
1.75	0.0400	0.9600
1.65	0.0500	0.9500
1.5	0.0668	0.9332
1.28	0.1000	0.9000
1.25	0.1056	0.8944
1.0	0.1586	0.8414
0.75	0.2266	0.7734
0.50	0.3085	0.6915
0.25	0.4013	0.5987
0	0.5000	0.5000

INDEX

DATE DUE

DATE DUE			
DEC 23 '80			
JUN 26 '84			